Introduction to ... ysics

Introduction to Environmental Physics

Planet Earth, Life and Climate

Nigel Mason
Department of Physics and Astronomy
University College, London, UK.

Peter Hughes
Kingsway College, London, UK.

with

Randall McMullan
Ross Reynolds
Lester Simmonds
John Twidell

with a foreword by

Sir John Houghton

London and New York

Cover: The weather station on the top of Germany's highest mountain, the Zugspitze, December 1999 (Source: P. Hughes)

First published 2001 by Taylor & Francis
11 New Fetter Lane, London EC4P 4EE

Simultaneously published in the USA and Canada
by Taylor & Francis Inc,
29 West 35th Street, New York, NY 10001

Taylor & Francis is an imprint of the Taylor & Francis Group

© 2001 Nigel Mason and Peter Hughes

Typeset in Goudy by Wearset, Boldon, Tyne and Wear
Printed and bound in Great Britain by TJ International, Padstow, Cornwall

British Library Cataloguing in Publication Data
A catalogue record for this book is available from the British Library

Library of Congress Cataloging in Publication Data
A catalog record for this book has been requested

ISBN 0-7484-0764-2 (hbk)
ISBN 0-7484-0765-0 (pbk)

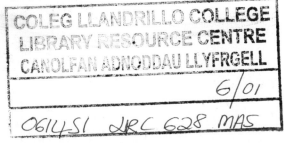

To Gaby and Jane

Contents

List of contributors		xv
Foreword		xvii
Preface		xviii
Acknowledgements		xx

Chapter 1   **Environmental physics: processes and issues**   **1**

1.1   Introduction   1

1.2   The environment: the science of the twenty-first century?   4

     1.2.1   Environmental concerns in the late twentieth century   4

1.3   What is environmental physics?   7

1.4   Physics in the environment   7

     1.4.1   Human environment   7

     1.4.2   Built environment   9

     1.4.3   Urban environment   10

     1.4.4   Global environment   11

     1.4.5   Biological environment   14

1.5   Environmental physics and the global environmental agenda   15

1.6   Summary   15

     References   15

Chapter 2   **The human environment**   **17**

2.1   Introduction   17

2.2   Laws of Thermodynamics   18

     2.2.1   First Law of Thermodynamics   18

     2.2.2   Second Law of Thermodynamics   18

     2.2.3   Entropy and the Third Law of Thermodynamics   19

2.3   Laws of Thermodynamics and the human body   21

     2.3.1   Energy and metabolism   21

     2.3.2   Thermodynamics and the human body   22

     2.3.3   First Law of Thermodynamics and the human body   23

		2.3.4	Second Law of Thermodynamics and the Gibbs free energy	24
	2.4		Energy transfer	26
		2.4.1	Conduction	27
		2.4.2	Convection	31
		2.4.3	Radiation	33
		2.4.4	Evaporation	38
		2.4.5	Energy budget equation	40
	2.5		Survival in the cold	42
		2.5.1	Thermal comfort and insulation	42
		2.5.2	Boundary layer	43
		2.5.3	Wind chill	44
		2.5.4	Hypothermia	46
	2.6		Survival in hot climates	47
		2.6.1	Effect of heat on the human body	47
	2.7		Taking risks, weather and survival	50
	2.8		Summary	50
			References	50
			Questions	52
Chapter 3			**The built environment**	**56**
	3.1		Introduction	56
	3.2		Thermal regulation in buildings	58
		3.2.1	Thermal insulation	58
		3.2.2	Thermal conduction effects	60
		3.2.3	Convection effects	62
		3.2.4	Radiation effects	62
		3.2.5	U-values	63
	3.3		Energy use in buildings	65
		3.3.1	Efficiency	65
		3.3.2	Energy losses	66
		3.3.3	Calculation of energy losses	67
		3.3.4	Energy gains	69
	3.4		Air regulation in buildings	71
		3.4.1	Ventilation requirements	71
		3.4.2	Ventilation installations	72
	3.5		Heat pumps	75
		3.5.1	Heat pump efficiency	75
	3.6		Condensation	76
		3.6.1	Water vapour	78
		3.6.2	Humidity	80
		3.6.3	Condensation in buildings	82
	3.7		Buildings of the future	84

	3.7.1	Checklist for a future house	84
	3.7.2	Energy use and carbon dioxide emissions	86
3.8	Summary		87
	References		88
	Questions		88

Chapter 4	**The urban environment**		**90**
4.1	Introduction		90
	4.1.1	Townscape	91
4.2	Energy in the city		93
	4.2.1	Electromagnetic induction	94
	4.2.2	Electrical power transmission	95
4.3	Transportation		96
	4.3.1	Energy efficiency in transport	97
4.4	Water for the urban environment		99
	4.4.1	Sewage	100
4.5	Lighting		102
	4.5.1	Sources of light	103
4.6	Urban pollution		105
	4.6.1	Urban pollutants	106
	4.6.2	Particulates	109
4.7	Smog		111
4.8	Acid rain		114
4.9	The car as an urban pollutant		116
	4.9.1	Internal combustion engine	117
	4.9.2	Efficiency of the car engine	119
	4.9.3	Reducing vehicle emissions	120
4.10	Noise pollution		122
	4.10.1	Human ear	123
	4.10.2	Sound levels	124
	4.10.3	Hearing loss	125
4.11	Summary		126
	References		126
	Questions		127

Chapter 5	**Energy for living**		**130**
5.1	Introduction		130
	5.1.1	World energy demand	131
	5.1.2	World energy supplies	132
5.2	Fossil fuels		133
5.3	Nuclear power		134
	5.3.1	Nuclear fission	134
	5.3.2	Nuclear reactors	135

	5.3.3	Nuclear fusion	139
5.4	Renewable energy		142
5.5	Solar energy		143
	5.5.1	Transferring solar energy	145
	5.5.2	Solar photovoltaic electricity	147
5.6	Wind power		152
	5.6.1	Average power of a moving mass of fluid	154
	5.6.2	Bernoulli's theorem and the aerofoil	155
	5.6.3	Forces acting on wind-turbine propeller blades	158
	5.6.4	Laminar and turbulent flow	162
5.7	Hydroelectric power		163
	5.7.1	Water moving through a cylindrical tube	164
5.8	Tidal power		164
5.9	Wave energy		166
	5.9.1	Mathematics of wave power	169
5.10	Biomass and biofuels		171
5.11	Geothermal power		174
5.12	Summary		176
	References		176
	Questions		177

Chapter 6   Revealing the planet   **180**

6.1	Introduction		180
6.2	Remote sensing		180
6.3	Orbits of satellites		182
6.4	Resolution of satellite images		185
	6.4.1	Image processing	186
6.5	Radar		187
6.6	Applications of remote sensing data		191
	6.6.1	Meteorological satellites	191
	6.6.2	Landsat	192
6.7	Summary		194
	References		194
	Questions		195

Chapter 7   The Sun and the atmosphere   **198**

7.1	Introduction		198
7.2	Solar energy		198
	7.2.1	Solar output	198
	7.2.2	Rhythm of the seasons	200
	7.2.3	Solar cycles and climate change	201
7.3	Structure and composition of the Earth's atmosphere		204
	7.3.1	Structure of the atmosphere	204

	7.3.2	Composition of the atmosphere	207
7.4		Atmospheric pressure	209
	7.4.1	Pressure and temperature as functions of altitude	209
	7.4.2	Escape velocity	210
7.5		Solar radiation	211
	7.5.1	Solar spectrum	211
	7.5.2	Earth's ionosphere	214
	7.5.3	The aurorae	215
	7.5.4	Solar photo-induced chemistry	216
7.6		Ozone	217
	7.6.1	The Earth's ultraviolet filter	217
	7.6.2	Ozone chemistry	219
	7.6.3	'Ozone hole'	220
	7.6.4	Ozone loss in the Antarctic polar region	222
	7.6.5	Ozone loss in the Arctic polar region	224
7.7		Terrestrial radiation	224
	7.7.1	Earth's energy balance	224
	7.7.2	Earth as a black body	226
	7.7.3	Greenhouse effect	227
7.8		Global warming	229
	7.8.1	Enhanced greenhouse effect	229
	7.8.2	Global warming: the evidence	231
	7.8.3	Global warming: the predictions	232
	7.8.4	Sea-level rise and global warming	234
7.9		Summary	236
		References	236
		Questions	237

Chapter 8		**Observing the Earth's weather**	**242**
8.1		Introduction	242
8.2		Observing the weather	242
	8.2.1	Air temperature	243
	8.2.2	Pressure measurement	245
	8.2.3	Wind measurement	246
	8.2.4	Humidity measurement	247
	8.2.5	Precipitation measurement	248
	8.2.6	Sunshine	249
	8.2.7	Visibility	251
8.3		Global weather monitoring network	253
	8.3.1	Surface network	253
	8.3.2	Upper atmosphere network	255

	8.4	Weather forecasting	257
		8.4.1 Folklore	257
		8.4.2 Computer modelling of weather	257
		8.4.3 Chaos in weather forecasting	258
	8.5	Cloud physics	262
		8.5.1 Water: the unique molecule	262
		8.5.2 Hydrosphere	264
		8.5.3 Types of clouds	265
	8.6	Physics of cloud formation	265
	8.7	Snow crystals	267
	8.8	Atmospheric electricity	269
	8.9	Summary	272
		References	272
		Questions	273

Chapter 9   Global weather patterns and climate   275

9.1	Introduction: atmospheric motion	275
	9.1.1 Air masses and weather fronts	275
9.2	Principal forces acting on a parcel of air in the atmosphere	276
	9.2.1 Gravitational force	276
	9.2.2 Pressure gradient force	277
	9.2.3 Coriolis force	278
	9.2.4 Frictional forces	279
9.3	Pressure gradients and winds	280
	9.3.1 Cyclonic motion	280
	9.3.2 Depressions and fronts	283
9.4	Thermal gradients and winds	284
9.5	Global convection	285
9.6	Global weather and climate patterns	287
	9.6.1 Global pressure field	287
	9.6.2 Global wind patterns	290
	9.6.3 Temperature fields	297
	9.6.4 Global humidity patterns	300
	9.6.5 Cloud patterns	307
	9.6.6 Precipitation	310
9.7	Summary	316
	References	316
	Questions	317

Chapter 10   Physics and soils   319

| 10.1 | Introduction | 319 |
| 10.2 | Soils | 319 |

	10.3	Water retention by soils	320
	10.4	Soil water suction	328
	10.5	Movement of water through soils	335
	10.6	Soil–water balance	342
	10.7	Leaching of solutes through soil profiles	344
	10.8	Evaporation from the land surface	347
		10.8.1 Energy requirement for evaporation	347
		10.8.2 Energy balance of wet and dry land surfaces	348
		10.8.3 Mechanisms for the transfer of latent and sensible heat away from the evaporating surface	351
		10.8.4 Potential evaporation and the Penman equation	353
		10.8.5 Evaporation from the land surface	357
	10.9	Summary	360
		References	360
		Questions	361
Chapter 11	**Vegetation growth and the carbon balance**		**363**
	11.1	Introduction	363
	11.2	Plant development	365
		11.2.1 Weather	365
		11.2.2 Rate of plant development	365
		11.2.3 Impact of global warming on crop distribution	369
	11.3	Plant growth	370
		11.3.1 Photosynthesis by individual leaves	371
		11.3.2 Photosynthesis by a vegetation canopy	376
		11.3.3 Respiration	382
		11.3.4 Allocation of new growth between the various plant parts	383
	11.4	Water stress and vegetation growth	383
	11.5	Carbon balance of the land surface	388
		11.5.1 Terrestrial carbon store	388
		11.5.2 Degradation of soil organic matter	390
		11.5.3 Modelling soil organic matter dynamics	392
	11.6	Summary	394
		References	394
		Questions	395
Chapter 12	**Environmental issues for the twenty-first century**		**397**
	12.1	Introduction	397

12.2	Demographic change	397
12.3	Urbanization	399
12.4	Sustainability	399
	12.4.1 Energy resources	401
12.5	Climate change, survival and health	402
12.6	Models, predictions and uncertainties	403
12.7	Environmental risk	405
	12.7.1 Risk benefit analysis	406
12.8	What is being done?	407
12.9	Summary: environmental physics as an enabling science	409

Appendices
1	Entropy	410
2	Mathematics behind Newton's law of cooling	412
3	Energy consumption self-assessment	412
4	Doppler effect	415
5	Pressure variation with altitude	418
6	Derivation of the lapse rate	420
7	Synoptic weather chart	421
8	Environmental risk and environment impact assessment of ozone-related disasters	423
9	Units and constants	425

Answers to numerical questions	426
Bibliography	429
Glossary	436
Index	455

Contributors

Peter Hughes
BSc (Liverpool University); Diploma of Education (London University); MEd (Hull University); MPhil (Bristol University).
Has taught in the Secondary, Further, Higher and Teacher-training sectors.
Lecturer in Physics, Mathematics and Statistics at Kingsway College London, Visiting Lecturer in Mathematics (Birkbeck College, London University), and Honorary Visiting Lecturer in the School of Engineering, City University.
Since 1990, first Chair of the Education Committee of the Institute of Physics Environmental Physics Group.

Randall McMullan
BSc (Mathematics and Physics), MSc (Physics and Geophysics): Victoria University of Wellington. Postgraduate research in geophysics. Chartered Physicist and Chartered Builder. Manufacturing Technologist at Kodak.
Senior Lecturer in Construction Science and Technology (College of NW London).
Construction training consultant with Taylor Woodrow and Kyle Stewart.
National Educational Advisor for UK Post-16 courses: for the National Council for Vocational Qualifications, Qualifications and Curriculum Authority.
Author and technical writer for book series and numerous publications on technology, the built environment and applied computing. Has written more than 14 books including *Practical AutoCAD*, *Noise Control in Buildings*, *Macmillan Dictionary of Building* and the best-selling *Environmental Science in Building*.

Nigel Mason
BSc and PhD (University College London).
1988–90: SERC Postdoctoral Fellowship. 1990–98: Royal Society University Research Fellow 1998–2001 Lecturer then Reader in Physics (UCL). Founder and Director of the Molecular Physics Laboratory (UCL), and co-founder of the Centre for Cosmic Chemistry and Physics (UCL). Current research interests: the study of the spectroscopy and collision dynamics of those molecules involved in the depletion of the terrestrial ozone layer. Fellow of the Institute of Physics, Honorary Secretary of the Division of Atomic, Molecular and Optical Physics of the Institute of Physics. Recorder of the Physics Section of the British Association. Chair of the European Physical Society Electron and Atomic Collisions Group. Visiting Professor of Physics at the University of Innsbruck, Austria 1999–2000.

Ross Reynolds

BA (Lancaster University); MA (Colorado University).

1972/73: Scientific Officer at the Institute of Oceanographic Sciences, Bidston Observatory. From 1973 he has been a Teaching Fellow in the Department of Meteorology, Reading University.

Teaches mainly meteorology courses and has produced many educational aids, some in collaboration with the Royal Meteorological Society, to promote weather science in schools. Served on the Education Committee of the Society for many years. Organizing and running INSET for 'A' Level teachers was instrumental in him winning the Society's Michael Hunt Award (1995), which recognizes major contributions to improving the public understanding of Meteorology. Recent activity involves the development of Computer-Aided Learning in Meteorology at tertiary level and spending the summer of 1998 as a Visiting Associate Professor in the School of Meteorology at the University of Oklahoma.

Lester Simmonds

Obtained 'A' Levels in Mathematics, Physics, Chemistry and Biology. Because he was keen to apply Science 'out of doors' he studied Agricultural Science at Leeds University. Excited by Environmental Physics (particularly in a soil context) he researched for a PhD at the University of Nottingham in Soil Physics. He then joined Professor John Monteith's Microclimatology of Tropical Crops research group there as a soil physicist, which involved work in India. Since 1985 he has been a lecturer, now Reader, in Soil Physics at the University of Reading. Current research includes work on the leaching of pesticides, remote sensing of soil water, with NASA, water conservation and crop production in the semi-arid tropics.

John Twidell

After taking undergraduate and postgraduate degrees at Oxford University (MA and DPhil), he worked in the Universities of Khartoum, Essex and, as Reader and Director of Energy Studies, at the University of Strathclyde. He has also been Head of Physics in the University of the South Pacific. Appointed as Professor to the Marmont Chair in Sustainable Energy Technology at De Montfort University in 1993. Other responsibilities include Editor of the *Journal of Wind Engineering*. With colleagues he has published about 100 research papers and books, mostly in energy policy, wind and solar energy. Joint author of the seminal book *Renewable Energy Resources*. Advisor to the House of Commons Select Committee on Energy.

Foreword

A prediction for the twenty-first century about which we can be rather certain is that the environment will be one of the century's dominant themes.

During recent decades three particular perspectives regarding the environment have begun to emerge. Firstly, we humans are beginning to realise that not only do we live on the Earth and use its resources at an increasingly accelerated rate, we also belong to the Earth and are a part of (not apart from) the total ecology of all living systems. Secondly, there are strong interactions between different components of the large and complex system that makes up our environment. The science of many of these interactions is as yet poorly understood, although we are beginning to appreciate some of the checks and balances, or in scientific language, the feedbacks that control them. Thirdly, with a population of 6 billion humans, probably rising during this century to over 10 billion, human impact on the environment is of great concern. Not only do we generate severe and damaging local pollution, all of us are responsible for the emission of gases into the atmosphere that cause pollution on a global scale. The consequences of such emissions are leading to damage of the ozone layer and to climate change, both of which are likely to result in serious consequences for human communities.

Because of these perspectives, we have begun to take a great deal more interest in and learn a lot more about our environment, how we interact with it and how it interacts with us. Knowledge of the underlying science is essential to aid this learning process and also to assist decision makers, those in industry, commerce and the concerned public to assess the likely consequences for the environment of human activities with their increasingly vast range and scale.

Aspects of environmental science are already being taught in our schools and there is a growing demand for environmental courses in our colleges and universities. Many different scientific disciplines are involved including the fundamental ones of physics, chemistry, biology and geology. As these sciences are applied to environmental processes, we find that phenomena in the environment provide a wealth of illustrations and evocative examples of many of the basic principles at the core of these fundamental disciplines.

I commend particularly the comprehensive coverage of the field of environmental physics provided by this textbook. One of its further strengths is that it begins to cross some of the barriers that have been built up with the other major areas of science involved. It will therefore encourage both teachers and students to take more of an integrated view of the science of the environment, a view that is essential if we as humans are to understand our environment more completely and protect it from the wanton destruction that large numbers of uninformed humans can so easily inflict.

Sir John Houghton
Co-chair, Scientific Assessment, Intergovernmental Panel on Climate Change

Preface

We can view with awe the breathtaking beauty and savagery of Planet Earth, and yet be aware that there are environmental issues and problems that affect us all; problems which have to be solved if humanity is to progress. The environment influences all of us, whether it is from the impact that we might have on it or how it in turn can affect us. In the past 30 years there has been an increasing interest and concern with our local and global environment – by people of all ages.

Humanity's ability to probe the mysteries of the Universe are boundless and profound, and Physics is central in understanding the global environment. Environmental Physics is concerned with the physics of environmental processes and issues, particularly where this focuses on the interaction between the atmosphere, the lithosphere, the hydrosphere and the biosphere. The aim of this book is to explore these processes, against a world background which is becoming smaller in terms of communication but in which the issues are becoming more complex.

H. L. Penman (Head of Physics at the Rothamsted Agricultural Research Station) and John Monteith, also a physicist (the first Professor of Environmental Physics in the British Isles at the University of Nottingham), were major pioneers in the development of Environmental Physics. In 1990 the Institute of Physics Environmental Physics Group was founded to advance the cause of Environmental Physics, especially in attracting committed people into this exciting branch of Physics. The Group was launched at the John Mason Lecture at the 1990 British Association's Annual Conference in Cardiff by Sir Herman Bondi. The Group's Education Committee has been instrumental in developing Environmental Physics in schools, colleges and universities, and young people's interest in the various aspects of Environment Physics has been matched by the take-up of the subject by several 'A' Level examination boards and certainly by the universities.

The book is aimed at the general reader, and, in particular, post-16 students and first-year undergraduates. The book is designed for students of Physics, Mathematics, Statistics, Chemistry, Biology, Geography, Geology, Environmental Science, Engineering, Construction, Computer Science and Ecology. In addition, it could prove useful to students of Medicine, Law, Sociology, Psychology, Politics, Business, Economics, Management, and those engaged in Outdoor Pursuits.

The structure of the book is as follows. It consists of two parts. The first part (Chapters 2–5) is concerned with the environment at the individual and urban level. The second part (Chapters 6–12) is concerned with the global issues that encompass our environment.

Chapter 1 introduces the main themes that underpin the environmental processes and issues that are to be examined. Chapters 2–11 will reveal the underlying physics of these processes, and Chapter 12 will draw the various strands together while providing glimpses into the future.

Chapter 2 explores the physical principles that underlie the energy exchanges that relate to human beings in cold and hot local environments. It will be subsequently seen that these principles link with a wide variety of environmental phenomena.

Chapter 3 will examine the application of physics, especially energy transfers, to the improvement of the design and operation of buildings.

Chapter 4 reviews various features of life in the urban environment. Rising environmental consciousness is inducing governments world-wide to move slowly towards the idea of sustainable cities, in which a major priority is the improvement of human health and the quality of life.

Chapter 5 will study various energy sources and energy production, with a focus on renewables.

In Chapter 6, the physics that underpins the operation of remote-sensing satellites and their application in studying various environmental contexts, including how human beings are affecting the planet and vice versa, will be reviewed.

Chapter 7 will focus on the importance of the Sun as the engine of the physical processes, including weather and climate, that form the framework for our global environment, and create the conditions for life itself. In examining the physics of the atmosphere there will also be discussion of global warming and the depletion of the ozone layer.

Chapter 8 observes the Earth's weather, examining how weather is monitored and the manner in which the formation of clouds is part of the wider hydrological cycle which supports life-forms.

In Chapter 9 we shall see how examining global weather patterns and climate can form the basis for weather forecasting. This will provide an introduction to some of the basic ideas of Meteorology.

Chapter 10 introduces those aspects of the biosphere that relate to the soil–water interface. In particular, it will emphasize how water is retained and moves through soils, and how the nature of the land surface influences evaporation.

Chapter 11 will explore the biosphere further by determining how the nature and properties of the land surface affect vegetation growth and the carbon balance.

Worked examples are included to illustrate physical principles and vignettes of scientists who have made a contribution towards this subject. At the end of each chapter there are questions for discussion, which may be suitable either for individual or group work, and numerical questions of increasing complexity.

At the outset of the twenty-first century there are many environmental challenges to be wrestled with, and though the environment is changing the underlying Physics does not. Though Environmental Physics is a young branch of Physics you will discover the contributions of several of the very greatest of scientists – Newton, Einstein, Boltzmann, Kepler, Planck and Faraday – to name but a few. Their perceptions have helped us to realize that the challenges can be met and that the solutions are not beyond our grasp. In essence, the environment belongs to all of us, and the knowledge and the understanding of its operation can be the pursuit of all. There is no doubt that Physics will continue to make a major contribution to that understanding, and it is hoped that this book will provide a foundation for new insights and inspired endeavours.

Nigel Mason and Peter Hughes
London 2001

Acknowledgements

Since its inception in 1990 the Environmental Physics Group and its Education Committee have been instrumental in developing Environmental Physics in British schools, colleges and universities. They have organized conferences, meetings, INSET for 'A' Level teachers and university lecturers. All those listed below have helped to create the climate in which Environmental Physics can germinate and develop.

Members of the Environmental Physics Group Committee (1990–2000):
Dr Peter Arnold, Dr Sian Bethan, Mr Richard Clarke, Dr Ian Colbeck, Dr Peter Dagley, Dr Barbara Gabrys, Dr John Garland, Professor John Harries, Dr Geoff Hassall, Dr Peter Hodgson, Mr Peter Hughes, Ms Susanna Lithiby, Dr Alastair McCartney, Dr David Pearson, Dr Douglas Peirson, Dr David Pugh, Dr Quintin Rayner, Dr Neil Roberts, Dr Derek Rose, Dr Michael Smithson, Dr Ranjeet Sokhi, Dr John Stewart, Professor Michael Unsworth, Dr Anne Wheldon, Ms Alexandra Wilson and Professor Edward Youngs.

Members of the Environmental Physics Group's Education Committee (1990–2000):

Mr Michael Chapple	Principal Examiner (London Board) and Middlesex University
Mr Ray Davies	Salford University
Dr Sally East	Atmospheric Physics Laboratory, UCL
Mr Peter Hughes (Chair)	Kingsway College London
Professor Colin Honeybourne	University of the West of England
Professor Alastair McArthur	University of the South Pacific
Dr Nigel Mason (Consultant)	University College London
Mr David Monteith	Toot Hill School, Nottinghamshire
Dr Jane Savage	London University Institute of Education
Dr Francisca Wheeler	Withington Girls School
Mrs Catherine Wilson, MBE	Institute of Physics

Others who have contributed:
Dr Jonathan Bamber, Dr Roger Brugge, Mr Alan Burgess, Mr Chris Butlin, Mr Michael Cook, Dr Linda Gray, Dr Martin Gregory, Professor Ron Hamilton, Dr David Hartley, Mr Bob Kibble, Sir James Lighthill, Dr Robin McIlveen, Sir John Mason, Professor Philip Meredith, Professor John Monteith, Mrs Susan Oldcorn, Dr Jane Ramage, Professor Michael Rycroft, Dr Jonathan Shanklin, Professor Keith Shine, Dr Craig Underwood, Dr Richard Wayne and Professor Roger Wootton.

We are particularly grateful for the contributions by:

Dr Jonathan Bamber (PhD Cambridge University). Formerly of the Mullard Space Science Laboratory (UCL), where his research involved the satellite remote-sensing of the polar regions. He is now Director of the Centre for Remote Sensing, University of Bristol.

Mr Jonathan Smith (MPhil Imperial College). Postgraduate research in particulates at Imperial College concerning the 'Deposition of radionuclides and their subsequent fate'. Lectured in Biology and Environmental Science at Kingsway College London and worked with the Ministry of Agriculture (MAFF) as a member of their Radiological Safety Division. He is now an Environmental Auditor with Wiltshire County Council.

We also thank Nykola Jones (BSc, PhD University College London) for her considerable assistance in the editing and preparation of much of the text in this book, Catherine Dandy and Sarah Nolan for the preparation of many diagrams in the text, and Roger Wesson for taking several of the photographs.

Acknowledgements

Fig 1.1a [ozone depletion] and plate 8 [Ozone hole] provided by courtesy of the Centre for Atmospheric Science, University of Cambridge, (www.atm.ch.cam.ac.uk/tour).

Fig 1.6 [hurricane] reproduced with permission of Dan Vietor, Unisys Corporation.

Fig 1.7 [lightning] Photograph by C. Clark, NOAA Photo Library , NOAA Central Library (www.photolib.noaa.gov).

Fig 2.1 [basal metabolic rate] J. H. Green, *An Introduction to Human Physiology*, 3rd edn. Oxford University Press, Oxford 1974, p. 119 and p. 121, © Oxford University Press.

Fig 2.2 [temperature distribution] and Fig 2.13 [clothing] from D. Emslie-Smith *et al.*, *Textbook of Physiology*, 11th edn. Churchill Livingstone, Edinburgh 1989, p. 511 and p. 527 respectively. With the permission of © Harcourt Publishers Ltd, 1989.

Fig 2.3 [energy transfers] R. N. Hardy, *Temperature and Animal Life*, Arnold, London 1979, p. 5. With the permission of Butterworth Heinemann.

Fig 2.7 [Schlieren optics] L. E. Mount, *Adaptation to Thermal Environment: Man and his productive animals*, Arnold, London 1979, p. 67. With the permission of Butterworth Heinemann.

Fig 2.7 [rabbit] J. Monteith and M. Unsworth, *Principles of Environmental Physics*, 2nd edn. Arnold, London 1990, p. 136. With the permission of Butterworth Heinemann, a division of Reed Educational and Professional Publishing Ltd.

Fig 2.12 [hypothalamus] from J. Simpkins, and J. I. Williams, *Advanced Biology*, 3rd edn. Unwin Hyman, London 1989, p. 350.

Fig 2.15 [Scott at the South Pole] © Scott Polar Research Institute, Lensfield Road, Cambridge, GB.

Fig 3.1b [African hut with solar panel] Dr Anne Wheldon.

Fig 3.8 [heat losses from a building] J. Houghton, *Global Warming: A Complete Briefing*, Cambridge University Press, Cambridge 1997, p. 198. With permission of Cambridge University Press.

Fig 4.1a [Parthenon] (http://www.providence.edu/dwc/grkhome.htm).

Fig 4.2 [Brasilia] courtesy of Augusto C. B. Areal (http://www.civila.com/brasilia).

Figs 4.4 [dynamo], Fig 4.5 [rotating magnet] and Fig 4.6 [three-phase] J. T. McMullan, R. Morgan, and R. B. Murray, *Energy Resources*, 2nd edn. Arnold, London 1983, pp. 50, 51 and 53 respectively. With the permission of Butterworth Heinemann.

Fig 4.12 [industrial pollution] by permission of the Natural Environment Research Council.

Figs 4.14 and 4.15 [London smog, 1952] *Interim Report of the Committee on Air Pollution, 1953* HMSO, London, Cmd 9011, pp. 18 and 19. © Crown copyright.

Fig 5.1 [growth in energy use] J. Houghton, *Global Warming; The Complete Briefing*, 2nd edn. Cambridge University Press, Cambridge 1997, p. 190. With permission of Cambridge University Press.

Fig 5.3 [Chernobyl plume] the Lawrence Livermore National Laboratory, USA, and based on R. Lange, M. H. Dickerson and P. H. Gudiksen, 'Dose Estimates from the Chernobyl Accident', *Nuclear Technology*.

Fig 5.5 [fusion reactor] reproduced with permission of the UK Atomic Energy Authority, Abingdon, Oxfordshire, UK.

Fig 5.6 J. W. Twidell and A. D. Weir, *Renewable Energy Resources*. E. and F. N. Spon, London 1996. Reproduced with permission by Taylor and Francis.

Fig 5.13 [solar cells] reproduced with the permission of Dulas Solar Medical, Dulas Ltd, Machynlleth, North Wales.

Fig 5.16b [Savonius rotor] R. Merrill, and T. Gage, eds. *Energy Primer: Solar, Water, Wind and Biofuels*, Delta Special/Seymour Lawrence, Dell New York 1978, p. 129. Reproduced with permission of Dell Publishing Company, Inc.

Fig 5.24 [wave energy contours] J. W. Twidell and A. D. Weir, *Renewable Energy Resources*, E. and F. N. Spon, London 1996. Reproduced with permission of Taylor and Francis.

Fig 5.28 [oscillating air column] reproduced with permission of the National Engineering Laboratory (NEL).

Fig 6.1 [Earth viewed from a satellite] the Space Photo Collection, NOAA Photo Library (www.photolib.noaa.gov).

Fig 6.10a [meteosat image] (http://www.nottingham.ac.uk/pub/sat-images/CTOT.jpg).

Fig 6.10b [meteosat image] (http://www.nottingham.ac.uk/pub/sat-images/ETOT.jpg).

Fig 6.11 [light pollution] © United States Air Force Defence.

Fig 7.5 [Earth's magnetic field] reproduced from *Scientific American*, Volume 200, No. 3 1959, p. 39. © The Estate of Irving Geis.

Fig 7.8 [solar spectrum] based on Fig 16.1, S. L. Valley, ed. *Handbook of Geophysics and Space Environments*, McGraw-Hill, New York 1965. With permission of McGraw-Hill Publishing Company.

Fig 7.9 [Fraunhofer lines] and Fig 8.17 [mechanisms of a thunderstorm] H. C. Ohanian, *Physics*, 2nd edn. 1988. With permission © W. W. Norton and Co. Inc, New York.

Fig 7.13 [absorption spectra] E. Boeker and R. van Grondelle, *Environmental Physics*, Wiley, Chichester 1995, p. 24. With permission by John Wiley and Sons Ltd.

Fig 7.15 [ozone hole] reproduced from *Physics World*, April 1994, p. 50. With the permission of Rolando Garcia.

Fig 7.17 [polar stratospheric clouds] photograph by Susan Solomon. With permission of *Physics World*, April 1994, p. 49.

Fig 7.20 [meteosat image] (http://www.nottingham.ac.uk/pub/sat-images/ETOT.JPG).

Fig 7.22 [global surface temperatures anomalies] with permission of the Meteorological Office, Hadley Centre for Climate Prediction and Research.

Fig 7.23 [Vostock ice core data] J. Houghton, *Global Warming: The Complete Briefing*, 2nd edn. Cambridge University Press, Cambridge 1997, p. 54. Reproduced with permission by Cambridge University Press.

Fig 8.10 [surface weather chart] *Weather*, Special Issue, The Storm of 15–16 October 1987. Vol. 43 (3), March 1988, p. 92. With permission of the Royal Meteorological Society.

Fig 8.12 [Lorenz attractor] E. Lorenz, *The Essence of Chaos*. © University of Washington Press 1993.

Fig 8.13 [global circulation model] K. McGuffie and A. Henderson-Sellers, *A Climate Modelling Primer*, 2nd edn. Wiley, Chichester 1997, p. 50. © Reproduced by permission of John Wiley and Sons Ltd.

Fig 9.6 [depression] S. Dunlop and F. Wilson, *Weather and Forecasting*, Chancellor, London 1983, p. 25. With the permission of © Chancellor/Philip's, London.

Figs 9.11–9.16, 9.19 (a and b) [global pressure fields, global wind fields, mid-tropospheric wind patterns, global temperature fields, mid-tropospheric temperatures, global humidity patterns], and Fig 9.19 [global precipitation patterns], with permission from the European Centre for Medium-Range Weather Forecasts.

Fig 9.17 [meteosat image northern winter] and Fig 9.18 [meteosat image northern summer] © 2000 EUMETSAT (http://www.eumetsat.de/).

Fig 12.1 [population growth] United Nations Population Division.

Fig 12.2 [rising global temperatures] reproduced by permission from *Nature*, Vol. 403: 756–7, 2000. © Macmillan Magazines Ltd.

Plate 2 [oceanic conveyor belt] David Webb and by courtesy of the Natural Environment Research Council (NERC).

Plate 3 [survival in the Arctic] with permission of the Telegraph Colour Library.

Plate 4 a and b [thermography] with permission of FLIR Systems Ltd UK.

Plate 5 [Aurora borealis] John Oldroyd. © With permission of the British Antarctic Survey, Cambridge, UK.

Plate 6 [Landsat image] with permission from the Space Sector, DRA, Farnborough, UK.

Plate 7 a and b [sea surface temperatures] and Plate 10 [global warming] reproduced by permission of the Hadley Centre for Climate Prediction and Research, UK Met Office.

Plate 9 [volcanic eruption] Science Photo Library. © I. and V. Krafft, Hoaqui.

Plate 11a [clouds (b) cumulonimbus] © Colin Harbottle [(c) Stratus] © Mike J.Dutton. [(d) Cirrus] © Alan Gair.

Plate 12 [hurricane] NOAA Photo Library, NOAA Central Library (www.photolib.noaa.gov).

Portraits

Sadi Carnot, from K. J. Laidler, *The World of Physical Chemistry*, Oxford University Press, Oxford 1995, p. 88. By permission of Oxford University Press.

Ludwig Boltzmann, reproduced with the permission of *Physics World*, Vol. 12 (4), April 1999, p. 39.

Joseph Fourier, from G. F. Simmons, *Calculus Gems: Brief Lives and Memorable Mathematics*, McGraw-Hill, New York 1992, p. 173. With permission of the McGraw-Hill Publishing Company.

Michael Faraday, from K. J. Laidler, *The World of Physical Chemistry*, Oxford University Press, Oxford 1995, p. 203. Reproduced with the permission of Oxford University Press.

Daniel Bernoulli, Johannes Kepler and Christian Doppler from H. C. Ohanian, *Physics*, 2nd edn. 1988. With permission © W. W. Norton and Co. Inc, New York.

Howard L. Penman, with the permission of Professor John Monteith.

Gro H. Brundtland, with permission from the World Health Organization/OMS.

The authors and publishers are grateful to the following companies for permission to reproduce examination questions: London Examinations (a division of EDXECEL), the Northern Examinations and Assessment Board (NEAB), Oxford and Cambridge and RSA Examinations (OCR).

The following photographs were taken by:
Peter Hughes
Fig 1.1c Urban pollution with a photochemical haze.
Fig 1.3 Population growth.
Fig 1.5 Energy efficient house. With the permission of Dr Susan Roaf.
Fig 3.1a English house.
Fig 3.4 Roof insulation.
Fig 3.12a Condensation.
Fig 3.12b Mould.
Fig 4.7 Energy efficient car.
Fig 4.10 Sewage plant.
St. Paul's Cathedral on page 84.
Fig 5.15 North German wind farm.
Fig 8.1 Mini weather station.
Fig 8.2 Stevenson screen. With permission of Frau Waltraud Gritsch.
Fig 8.3a and b Barometers.
Fig 8.6 Physicist operating a rain gauge.
Fig 8.7 Sunshine recorder. With permission of Dr Nicola de Podesta.
Fig 11.13 Rothamsted Experimental Station.
Plate 1 Components of Environmental Physics. East Devon coastline, GB.
Plate 11a Cumulus clouds over Vent in the Ötz Valley, Austrian Tirol.
Plate 13 Soil profile (with teachers and students) in a gravel pit near Stockholm, Sweden.

Nigel Mason
Fig 1.1b Sinai Desert, Egypt.
Fig 4.9 Fatehpur Sikri, India.

Roger Wesson
Fig 3.3 Canary Wharf.
Fig 3.9 Stairwell.
Fig 3.16 Building Research Establishment, Watford.
Fig 3.17 The Millennium Dome.
Fig 4.1b Modern city.
Fig 4.16 Acid rain.
Fig 4.17 Model T Ford.
Fig 5.7 Solar wall.
Fig 8.4 Cup anemometer.
Fig 8.5 Wet and dry bulb thermometer.
Fig 8.11 The Meteorological Office.

Chapter 1

Environmental physics: processes and issues

1.1 Introduction

Our Earth is a marvellous and awesome place. Since its formation 4.6 billion years ago both living and non-living entities have developed. In a global environment that is structured within the relationship between the land, the air, the oceans and the biosphere. However, to appreciate our environment it is necessary to understand the basic physical science that regulates its development.

In the past few decades the possible detrimental impact humanity is having on the planet has caused increasing concern. As humanity has sought to improve its prosperity, it has often done so by exploiting the Earth's abundant natural resources. At first such exploitation was controllable and it seemed that the Earth would provide limitless resources for human development, but with the Industrial Revolution and an increasing global population it has become apparent that an era will be reached where some natural resources are likely to be depleted. By our own actions we are changing, perhaps irrevocably, the natural balances established over billions of years.

The discovery of the ozone hole, the first signs of industrially induced global warming, the widespread phenomenon of acid rain and the growing evidence of health problems caused by urban pollution (Figure 1.1), have attracted world-wide attention from both social and political commentators. Substantial, and often heated, debates have taken place, both in the scientific and political communities, about the actual evidence for such phenomena and what actions should be taken to alleviate such impacts.

To understand and assess the possible dangers to the Earth caused by the exploitation of its resources and the development of industry, a new branch of science has evolved in the past 30 years. It is dedicated to the study of 'Environmental Issues', in which the science underpinning the regulation of the global environment is studied, and the consequences of current policies and actions are examined in relation to sustaining ecological balance on the Earth.

The environmental sciences embrace many areas of 'traditional physics and chemistry', particularly through the adoption of techniques developed in well-established fields that may be adapted for the study of specific environmental issues. Many mathematical modelling techniques developed for 'pure research' have been utilized to model environmental phenomena. For example, models of fluid dynamics have been adapted to study water flow through the soil, and atomic and molecular codes developed for physical chemistry have also been used to determine spectroscopic parameters for remote sensing within the Earth's atmosphere. Similarly, experimental apparatus designed in the laboratory to study

Figure 1.1 Modern environmental issues: (a) the ozone 'hole' observed over the Antarctic by TOMS satellite (b) global warming leading to increased desertification and (c) urban pollution, a risk to human health.

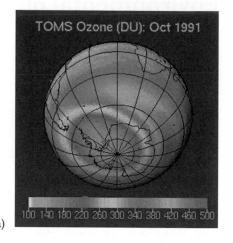

(a)

(b)

(c)

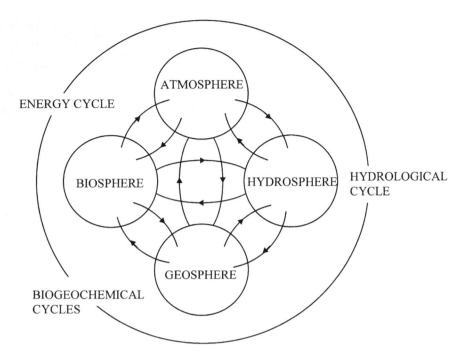

Figure 1.2 Principal interlocking themes that comprise environmental physics.

fundamental properties of matter have been adapted to make field measurements of atmospheric pollutants, the most common examples being the design of high-resolution laser systems for probing tropospheric pollution and the development of spectroscopic methods to study the Earth from satellites.

Behind all these advances has been the development of the understanding of the physics underpinning environmental processes. The aim of this book is to introduce the fundamental ideas of environmental physics and demonstrate how the laws and concepts of physics play a pivotal role in the regulation of the Earth's environment.

Environmental physics is, therefore, a broad and interdisciplinary subject (Figure 1.2) that integrates the physical processes in the following disciplines:

- The atmosphere
- The biosphere
- The hydrosphere
- The geosphere.

In this first chapter the parameters of the subject will be defined by 'setting the scene' and show how this new scientific discipline has evolved in the last few decades of the twentieth century. The environmental processes and problems involved in understanding the major issues likely to concern scientists, politicians and the general public in the twenty-first century will be presented and outlined.

1.2   The environment: the science of the twenty-first century?

The Earth is a small planet in orbit around a rather ordinary star (the Sun). It is, as far as is known, the only site in the Universe where life has evolved in all its exquisiteness and complexity. The Earth has infinite variety; in every part of the globe, even in regions with the most extreme climates, in the polar wastes or among the subterranean vents on the deepest ocean floors, life in thousands of diverse forms is found, each adapting to its own local environment, each having evolved to exploit the Earth's natural resources to flourish and grow. Yet only one life form has significantly altered the ecology of the planet and only one has evolved to such an extent that it can now dictate the future of that life form, i.e. *Homo sapiens*.

At first, humanity evolved in response to particular environmental conditions, adapting its behaviour to the world around it, surviving by hunting animals and by gathering plants. Yet even as they roamed large areas, the earliest humans were beginning to shape the development of the world in which they lived. They could anticipate when and where migrating herds could be found, and where and when particular plants would flower and provide fruit. Gradually humanity began to exploit the Earth's resources and harness them for its own needs. In clearing large tracts of land for agriculture, humanity began to change the landscape and thus fashion the environment to its wishes.

With the emergence of civilization for the first time humanity could begin to shape the world. The establishment of agriculture with its attendant fields, dams and irrigation; the building of cities and roads, canals, and railways; and the mining of the Earth's minerals have all led to the shape of the world in which we live today. Humanity's understanding and exploitation of science has led to our being able to determine the future evolution of the planet and all life forms on it. Through the understanding of nuclear physics, humanity has the ultimate capability to destroy the planet, but it also has the opportunity to use its knowledge of science not only to sustain the Earth and all its species, but also to develop its resources to the benefit of all.

1.2.1   Environmental concerns in the late twentieth century

The future of the environment and the stability and sustainability of the Earth's physical and biological systems have become major concerns only during the past 30 years. The early twentieth century was so dominated by the devastating effects of the two world wars, a global economic depression, the huge costs of reconstruction, and the spectre of nuclear war that there was little intellectual or political incentive to think about potential new threats to the global environment or to invest in studies of environmental risks, particularly as such dangers seemed likely to occur on decadal or longer time-scales.

In the 1950s the major environmental issue was perceived to be the dramatic growth of world population (which doubled between 1950 and 1987). Concern with demographic trends was not new. In 1798, the Anglican minister and political economist Thomas Malthus had written *An Essay on the Principle of Population*. He perceived that, if it was unrestricted, a population would increase geometrically, whereas food would only increase arithmetically. Malthus believed that diminishing food supplies would constrain population growth, possibly through famine, and ensure that a natural balance would be maintained. However, his warnings were largely ignored as the 'opening up' of new farming

Figure 1.3 Global population growth remains a major environmental issue.

territories throughout the world, together with the impact of the Industrial and Agricultural Revolutions, dramatically boosted world food production. Science and humanity's
unquenchable ability to improvise seemed to suggest that humanity would overcome all
obstacles. Britain was the first country to experience the *demographic transition*, which
involved the change from high birth- and death-rates in the eighteenth century to low
birth- and death-rates by the mid-twentieth century. Major factors in this transition were
the improvement in living standards and the medical advances which addressed issues of
environmental and public health. Indeed, by the mid twentieth century these advances
had facilitated the *epidemiological transition*, i.e. the eradication in many countries of those
terrible diseases which had previously generated high mortality rates.

The droughts, harvest failures and reductions in global food reserves in the 1960s came
as an unpleasant shock to many. These issues, coupled with the growing urbanization of
the world's population, led to a heightening of awareness that personal living standards
and health issues were increasingly affected by population growth.

The implications of an ever-increasing world population with finite resources and the
unrelenting drift towards urbanization are clear (Figure 1.3). In 1990, the world's population was about 5.3 billion. It was over 6 billion in 2000, and may well be in excess of 9
billion by 2030. Of the population increase, 95% will occur in the Third World. The
1990s also witnessed a move towards a global environmental agenda. This was the culmi-
nation of developments initiated in the early 1970s when the Club of Rome was founded.

In the 1970s, as a consequence of the Arab–Israeli War, the major industrialized countries were reminded of their reliance on fossil fuels, particularly oil, for the operation of
the modern state infrastructure. For the first time it was apparent that the Earth could not
provide limitless resources and that humanity would have to learn to husband its natural

resources and develop new energy sources if the world economy was to continue to be technologically driven.

As scientists investigated these new environmental issues, concern was expressed that the world's climate, hitherto regarded as essentially unchanging, might now be influenced by increasing emissions of carbon dioxide and other pollutants from global industry. These issues were discussed at the First World Conference on the Environment in Stockholm in 1972 and subsequent developments led to the organization of huge international research programmes. These included the ongoing World Climate Research Programme, the International Geosphere–Biosphere Programme, the 10-year World Ocean Circulation Programme and many smaller scale experiments to study specific environmental phenomena and mechanisms.

Thus, at the beginning of the twenty-first century there is a growing appreciation of the need to conserve the Earth's environment and to be responsible for the continuing development of the global ecosystem. Some of the present issues and problems being discussed and researched today include:

Global ecosystem:
- Changing global environment
- Earth as an ecosystem
- Biodiversity
- Deforestation
- Land degradation and soil erosion
- Desertification and drought.

Human dimension:
- Survival in various climates
- Demographic trends and the long-term effect of continuing population growth
- Urbanization
- Humanity affecting the environment, and vice versa
- Agricultural resources for human development, particularly water and food supplies.

Environmental risk:
- Pollution
- Ozone depletion
- Global warming
- Environmental risk and the quality of life
- Aerosols, particulates and ionizing radiations.

Sustainability:
- Sustainability and the future
- Limits to growth
- Developing the technology required to harness renewable energy sources
- Search for domestic and industrial energy efficiency.

To understand these issues and provide rational solutions to the problems they raise, it is necessary to understand the science that governs the Earth's environment. Mathematics, physics, chemistry, biology, geology and ecology all play a role in the environmental sciences. In this book, the focus will be on how the principles of physics underpin many of the environmental processes affecting the world today; applications of physics now often summarized as a new area of physics – *Environmental physics*.

1.3   What is environmental physics?

Environmental physics can be defined as the response of living organisms to their environment within the framework of the physics of environmental processes and issues. It is structured within the relationship between the atmosphere, the oceans (hydrosphere), land, (lithosphere), soils and vegetation (biosphere) (Plate 1). It embraces the following themes:

- Human environment and survival physics
- Built environment
- Urban environment
- Renewable energy
- Remote sensing
- Weather
- Climate and climate change
- Environmental health.

To understand how any specific environmental process evolves, it is necessary to appreciate that all these processes are interdependent. The formation and mobility of clouds, for example, illustrate just one aspect of a number of global environmental processes (Figure 1.4) and require the study of:

- Solar radiation transformations and the radiation balance
- Phase changes in the water cycle
- Monitoring physical phenomena
- Exchanges between the Earth, the oceans, the atmosphere and the biosphere
- Transport phenomena, especially mass and thermal energy transfer.

However, it is important to appreciate that the principles and laws of physics are in evidence in many different environments and govern how all species live on the Earth.

1.4   Physics in the environment

The environment may be defined as the medium in which any entity finds itself. For example, for a cloud, its environment may be the region of the atmosphere in which it is formed, while for a plant, it is the field in which it lies, and for a whale it is the sea in which it swims. Thus, it is informative to discuss environmental issues within the context of the surroundings in which an object finds itself.

1.4.1   Human environment

Living organisms have to adapt and survive in a variety of environmental conditions, including hot and cold climates. The basic necessities for the sustaining of human life, with a modest state of comfort and security requires the provision of food, clothing and shelter. Living organisms are thermodynamic entities characterized by energy flows both within the body, and between the body and its environment. For people to

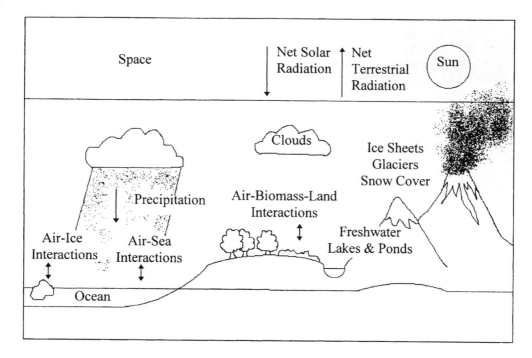

Figure 1.4 Global environmental processes.

survive, the core body temperature has to be maintained within a narrow temperature range of 35–40°C. In cold weather, without adequate insulation and heating, the core body temperature may fall to below 35°C. Hypothermia may result. In contrast, in very hot climates dehydration might render the body vulnerable to heat stress. Control of the body's temperature is maintained through the body's metabolism, which is essentially a series of energy processes within the body. The rate of these energy transfers and the mechanism of thermoregulation are governed by the following laws and concepts of Physics:

- Laws of thermodynamics
- Principles of entropy, enthalpy and the Gibbs free energy
- Principles of conduction, convection, radiation and evaporation
- Newton's law of cooling
- Stefan–Boltzmann radiation law.

Understanding these physical principles and applying the laws allows us to understand many of the mechanisms by which the human body regulates itself. They can also be applied to select the most suitable clothing to wear in particular climates and how to reduce the risk of injury and improve human health. The reason why human beings can survive is because of an energy balance and this has very important implications for the energy exchanges between houses and their environment, and the globe and its environment. Human survival, will be discussed in Chapter 2.

Figure 1.5 A modern, energy-efficient, purpose-designed house in Oxford, UK.

1.4.2 Built environment

From the earliest times, people have adapted their habitat to give shelter from the weather. Today, the designs of houses, offices, schools and factories are regulated by strict Health and Safety legislation to ensure comfort and security for the occupants. The designs of all buildings, and hence the regulation that governs them, must be compatible with the laws of physics.

Obviously, it is redundant to construct any building without allowing for the forces that it will be subjected to. The risk of strong winds, storms, floods and/or earthquakes will determine the type of materials that should be used in a building's construction and might determine its height and the depth of the foundations. However, more recently, great emphasis has been placed on the environmental consequences of building construction. The heating and lighting of buildings is one of the major energy sinks of global energy production. As the world's traditional fuel reserves are depleted, so the need to introduce energy-saving systems in buildings is emphasized. This may be achieved by the introduction of thermal insulation, which can reduce the amount of energy needed to establish comfortable living conditions. However, a fully insulated building may not be a healthy one since ventilation may be limited. Thus, physics is required not only to formulate the best construction building materials, but also to evaluate the best design incorporating insulation with ventilation and elegance (Figure 1.5).

The physics encountered in the development of the built environment includes the laws encountered in Chapter 2 and, in addition, the principles of heat pumps and heat engines and the properties of water vapour. The application of these physical principles to building design are reviewed in Chapter 3.

1.4.3   Urban environment

Humans are social and choose to live in larger groups so that they may share resources and expertise. If human beings can affect the planet, then population growth and population density will have an important bearing on an environment, especially where this interacts with ill-planned urban development. A city must provide:

- Sufficient food supplies for its population
- Sufficient energy supplies for heating and lighting
- Sufficient water supply for domestic and industrial use
- Adequate sewage to allow waste disposal
- Transport links to/from the city
- Housing to provide adequate safe shelter and comfort.

The Industrial Revolution facilitated the rise of great cities and conurbations. In 1850, there were four cities in the world with populations in excess of 1 million, this increased to 19 cities by 1900, while in 1960 there were 141. Increasingly the twenty-first century will see the emergence of the *mega-city*. For the first time in history the 1990s saw the numbers of people living in urban areas exceed those living in rural areas. In the twenty-first century the world will become increasingly urban. Many 'developed' societies now have the problems and features of post-industrial societies – intensive agriculture, expanding and polluted cities, and ageing populations. Urbanized societies are characterized by built environments with cities as exciting centres of commerce and culture, but also having stress, noise and air pollution. It took the Great London Smog of 1952, for example, which produced an excess mortality rate of 4000, to force Parliament to pass the Clean Air Act.

Thus, the growth of conurbations has immediate environmental impact, not only on the land on which they are constructed, but also on their citizens. Several of the major environmental issues under discussion today arise from urbanization, and industrial development, with its heavy use of fossil fuels and marked population growth, has generated a range of deleterious effects. Urban energy requirements have led to the construction of large power stations, the emissions from which may lead to air pollution. For example, sulphur dioxide emissions from coal-fired power stations lead to the formation of acid rain, which is detrimental to the fabric of urban buildings, the quality of crop yields and the health of forests in the surrounding countryside.

Much urban air pollution is due to the transportation of people and goods. Nitrogen oxide, carbon monoxide and particulate matter emissions from car engines are harmful to human health and in the most severe cases can lead to increased urban mortality rate levels. The removal of such air pollutants, the monitoring of pollution levels, and the design of less polluting vehicles and industrial plants all require a basic understanding of the underlying science. Chapter 4 discusses the urban environment, its problems and possible solutions; Chapter 5 describes the options for the new energy sources that may provide renewable and clean energy for the cities of the future. A crucial factor, encompassing demographic and urban growth, that determines global economic and industrial development, is how global energy requirements are to be met. The exploration of improved domestic and industrial energy efficiency is involving physicists and engineers in the development of the technology required to exploit such renewable energy sources as solar energy, wind power, and tidal and wave power. Already, for example, hydrogen fuel cells and photovoltaics are at an advanced stage of development. How economies plan to evolve, and in particular develop urban transport policies, will have critical implications for their energy requirements.

1.4.4 Global environment

Environmental issues are not just local or even national in character. Many of the major environmental issues facing the world today have global consequences. The risk of an emerging hole in the Earth's ozone layer, allowing biologically harmful solar ultraviolet radiation to reach the Earth's surface, cannot be solved by individual citizens or even by national action. Ozone depletion is caused by global emissions of chemicals, such as the chlorofluorocarbons (CFCs), and can only be tackled by global agreement.

The prospect of a warming of the Earth's surface temperature, through global warming, and subsequent changes in the Earth's climate arises from global emissions of carbon dioxide (and other 'greenhouse gases') caused by the burning of fossil fuels, mainly by industrialized countries. Thus, although the consequences may seriously affect only limited regions of the world (e.g. desertification in North Africa or increased flooding in Bangladesh), the solution often does not lie in the country being most affected but depends on the global community taking action to change its current practices.

In the latter part of the twentieth century it had been recognized that to understand global issues it is necessary to study the global environment. This has been immensely aided by the rapid development of technology and, in particular, through the birth of the 'Space Age'. It is now possible to study nearly all major environmental processes from space satellites, through the science of remote sensing (Chapter 6). The use and application of such technology requires an understanding of:

- Newton's laws of motion and gravitation
- Kepler's laws of planetary motion
- Physical principles of circular motion
- Principles of radar and the Doppler effect
- Application of optical imaging and spectroscopic techniques
- Development of high-performance scientific computing.

The environmental processes that form the fabric of environmental physics are mediated through the integration of the four components of the land, oceans, air and biosphere. The surface of the globe consists of the upper part of the geosphere (the lithosphere), the hydrosphere (which includes the clouds and water vapour that envelop the Earth) and the biosphere. Seen from space, the Earth's surface is covered with about 70% water, and the oceans, as part of the hydrological cycle, play an extremely important role in the development and maintenance of the global climate, through the transport of mass and thermal energy (Plate 2). Those parts of the land-masses consisting of deserts and the polar regions of the cryosphere also have a role in this energy distribution. In addition, the lithosphere,

Figure 1.6 Satellite image of a hurricane over the Caribbean.

which consists of the rocks and soils that comprise the Earth's surface, has a crucial role in biogeochemical cycles and in the migration of the tectonic plates that cause earthquakes and catalyse the volcanic activity that results in the atmospheric dispersion of particulates that can affect the weather.

The structure and composition of the atmosphere reveal its thermal nature and it too is essential for the conditions for life. The lowest layer of the atmosphere, the troposphere, is where most of the 'weather' occurs, and the impact of solar energy on the Earth and the atmosphere, together with the effect of a rotating Earth, generate the characteristic wind patterns responsible for the different global weather systems, such as the Pacific's El Niño, the North Atlantic circulation and the development of hurricanes (Figure 1.6). The complexities of the global climate system are mediated through coupling systems, such as that between the atmosphere and the oceans. Weather is concerned with the daily fluctuation of a number of parameters – rainfall, solar insolation, pressure, temperature, wind, humidity and cloud cover. Climate is the average weather conditions that prevail in a given geographical locality over a long period.

Life on Earth is tenuous; if the Earth had been a little further away from the Sun's surface, temperatures would be too low to sustain life as we know it (since water would remain frozen), while if the Earth were a little closer to the Sun's surface, temperatures would be too high for water to have formed the oceans in which life originally evolved. The Sun, and the Earth's proximity to it, therefore controls much of the global environment. Radiation from the Sun sustains the Earth's surface temperature and leads to the production of clouds that carry the salt-free water needed by most plants and animals to live and grow. Energy from the Sun establishes the dynamics within the atmosphere that leads to winds that transport the clouds. Radiation reaching the Earth's surface is used by plants in photosynthesis to produce the oxygen that all mammalian life needs to exist and which has, in turn, led to the establishment of an atmosphere around the planet.

Therefore, if we are to understand global environmental processes it is necessary to understand how solar energy drives many of the processes in the atmosphere. An understanding is required of:

- Laws of radiation and thermodynamics
- Physical process of evaporation and the principle of latent heat
- Properties of ice and water vapour
- Physical concepts of humidity and saturation vapour pressure.

Figure 1.7 A thunderstorm: the physical laws of electricity and cloud physics are required to understand this dramatic weather event.

In addition, an understanding of the following is required:

- Properties of fluids and the physical laws of fluid dynamics
- Principles and laws of electricity (e.g. involved in thunderstorms) (Figure 1.7) and magnetism (e.g. involved in the aurorae).

The role of solar radiation on the Earth's surface and atmosphere is examined in Chapter 7, where the major environmental issues of global warming and the seasonal depletion of the ozone layer over the Antarctic – the ozone hole – are also discussed.

As humanity has studied global atmospheric processes, so it has sought to predict their future consequences. On a daily basis, an understanding of the dynamics within the Earth's atmosphere has allowed the prediction of the weather across the globe to the benefit of all its citizens, while an understanding of how global weather patterns are influenced by changes in solar irradiation, as part of annual cycles within a geographical framework, have had dramatic consequences for the development of world-wide agriculture. Chapters 8 and 9 discuss the fundamental physics and concepts necessary for understanding the Earth's weather and climate, and describe how these are monitored and how they provide predictions.

Nowhere is the public interest in the weather greater than when discussing the causes and consequences of climate change. Looking back over geological time it is evident that climatic variations have been marked, and it has been suggested that climate change may have been instrumental in species mass extinctions and in the demise of some ancient civilizations. Palaeoclimatology provides exciting clues to past climates through analysis using ice cores (such as the Greenland ice core, which is the longest at 3000 m and which represents a timescale of about 500 000 years), oxygen isotopic ratios, pollen evidence and tree rings. The Croll–Milankovitch astronomical theory of ice ages also provides a means of explaining how climate change might occur. What is still being debated at the outset of the twenty-first century is the extent to which anthropogenic inputs, through the increasing emissions of greenhouse gases, are facilitating the enhanced greenhouse effect, and are possibly contributing to rising global temperatures.

The environmental physics research that is now taking place around the world is at the frontier of knowledge. Mathematics and physics lie at the heart of environmental modelling, and the development of the mathematical basis for climate modelling and weather forecasting provides a marvellous example of how increasingly sophisticated methods have evolved; from the numerical methods of L. F. Richardson through Global Circulation Models (which embrace the physics of thermodynamics and fluid dynamics) to the state-of-the-art ensemble forecasting which uses chaos theory.

1.4.5   Biological environment

Ultimately, the future of the Earth's environment lies with its citizens and how they respond to the environmental challenges facing them. The greatest problem facing the globe is the ever-increasing number of people seeking to exploit and farm its resources. Thus, Chapters 10 and 11 explore the soil–water–vegetation–atmosphere continuum with emphasis on the physics that underpins the sources of food and water that humanity needs to survive.

Central to this is again the energy from the Sun. The key process is photosynthesis, the means by which carbon dioxide and water combine, under the mediation of sunlight and the chlorophyll in plants, to produce carbohydrates and oxygen. The efficiency mechanism by which solar radiation is absorbed by any plant is similar to that by which solar radiation is transported through the Earth's atmosphere, and can be studied using spectroscopic methods.

Interconnections lie at the heart of environmental physics. One can see this even in the origins of agriculture. It is not coincidence that the Neolithic Revolution, which saw the organization of the cultivation of crops and domestication of animals, was accompanied by the climate change prevalent at the time of the recession of the last glacial ice-sheets approximately 10 000 years ago. Reference to physics principles can explain why certain types of plant grow in certain ways, why some have long, tall leaves, while others have small leaves spread over a wide area. For crops to thrive, the nature of the soil in which they are planted is paramount and this is reflected in the underlying geology and the biogeochemical cycles that provide the sources, sinks, routes and feedbacks for such elements as nitrogen, carbon, sulphur and phosphorus. In linking with the lithosphere (which includes besides the rocks and soils the physico-chemical interactions that provide nutrients for plants) and the biosphere they generate the conditions for germination and subsequent plant development. The rate at which solar energy is transported through the soil can be explained by considering the principles of thermodynamics.

Physics also explains the means by which water transport allows vegetation to thrive. The percolation of water from the soil to the plant, through it and out from it can only occur when water rises due to a difference in water potential. Physical principles, involving gravity, capillarity and osmosis, can therefore explain how water is drawn through the soil into the plants and how it is released once again into the atmosphere through evaporation and transpiration from the leaves.

There is a widespread feeling that there is a limit to global growth and that there is a global-carrying capacity for food and other utilities. Nevertheless, as the global population increases a major question arises: how can more land, and the oceans, be brought into production, and how can agricultural productivity be increased without destroying the very soil that generates this production?

1.5   Environmental physics and the global environmental agenda

Gradually, concern with local, national and global environments has grown, and the consequences of many environmental issues can no longer be ignored. The concerns of local environmental groups have coalesced into global commitments and these have become a conspicuous part of national and international political agendas. There has been an undeniable growth in environmental consciousness, and the implication that this might have for the relationship between our environment and the quality of life that we wish to enjoy and pass on to our children. It shall be shown in Chapter 12 that at the international level action is being taken as the gathering momentum of global involvement with environmental issues becomes plainly visible.

A widespread sense of environmental crisis, and the problems that generate them, also present challenges. Physics is concerned with providing solutions to problems, and because it is a science, if not *the* science, it has formulated the means and the power to predict and facilitate the transformation of our world. We need to be aware, for example, if one continues to pollute the planet what the consequences will be. We need to know what contingencies have to be made in the event of the depletion of water supplies, food stocks and sources of energy. Through measurement, monitoring and analysis, and by reducing the uncertainty in modelling, environmental physics provides an understanding of environmental processes, especially where global environmental change is concerned, and this has implications for environmental impact, risk and science policy. Environmental physics can therefore make a contribution to how certain features of our society may be organized.

1.6   Summary

This introductory chapter has provided an overview of the nature, scope and themes of environmental physics. In each of the following chapters the applications of the principles of physics to environmental processes and problems will be discussed and put in the context of current environmental issues.

References

Environmental physics

Boeker, E. and van Grondelle, R., *Environmental Physics*. Chichester: Wiley, 1995.

Guyot, G., *Physics of the Environment and Climate*. Chichester: Wiley, 1998.

Monteith, J. L. and Unsworth, M. L., *Principles of Environmental Physics*, 2nd edn. London: Arnold, 1990.

Environment and development

Briggs, D., Smithson, P., Addison, K. and Atkinson, K., *Fundamentals of the Physical Environment*, 2nd edn. London: Routledge, 1997.

Demeny, P. and McNicoll, G., eds, *Population and Development*. London: Earthscan, 1998.

McMichael, A. J., *Planetary Overload: Global Environmental Change and the Health of the Human Species*. Cambridge: Cambridge University Press, 1995.

Park, C., *The Environment: Principles and Applications*. London: Routledge, 1997.

Pickering, K. T. and Owen, L. A., *An Introduction to Global Environmental Issues*, 2nd edn. London: Routledge, 1997.

Roberts, N., ed., *The Changing Global Environment*. Oxford: Blackwell, 1994.

Turner, B. L., Clark, W. C., Kates, R. W., Richards, J. F., Mathews, J. T. and Meyer, W. B., eds, *The Earth as Transformed by Human Action: Global and Regional Changes in the Biosphere over the Past 300 years*. Cambridge: Cambridge University Press, 1993.

UN World Commission on Environment and Development, *Our Common Future: the Brundtland Report*. Oxford: Oxford University Press, 1987.

White, I. D., Mottershead, D. N. and Harrison, S. J., *Environmental Systems*. London: Chapman and Hall, 1994.

Chapter 2

The human environment

2.1 Introduction

Living organisms survive, indeed often adapt to and thrive in, a multitude of different habitats: people in urban areas, plants on mountains and in the desert, and animals in the air, the oceans or sometimes in the world's most inhospitable regions. Human beings have managed to live in all the different environments present throughout the Earth: from the wastes of the Arctic (Plate 3) to the deserts of Mongolia, from the jungles of Africa to the coral islands of the Pacific.

At the macroscopic level the Earth is a thermodynamic entity. Life is sustained on it because of the temperature gradient maintained by the net energy balance between the energy that comes from the Sun and the energy that leaves the Earth. At the microscopic level, i.e. at the level of the individual person or animal, living organisms are also thermodynamic structures exhibiting energy exchanges between themselves and the environment.

Mammals, including humans, have the remarkable ability to maintain a constant body temperature, in spite of dramatic changes in environmental conditions. They are called *homeotherms*. They sustain their body temperatures by adjusting the rate of energy transfer and energy production. In birds, this is achieved through the use of feathers; in some animals through fur and fleece insulation.

In contrast, certain animal species, such as reptiles and amphibians, have core body temperatures that respond to environmental temperatures. Such animals are called *poikilotherms*. Both homeotherms and poikilotherms respond to conditions in a variety of physiological and behavioural mechanisms. In cold weather we put on 'warmer' clothing, while bears have fur. In hot weather we wear thinner clothing; animals with fur may pant and hippopotami will lie in rivers to dissipate excess heat. The transfer of heat and the maintenance of a stable equilibrium within any animal body, through biological processes, may be understood by the study of the laws of physics.

Planet Earth provides many environmental and ecological contexts for living things to survive and develop. For life to be sustained we should not only be concerned with the chemistry and biochemistry of metabolic reactions, but also with the physics of thermal processes. This chapter will focus primarily on the physical principles underlying the energy transfers to and from human beings in a variety of environmental contexts. In examining these principles it is necessary to discuss the First and Second Laws of Thermodynamics to see how they apply to the body's energy metabolism. The mechanisms by which thermal energy can be transferred from a human being and an energy balance maintained will then be examined and how these principles apply to survival in cold and hot environments will be investigated.

2.2   Laws of Thermodynamics

We will first review the Laws of Thermodynamics and then see how the thermal processes they describe relate to the human body.

2.2.1   First Law of Thermodynamics

The general formulation of the First Law of Thermodynamics for an ideal gas is that

$$dQ = dU + dW \qquad (2.1)$$

where dQ is the energy supplied to or extracted from a closed system, dU is the change in the internal energy of the system and dW is the work done by the system against external forces $(+)$ or done on the system by external forces $(-)$.

The First Law is an expression of the *principle of the conservation of energy*, and the *internal energy* refers to the total kinetic energy of all the atoms and molecules comprising the gas. Several great scientists were involved in the development of the First Law, including the Salford brewmaster J. P. Joule (1818–89) and the Bavarian physician and physicist J. R. Mayer (1814–78) (see Chapter 11).

Another idea useful at this stage is that of enthalpy. *Enthalpy* (H) is the heat content of a system and is a *thermodynamic state function*, which is related to U, P and V in the form:

$$H = U + PV, \qquad (2.2)$$

where PV is the product of the pressure of a system and its volume. A thermodynamic state function is characteristic and descriptive of the thermodynamic state of a system. Examples include internal energy (U), temperature (T) and entropy (S). For such functions the pathway is not relevant, just the initial and final states.

Often it is more useful to speak of the enthalpy change, dH, of a chemical reaction. In the situation where no external work is achieved, $dW = 0$. Thus, $dH = dQ$. This enthalpy change can be assessed by the amount of energy generated (or absorbed) in a reaction. For this reason, it is sometimes referred to as the '*heat of reaction*'. By convention, a reaction in which there is an absorption of energy implies that dH is positive, and if energy is dissipated, then dH is negative. The two reactions are called *endothermic* and *exothermic* respectively.

2.2.2   Second Law of Thermodynamics

An internal combustion engine and the human body have similarities in that they function as *heat engines*. A heat engine is a means of extracting useful mechanical work from a system with a temperature difference between its interior and it environment. The heat engine is, therefore, a useful analogy for our bodies. The operation of any heat engine is governed by the Second Law of Thermodynamics, originally stated by the French physicist Sadi Carnot. Carnot proposed that in a heat engine the 'motive power', or work done, by a system is obtained from the energy transferred between one body at a higher temperature to another at a lower temperature or, as the Prussian physicist R. Clausius suggested in 1850, that energy flows from a higher to a lower level, and that it cannot of itself go in the opposite direction unless acted upon by an external agency. Mathematically, it is often expressed in terms of efficiency:

$$\text{Efficiency, } \eta = \frac{\text{power}_{out}}{\text{power}_{in}} \times 100\%$$

$$= \left(\frac{T_1 - T_2}{T_1}\right) \times 100\% \tag{2.3}$$

where T_1 is the higher temperature and T_2 is the lower temperature. Expressed in this way, the Second Law argues that in a cycle of operations, such as in an internal combustion engine, there will be a limit on the efficiency of the process.

Equation (2.3) may be applied to internal combustion engines, power stations and the human body since the body can do work by virtue of the temperature difference that exists between it and its environment. The importance of the Second Law is that it defines the *direction* in which thermal energy will flow.

Sadi Carnot (1796–1832). Son of the French revolutionary general Lazare Carnot, Sadi became a military engineer in the French Army and then a research scientist. He became interested in thermal processes and engines, and in his magnus opus *Reflexions sur la Puissance Motrice du Feu et sur les Machines propres a developper cette Puissance* (1824) he discussed the efficiency of heat engines. He died in the cholera pandemic of 1832 at the early age of 36.

2.2.3   Entropy and the Third Law of Thermodynamics

If a cup of tea at 60°C is left in a room at 20°C, it will gradually cool. The temperature of the tea will decrease from a higher to a lower level. Without any external input, it is not possible for its temperature to rise. That is, the process is *irreversible*. This is a simple example of the Second Law of Thermodynamics. Similarly, for a human, without the external agency of food as a source of chemical energy and the impact of solar radiation, the body's temperature would fall, and with starvation, death would result. The temperature difference between our bodies and the local environment not only sustains us, but also allows us to produce useful mechanical work. Since the temperature of the body is usually greater than that of the surroundings, energy flows out of the body into the environment.

The process is irreversible, and the environment gains energy (Q) at this environmental temperature (T). This provides us with a definition of *entropy* (S),

$$S = Q/T \tag{2.4}$$

Therefore, for an energy change dQ there will be an entropy gain (dS) of dQ/T implying that dS is the entropy change when energy is added to a system at a temperature T. Thus, if one takes the body and its surroundings (i.e. the Universe), there will be an entropy gain, dS, in the entire system. This can be expressed in the form $dS_{body} + dS_{environment} > 0$ and it implies that even if dS_{body} decreases, a more positive entropy gain by its surroundings will result in net entropy gain in the Universe.

Ludwig Boltzmann also defined entropy in terms of the probability (W) of the number of ways in which energy distributions can be generated:

$$S = k \ln W \tag{2.5}$$

where k is Boltzmann's constant $= 1.38 \times 10^{-23} \, \mathrm{J \, K^{-1}}$.

W tells us that the probability of obtaining certain outcomes in a particular energy distribution depends on the number of ways it can be distributed. There is a direct relationship between probability (W) and 'disorder', and this implies that as W increases, disorder increases. For more details on entropy, see Appendix 1.

Ludwig Boltzmann (1844–1906). An Austrian mathematical physicist who held professorships in Austria and Germany. He is considered the father of the branch of physics called Statistical Mechanics. He amplified Maxwell's work on the distribution of molecular velocities and made major contributions to the kinetic theory and the Second Law of Thermodynamics, which he approached from a statistical viewpoint. The inscription on his Viennese tombstone reads: $S = k Ln \Omega$

Boltzmann's equation helps one understand why the entropy increases in a system when it is warmed. For example, the phase changes from solid to liquid to gas are accompanied by entropy increases. At absolute zero, the entropy is zero, since all the atoms are fixed in the crystal lattice and not oscillating with simple harmonic motion. As the temperature of the material increases, the 'disorder' increases. The absolute entropy (S) of a system can be determined if use is made of the *Third Law of Thermodynamics*, which assumes that at absolute zero ($0 \, \mathrm{K}$), entropy is zero. By summing the incremental changes of the energy supplied (dQ) to the system for incremental changes in temperature, between absolute zero and the temperature under consideration the total entropy of a system can be found. The absolute entropies of 1 mole of pure water as ice at $0 \, °\mathrm{C}$ and as liquid at $0 \, °\mathrm{C}$ are 41 and $63 \, \mathrm{J \, K^{-1}}$ respectively.

However, this approach is time-consuming and unnecessary because it is easier to define entropy in terms of the entropy change, $dS = dQ/T$. The entropy change accompanying the phase change (i.e. from solid to liquid) of 1 mole of ice into water, under *isothermal* conditions (i.e. under conditions of constant temperature), is then $22\,J\,K^{-1}$; the difference between the absolute entropies of water in the two states.

This is precisely the quantity of energy (dQ), extracted from the surroundings, that brings about the change of phase. Since $dQ = mL_f$, where m is the mass of 1 mole of water $= 0.018\,kg$, and L_f is the latent heat of fusion $= 336\,000\,J\,kg^{-1}$, then the entropy change will be

$$dS = dQ/T = 0.018 \times 336\,000/273 = 22\,J\,K^{-1}$$

In a physical sense, entropy is, therefore, a measure of the 'disorder' of a system. Most natural processes, like the cooling tea or the decreasing radioactivity resulting from a radioactive source, are irreversible. If a process goes in its 'normal' manner, the entropy of the system increases; if it proceeds in the opposite direction, the entropy decreases.

2.3   Laws of Thermodynamics and the human body

The Second Law governs changes that act in the direction in which entropy increases. We will now see through a more detailed examination how the laws of thermodynamics relate to the energetics of the body.

2.3.1   Energy and metabolism

Metabolism is the total of all the chemical processes that occur in the cells of a body. It consists of *anabolism* in which molecules are built-up and *catabolism* in which enzymes break down the food consumed through hydrolysis, and at the cellular level involves the process of *phosphorolysis*.

The *basal metabolic rate* (BMR) is the rate at which a fasting, sedentary body generates sufficient energy to achieve the vital functions of respiration, maintaining the body's temperature, the heart beat and production of tissue. BMR is approximately equal to the metabolic rate while sleeping, and while resting most of the energy is dissipated as thermal energy. BMR can be calculated using direct calorimetry or by use of a spirometer, which measures the oxygen consumption per unit time. In the calorimetric method, a person is placed in a chamber through which there are pipes carrying water (Figure 2.1a). The amount of energy produced can be determined from the energy gained by the water passing through the pipes. In spirometry (or indirect calorimetry), the energy generated is related to the amount of oxygen taken in during respiration (Figure 2.1b), and thus the metabolic rate measured.

For a man, BMR is about $170\,kJ\,m^{-2}\,h^{-1}$, and is $155\,kJ\,m^{-2}\,h^{-1}$ for women. Thus, for a man of about $1.8\,m^2$ surface area, this would make $7200\,kJ\,day^{-1}$ or about $80\,W$. During the day, in addition to the basal requirements, energy will be required for mechanical work and physical exercise. Typical energy dissipations are:

- Sleeping: $75\,W$
- Sitting: $80–100\,W$

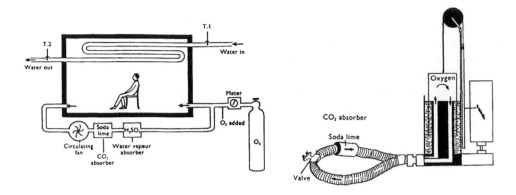

Figure 2.1 Measurement of the basal metabolic rate using (a) direct calorimetry and (b) spirometry.

- Walking: 150–450 W
- Running hard: 400–1500 W.

The average person needs an additional 4200 kJ for a 'normal' working day; thus making a total requirement of about 12 000 kJ day^{-1}. Since carbohydrates provide about 17 kJ g^{-1}, proteins 38 kJ g^{-1} and fats 17 kJ g^{-1}, by adjusting the various amounts this figure can be attained. A Mars bar, for example, provides 1225 kJ (comprised of 2.7 g protein, 44.7 g of carbohydrate, 11.3 g fat).

Metabolism involves the chemical processes in the body in which energy is transferred between various chemical compounds and in which thermal energy is generated. If the rate of metabolic reactions increases, then the rate of energy generation also increases. People require certain amounts of energy to achieve certain tasks. This has implications, for example, for athletic performance and survival. A sedentary man can produce energy of the order of 0.07 kJ kg min^{-1} which is about 80 W. In contrast, a marathon runner would generate about 1.1 kJ kg min^{-1}.

2.3.2 Thermodynamics and the human body

Humans breathe in oxygen and eat food, which is composed of carbohydrates, fats, oils and proteins. The carbohydrates are converted into glucose, the proteins into amino acids, and the fats into fatty acids. The blood then transports these, together with oxygen, to the cells, where enzymes, which are biological catalysts, convert the glucose into pyruvic acid, through the process of glycolysis. The fatty and most of the amino acids are converted into acetoacetic acid. These are changed into acetyl Co-A, and with further oxidation, produce adenosine triphosphate (ATP), carbon dioxide and water. This entire process is called the *Krebs Cycle*.

ATP generates the energy that could be potentially used by the cells. The energy is stored in the phosphate bond when adenosine diphosphate (ADP) is transformed to adenosine triphosphate, and is dissipated when ATP is converted into ADP. When the energy is released it takes the form of heat, and this is transported by the blood,

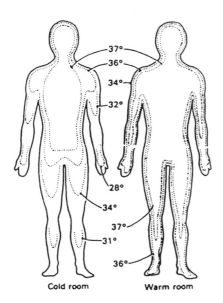

37°
36°
34°
32°
28°
34°
37°
31°
36°

Cold room   Warm room

Figure 2.2 Internal cross-sectional temperature distribution of the body.

around the body. Energy is also transferred from the cells to their surroundings by conduction because of the thermal gradient created between the cells and their environment.

Thermal energy loss from the body is achieved through conduction, convection, radiation and evaporation (from the skin and through respiration). In humans energy is transferred to the surroundings at the skin's interface with the air outside. Since cooling results, this implies that a temperature gradient exists between the body's core and the skin's surface. This body temperature is stable as long as the production of energy equals the energy loss.

Living organisms are also thermodynamic entities, in which thermal processes are characterized by energy flows and fluxes both within the body, and between the body and its environment. For people to survive, the core body temperature has to be maintained within a narrow temperature range of 35–40°C. The normal body temperature is 37°C. However, this is the core temperature, and as Figure 2.2 shows, there is a temperature gradient as one moves away from the core. Hence, not only is there a temperature drop between a person and their external environment, but also there is one within the body. Figure 2.2 also illustrates the temperature distribution across various parts of the body.

What is the relevance of physics in a discussion on energy and metabolism? Physics underpins the biochemical processes that provide us with energy. Although this chapter is not concerned with biochemical processes, we will look at how physics, through the First and Second Laws of Thermodynamics, relates to metabolic processes.

2.3.3 *First Law of Thermodynamics and the human body*

For an energy balance, under steady-state conditions where the core body temperature and the ambient temperature remain constant, the quantity of energy produced will equal the quantity of energy dissipated. Hence, it is possible to invoke the First Law of Thermodynamics to the body.

The total energy produced in the body is called the metabolic rate (dM). It is related to the total metabolic energy production of the body (dH), and the external work done by the body (dW), by the expression:

$$dM = dH + dW \qquad (2.6)$$

There is an obvious analogy if this is compared with the expression for the First Law (equation 2.1). dH, measured in watts per square metre ($W\,m^{-2}$), varies from one person to another, and depends on the activity engaged in, and on, the body's surface area. On average, the body's surface area is about $1.84\,m^2$, the average male mass is 65–70 kg and the average female mass 55 kg. For a sedentary (i.e. sitting) person the metabolic rate is about 100 W, and is 400 W for a person engaged in heavy physical work. Energy transfers in metabolic processes are governed by the First Law of Thermodynamics, and the law can be applied to determine the quantity of energy that can be generated. If no mechanical work is done (dW = 0), then the chemical energy input will be transferred as thermal energy, i.e. dH = dU is the energy produced by the oxidation of the chemicals and the total dH is the mass of the chemicals oxidized (kg) × dH ($kJ\,kg^{-1}$).

2.3.4   Second Law of Thermodynamics and the Gibbs free energy

In applying the First Law of Thermodynamics and looking at changes in enthalpy the mechanism or metabolic pathway does not have to be included, just the initial and final states, but when looking at the Second Law of Thermodynamics one has to address the question of *directionality*. Are all chemical reactions reversible?

If a metabolic process occurs in a particular direction, does it also occur in the reverse manner? The Second Law helps to explain both the direction and attainment of equilibrium in metabolic processes, and now it can be seen that entropy and entropy change can assist in the understanding of the direction that a metabolic process will take. It also tells us whether that particular process will occur.

In the oxidation of glucose a certain amount of energy is 'wasted'. Thus, the process is not 100% efficient. The 'waste' is the production of energy as heat – a prerequisite for maintaining the core body temperature. This 'wasted' energy is the driving force for the direction in which a metabolic process should go. The idea of potential energy is a useful starting point for trying to predict whether a biochemical reaction will occur and in which direction it will occur. If one drops a mass, its potential energy is transformed into kinetic energy and then into heat, sound and, possibly, light. As a result, the entropy of the surroundings (i.e. the Universe) will increase. The change in entropy is a function of the energy transferred from the body. Thus, equation (2.4) can be applied in the form:

$$dS_{environment} = -dQ_{body}/T$$

where the conditions, at the cellular level, are assumed to be isothermal. This equation implies that if energy is lost from a body, indicated by the negative sign, then $dS_{environment}$ will increase.

If the entropy S tells us the direction of a spontaneous change, it would be useful to develop the criteria, from energetic considerations, for the propensity of a system to provide 'free energy' to do useful work. The criteria is provided by the idea of the *Gibbs free energy* (G), named after the American mathematical physicist Josiah Willard Gibbs (1839–1903).

Since the First Law of Thermodynamics can be represented as:

$$dQ = dU + P.dV$$

where $P.dV$ is the product of the pressure and the change in volume, and the Second Law by:

$$dS = dQ/T$$

then

$$dQ = T.dS$$

where dS is the change in entropy related to a change in energy, dQ. Thus

$$T.dS = dU + P.dV$$

Therefore, the change in internal energy is

$$dU = T.dS - P.dV \tag{2.7}$$

This is the *Gibbs equation* (1873) and it is important equation because, by incorporating the idea of temperature, it embraces the *Zeroth law*, and the First and Second laws of Thermodynamics. The *Zeroth law*, suggested by Maxwell (1872) as the 'law of equal temperatures', says that if two bodies are in thermal equilibrium with a third body, then they are in thermal equilibrium with one another. The law provides a working reference for the idea of temperature, and it implies that temperature is a central characteristic of a thermodynamic system. It can be applied to physical systems, and it shall be shown that it can also be applied to the biophysical system that is the human body.

Using the definition of *enthalpy* (equation 2.2) $H = U + PV$ then

$$dH = dU + P.dV + V.dP$$

From equation (2.7):

$$dH = T.dS + V.dP$$

Now, the Gibbs free energy, G, is defined as

$$G = U - TS + PV \tag{2.8}$$

Thus, the change in the Gibbs free energy (dG)

$$dG = dH - T.dS \tag{2.9}$$

Equation (2.9) defines the maximum possibility of a process achieving work. In the equation dH is energy, while $T.dS$ is not, because entropy is not conserved, i.e. there is not one single sum of entropy. The requirement is that it has to increase. G is not free energy in the sense that it comes from nothing. It implies that it is the energy *available* for work. dG influences the possible direction of a metabolic process. If it is negative, then free energy is released and the process will occur. If it is positive, it will not. This will be referred to in the discussion on photosynthesis in Chapter 11.

2.3.4.1   Gibbs free energy and the human body

How can this physical concept be applied to biochemical reactions? The Gibbs free energy is a function of potential energy (PE) and gives an indication of the direction of a physical change or a chemical change that underlies a metabolic reaction. It also reveals the maximum quantity of work that can be made available from a given system. In chemical reactions, dG can be affected by factors, which include:

- Types of chemical compound involved
- Concentrations of the compounds
- Quantity, expressed as the number of moles, n, of chemicals in the reaction.

These can be combined in the form:

$$dG = n.dG^0 + nRT \ln \frac{\pi c_p}{\pi c_R}$$ (2.10)

where π is the sum of the products ($c_1 \times c_2 \times \ldots$), c_p is the concentration of products, c_R is the concentration of reactants and dG^0 is the change in free energy when 1 mole of chemical reactant is converted into a product under specific standard conditions.

If $dG = 0$, it implies that the system is in equilibrium, i.e. $A \leftrightarrow B$. Both the magnitude and sign of dG depends on dG^0 and the molar concentrations of chemical reactants and products.

How can this be linked to a metabolic reaction? As an example, consider the hydrolysis of ATP to ADP:

$$ATP + H_2O \rightarrow ADP + P_i$$

At the cellular level, where P_i is an inorganic phosphate the concentrations are approximately ATP 10 mM, ADP 0.1 mM, P_i 10 mM (mM is a milli Mole) and water 55 M. This generates a value of $dG = -55 \, kJ \, mol^{-1}$. This tells us that the reaction will occur.

One can see here how relevant is an appreciation of how the physical principles of thermodynamics can be applied to biological processes within the human body. The discussion shall be broadened to an examination of the transfer of energy between that body and its environment.

2.4   Energy transfer

Why is energy transfer important? An appreciation of thermal processes can help make sensible decisions about how buildings can be efficiently heated (Chapter 3). It can help us to understand why the Earth's average surface temperature is 15°C, and in determining how the human body regulates its temperature.

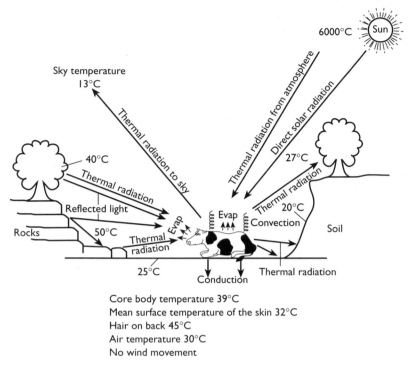

Figure 2.3 Energy transfers between an animal and its environment.

To feel warm, whether one is in the house or walking outside, is a question of energy conservation, but the underlying principle is that of an *energy balance*, and for this to be achieved energy exchange is necessary. Figure 2.3 shows the various modes of energy exchange between a homeothermic animal and its local environment. Energy can be transferred from one point to another by the following mechanisms: conduction, convection, radiation and evaporation. The physics of each of these mechanisms will be discussed in turn.

2.4.1   Conduction

Thermal conduction is the process by which energy can be transferred between two points in a material at different temperatures. In solids this is achieved in two ways: (a) through molecular vibrations transferring energy through the crystal lattice, and (b) through the mobility of free conduction electrons throughout the lattice.

In semiconductors both components contribute to energy transfer because there are less free electrons but in insulators (a) predominates. These lattice vibrations, which are called *phonons*, generate elastic acoustic standing waves which pass through the material at the speed of sound for that material.

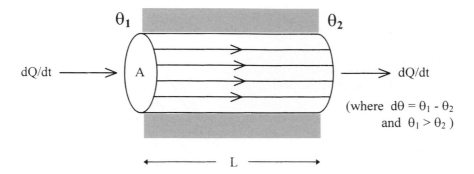

Figure 2.4 Thermal energy flowing through a lagged rod.

The French mathematician J. Fourier had discovered in 1822 that the rate of flow of thermal energy (Q/t) through a material depends on the cross-sectional area (A), the length or thickness of the material (L), and the difference in temperature between the two sides ($d\theta = \theta_1 - \theta_2$) (Figure 2.4).

This can be expressed as:

$$dQ/dt \propto A$$
$$\propto d\theta/L$$

Therefore

$$dQ/dt = -kA.d\theta/L \qquad (2.11)$$

Equation (2.11) is *Fourier's law of thermal conduction*. k is the *thermal conductivity*, defined as the rate of thermal energy flow per unit area of a material per unit temperature gradient (it has units $W\,m^{-1}\,K^{-1}$). The effectiveness of a material as an insulator can be determined

Joseph Fourier (1768–1830). Having attended a military school in France, Fourier became a mathematics teacher at 16. He subsequently accompanied Napoleon Buonaparte's expedition to Egypt in 1798 and was appointed Governor of Lower Egypt. In 1816 he succeeded Laplace as Professor at the Polytechnic Institute in Paris. His later investigations included the mathematics underpinning thermal conduction in solids. He also studied waveforms and this provided the basis of Fourier analysis, a technique widely used in telecommunications and which involves the expression of a signal as a series of sine and cosine functions.

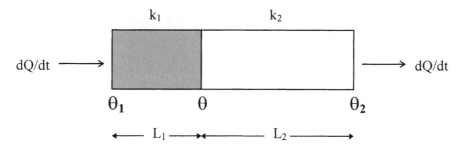

Figure 2.5 Energy flow in a composite slab.

by measuring its thermal conductivity. Good thermal conductors like copper have a high thermal conductivity (e.g. $380\,\mathrm{W\,m^{-1}K^{-1}}$), while poor conductors, like water, have a low thermal conductivity (i.e. $0.59\,\mathrm{W\,m^{-1}K^{-1}}$).

The ratio of the temperature difference divided by the length is called the *temperature gradient*. It is this that is the driving force for the flow of thermal energy, and is analogous to the potential gradient (dV/L), which provides the flow of electric charge (i.e. current) in a conductor.

The minus sign in equation (2.11) is significant. It shows that the flow of energy is from the region at the higher temperature to that at the lower temperature, i.e. it flows along the temperature gradient. It implies that energy flow is unidirectional. The equation is true for steady-state conditions, i.e. when the two temperatures are stable and that the thermal energy input equals the thermal energy out. For the rod in question, it implies that there are no lateral heat losses, that the thermal energy flow-lines are laminar, and that dQ/dt is constant.

Fourier's equation can be applied to the situation which involves *composite materials*, i.e. materials constructed out of two or more layers.

In Figure 2.5, energy is flowing from left to right and with fixed θ_1 and θ_2. It is assumed, with steady-state conditions, that the rate of energy flow is the same for both materials. This means that Fourier's equation can be applied for both materials and that the interface temperature (θ) can be obtained from equation (2.12). From this the temperature gradients for each material can be derived:

$$\frac{dQ}{dt} = -k_1 A \frac{\theta_1 - \theta}{L_1} = -k_2 A \frac{\theta - \theta_2}{L_2} \tag{2.12}$$

where 1 refers to one material and 2 refers to another, and A is assumed to be the same for both. As is the case for the lagging round a boiler, the temperature drop across the material of which the boiler is made will be much smaller compared with the drop across the insulating material.

Comparing the human body and a furry animal (Figure 2.6), as in the shape of a cylinder, one can apply Fourier's law to the energy flow from the body's core. This model can be extended to include the energy flow through the clothes we wear and which maintain body warmth.

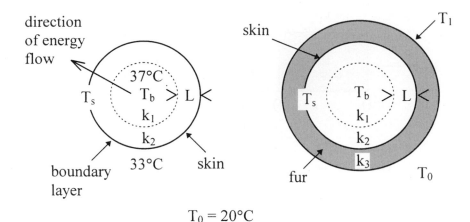

Figure 2.6 Plan of the direction of energy flow from (a) the human body and (b) a furry animal, where T_b is core body temperature, T_s is the surface temperature, L is the thickness of the insulating layer, T_0 is the ambient temperature, T_1 is the temperature of the surface of the fleece or fur, k_1 is the thermal conductivity of the core of the body, k_2 is the thermal conductivity of the outer part of the body and k_3 is the thermal conductivity of the fleece or fur.

Worked example 2.1   Fourier's law of thermal conduction.

A rambler is walking up a steep hillside in January. He is wearing clothing 1 cm thick, his skin temperature is 34°C and the exterior surface is close to freezing at 0°C. Determine the rate of flow of energy outwards from his body, through thermal conduction, when:

(a) It is fine and dry. Assume that the thermal conductivity for clothing, under dry conditions, is $0.042\,W\,m^{-1}K^{-1}$.
(b) It has been raining heavily and the rambler is soaked. The thermal conductivity is now $0.64\,W\,m^{-1}K^{-1}$. Assume that the walker has a surface area of $1.84\,m^2$.

Comment on the difference between the rates.

Solution
(a) Apply Fourier's law of thermal conduction. The rate of energy flow outwards from the body, $dQ/dt = kA.d\theta/L = 0.042 \times 1.84 \times 34/0.01 = 263\,W$.
(b) Applying Fourier's law again, $dQ/dt = 4004\,W$.

The rate of energy loss increases by a factor of more than 15. When clothing becomes wet, it becomes a better conductor for the outward dissipation of energy because water has a higher thermal conductivity than dry clothing. This is why jeans are inappropriate trouserware for strenuous outdoor pursuits in wet weather.

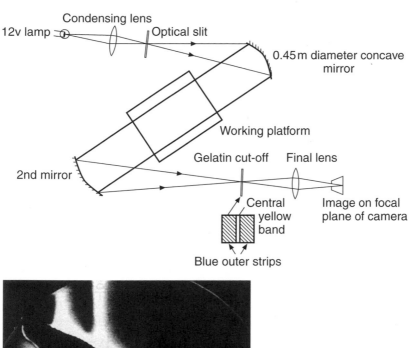

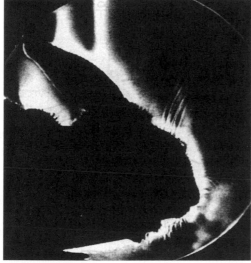

Figure 2.7 Experimental setup for studying convection currents using Schlieren optics with typical results around a rabbit. The lighter parts indicate warmer regions.

2.4.2 Convection

Convection occurs when thermal energy is transferred by the motion of a fluid. The fluid can be either a liquid or a gas. The air in close proximity to any living body will warm up due to heat radiating from the body and expand, so becoming less dense and rise. Colder and more dense air will take its place, and a convection current will be set up. A similar process occurs but on much larger scale in the Earth's atmosphere (see Chapters 8 and 9).

Convection from heated objects can be studied using Schlieren optical techniques (Figure 2.7). These work on the principle that convection currents consist of areas with different densities, and therefore with different refractive indices, which can be monitored. Schlieren optics were first developed to study the shock waves produced by bullets and jet airfoils when travelling at supersonic velocities.

There are two types of convection: *natural* (when fluids move without forcing) and *forced* (when the fluid flow is forced, such as blowing over a hot cup of tea). Of particular importance for the human environment is the latter case and Newton's law of cooling provides a mathematical model of forced convection.

2.4.2.1 Newton's law of cooling

A number of factors will influence the rate of convection from an object in a fluid, including the temperature of the object, the shape, the size, the temperature of the fluid and the type of flow relative to the object. Newton determined that the rate at which energy is lost from a body dQ/dt is directly proportional to the difference between the body's temperature (T) and the ambient, or environmental temperature (T_0), i.e.

$$dQ/dT = -kA(T - T_0) \qquad (2.13)$$

Equation (2.13) represents Newton's law of cooling, where k is a proportionality constant whose magnitude depends upon the nature and surface area of the body, and is called the *convective energy transfer coefficient*. k for a plate in still air is $4.5\,\mathrm{W\,m^{-2}K^{-1}}$ and is about $12\,\mathrm{W\,m^{-2}K^{-1}}$ when air flows over it at $2\,\mathrm{m\,s^{-1}}$.

Strictly, this law applies to objects cooling in a draught, such as blowing over a cup of tea or coffee. It does not apply to humans as the process of metabolism attempts to maintain the body temperature at a constant value. However, walking against a strong wind or in a wind tunnel is a reasonably good approximation of this law.

When energy (dQ) is supplied or extracted from a system, then the energy gained or lost is $mc\theta$, where θ (or dT) is the temperature change and c is the specific heat capacity of the body. Assuming that the heat capacity (mc) does not change with temperature

$$dQ = mc.dT \text{ or } dQ/dt = mc.dT/dt$$

and substituting into equation (2.13)

$$dT/dt = -\frac{kA}{mc}(T - T_0) \qquad (2.14)$$

Equation (2.14) gives a relationship between temperature and time. For a full mathematical treatment, see Appendix 2. The negative sign implies that the temperature decreases with time. If one assumes the human body is cylindrical in shape, of diameter D, becoming progressively colder in a wind velocity of V, then the convection coefficient becomes:

$$k = 9\sqrt[3]{\frac{V}{D^2}} \qquad (2.15)$$

Worked example 2.2   Fluid flow and convection.

A student volunteers to take part in the following simulations of convective energy loss:

(a) He is placed in a wind tunnel in which air, at $-2°C$, blows through at $40 \, \text{km h}^{-1}$.
(b) He is now placed in a flow of water, up to the neck, which is at $12°C$. The velocity of flow is $0.5 \, \text{m s}^{-1}$.

In each case calculate (a) the convective energy transfer coefficient (k) in $(\text{W m}^{-2} \text{K}^{-1})$ and (b) the convective energy transfer flux (dQ/dt). Assume $A = 1.8 \, \text{m}^2$ and the skin temperature is $31°C$.

   How do the results compare with the energy lost through convection under normal sedentary conditions (i.e. $25–40 \, \text{W m}^{-2}$)?

Solution
Apply equation (2.15) to determine k in each case. Convert km h^{-1} into m s^{-1}. $k = 44.8$ in the first case and 34.4 for the second. Now, apply Newton's law of cooling:

$$dQ/dt = -kA(T - T_0).$$

For (a) $dQ/dt = 44.8 \times 1.8 \times (31 + 2) = 2661 \, \text{W}$.
   For (b) you should find that it is $1176 \, \text{W}$.
   These figures are very high compared with sedentary conditions, and reinforce the idea that immersion in a moving fluid, especially if it is turbulent, facilitates energy loss.

2.4.3   Radiation

Radiation plays an important role in the energy balance of human beings. It is the process in which energy can be transferred in the form of electromagnetic waves from one point to another through a vacuum. Human beings emit radiation in the infrared band with wavelengths between 90 and 200 μm. Both solar and thermal radiation are part of the wider electromagnetic spectrum (Figure 2.8).

   A useful means of studying the variation of human and animal surface temperatures is that of *thermography*. It is used in medical imaging and in studying energy transfer from people, animals and buildings. Plate 4 shows how energy losses and temperature distributions can be studied with a colour-coded thermogram for (a) a building, and (b) a man.

   All objects above absolute zero release energy in the form of electromagnetic waves. The best absorbers usually make the best emitters of radiation and these are called *black-body* or *full-body radiators*. They absorb all the radiation falling on them, i.e. they do not reflect or transmit it, and can re-radiate.

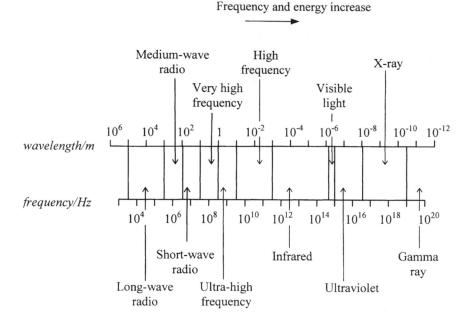

Figure 2.8 The electromagnetic spectrum.

Case study 2.1   Quantum theory and black-body radiation.

In 1899 Lümmer and Pringsheim investigated how the energy from a black-body radiator was distributed. The shape of the resulting spectral lines (Figure 2.9) was not easy to explain. According to the prevailing classical theory, which argued that radiation was emitted continuously, the energy radiated at each wavelength (E_λ) should vary with wavelength (λ) (Figure 2.10).

The implication of Figure 2.10 is that as wavelength (λ) decreases, energy (E_λ) should increase towards infinity. In fact this does not happen. How could this discrepancy between theory and experiment be resolved?

We now enter one of the great paradigm-shifts in science. The German physicist Max Planck, one of the titans of twentieth-century physics, made a revolutionary suggestion in 1900. He argued that the radiation was released not continuously but in discrete packets of energy called *quanta*. Thus, the *quantum theory* was conceived.

He also derived an expression that modelled the curves in Figure 2.9:

$$E_{\lambda\max} = \frac{8\pi hc}{\lambda^5\left[e_{xp}\left(\dfrac{hc}{k\,\lambda\max\,T^{-1}}\right)\right]} \tag{2.16}$$

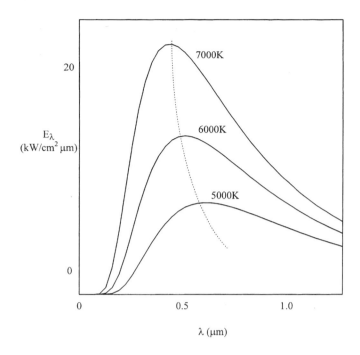

Figure 2.9 Black-body curves.

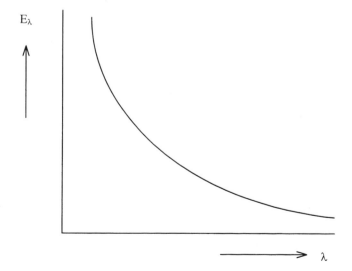

Figure 2.10 Expected black-body curve according to classical theory.

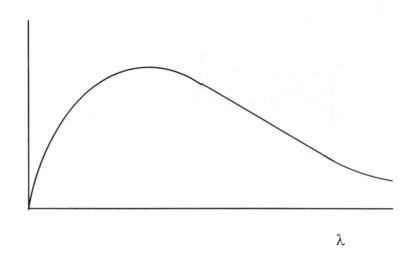

Figure 2.11 Black-body curve modelled on Planck's quantum theory.

where h is Planck's constant $= 6.63 \times 10^{-34}$ Js, k is Boltzmann's constant $= 1.38 \times 10^{-23}$ JK^{-1}, T is absolute temperature and c is the velocity of light $(3 \times 10^8 \, \mathrm{m \, s^{-1}})$, which generated the spectral curve in Figure 2.11.

By varying the temperature of the object, a series of spectral curves is produced, as in Figure (2.9). Note that for a given temperature (T), the energy (E_λ) against wavelength (λ) graph varies in such a manner that E_λ reaches a maximum for that temperature. This shall be called $E_{\lambda_{max}}$. As the absolute temperature (T) increases, the curves are shifted to the left, i.e. towards shorter wavelengths. The dotted line in Figure 2.9 denotes the locus of the maximum E_λ values. As $E_{\lambda_{max}}$ increases, λ_{max} decreases. This means that as T increases, λ_{max} decreases. This is summarized in *Wien's law*:

$$\lambda_{max} \times T = \text{a constant} = 3 \times 10^{-3} \, \mathrm{m \, K} \tag{2.17}$$

where m K means metres Kelvin.

We will see that the quantum theory will have applications to solar cells (see Chapter 5) and in photosynthesis (see Chapter 11).

2.4.3.1   Stefan–Boltzmann law

The total amount of radiation (E), or emissive power, released per metre squared per second $(\mathrm{m^{-2} \, s^{-1}})$ from a full- or black-body radiator was discovered to be proportional to the fourth power of the absolute temperature, and the law can be expressed as:

$$E = \sigma T^4 \tag{2.18}$$

or as power

$$P = A\sigma T^4 \tag{2.19}$$

where σ is Stefan's constant $= 5.7 \times 10^{-8} \, \mathrm{W \, m^{-2} K^{-4}}$ and A is the area of the surface emitting the radiation.

This law can be applied to clouds and the Earth's surface (see Chapter 7), buildings (see Chapter 3) and people, since these may all be treated as black-body radiators. Chapters 6 and 7 describe how this law can be applied in determining how a space satellite can 'sense' temperatures on Earth and in examining the Earth's radiation balance.

If the environmental temperature is to be included, and thereby make the calculations for radiation more accurate two modifications have to be made to equation (2.18).

No real body is a perfect black-body radiator, which absorbs all the wavelengths of the radiation incident on it. To distinguish between a perfect black body and real bodies the idea of *emissivity* (ϵ) has to be introduced. This is defined as the amount of radiation emitted per unit area when compared with a black body at the same temperature. Equation (2.18) then becomes:

$$E = \epsilon \sigma T^4 \tag{2.20}$$

and equation (2.19) becomes

$$P = \epsilon A \sigma T^4 \tag{2.21}$$

For a non-perfect black-body radiator, the emissivity is less than one ($\epsilon < 1$), whereas for a perfect black-body radiator $\epsilon = 1$.

A second modification of the Stefan–Boltzmann law relates to the effect of the local environment. Equation (2.20) is true for an isolated object emitting energy. In reality, most objects release energy in an environment which itself is radiating energy. If the object is in an environment at an ambient temperature of T_0, then it will absorb energy at a rate of

$$E_{\mathrm{absorbed}} = \sigma T_0^4 \tag{2.22}$$

This means that the net rate of radiation (E) emitted is the difference between what is emitted by the object and what it absorbs from its surroundings:

$$E = e\sigma T^4 - e\sigma T_0^4 = e\sigma(T^4 - T_0^4) \tag{2.23}$$

Equation (2.23) is another form of the Stefan–Boltzmann equation. J. Stefan conducted the experimental aspects of the problem while Boltzmann analysed the process mathematically.

Worked example 2.3   Radiation and the Stefan–Boltzmann law.

A person sitting reading a book releases radiant energy of between 70 and 100 W. Calculate how much energy the person is radiating. Assuming that the emissivity for the human body is 0.5, with an average surface temperature of 35°C, that the room temperature is 20°C, and the body surface area is 1.8 m².

Solution
Applying the Stefan–Boltzmann law, $E = 0.5 \times 5.7 \times 10^{-8} \times (308^4 - 293^4) = 46.43 \, W \, m^{-2}$. Thus, for a person with a surface area of 1.8 m², the energy radiated would be $46.43 \times 1.8 = 83.6 \, W$.

2.4.4   Evaporation

Anyone engaged in strenuous physical activities increases their metabolic rate, and to maintain a constant core body temperature losing energy by conduction, convection and radiation is insufficient. Evaporation constitutes a fourth method for obtaining an energy balance for energy transfer. Evaporation is important in understanding weather and climate generally and clouds in particular (Chapters 8 and 9).

Evaporation is the process whereby a liquid can be transformed into a vapour. This implies a *phase change*, or change of state, and it is an example of a latent heat change, in which the evaporative energy loss depends on the mass of the liquid and the energy required to vaporize the liquid:

$$Q = mL \qquad\qquad (2.24)$$

where Q is the energy (in joules) extracted or supplied to bring about a phase change, m is the mass of liquid to be vaporized (kg) and L is the specific latent heat of vaporization ($J \, kg^{-1}$). For pure water, 2.25 kJ is needed to vaporize 1 g of pure water, but sweat, which is

Worked example 2.4   Sweat and evaporation.

A hiker is sweating at 2 litres h⁻¹. Determine how much energy is required to evaporate all the sweat.

Solution
It requires $2.25 \times 10^6 \, J$ to vaporize 1 kg sweat, which is taken here to be water. Now, 2 litres sweat has a volume of $2 \times 10^{-3} \, m^3$ and mass = density × volume $= 1000 \times 2 \times 10^{-3} \, m^3 = 2 \, kg$.

Therefore, applying $Q = mL = 2 \times 2.25 \times 10^6 \, J$ for 2 kg in 1 h. If this occurs over the surface area of the body, which can be assumed to be 1.84 m², then the energy dissipated is $4.5 \times 10^6 / 1.84 = 2.45 \, MJ \, m^{-2} h^{-1} = 681 \, W \, m^{-2}$.

99% water with sodium chloride as solute, is an electrolyte with $L = 2.43 \times 10^6 \, \text{J kg}^{-1}$. For humans and animals, L depends on temperature.

As sweat vaporizes, the energy to achieve this has come from the hotter body. The net result is a cooling effect, and the surface temperature of the body decreases.

The rate of evaporation depends on the surface area, the temperature difference, the humidity and, therefore, the difference in vapour pressure, the rate of sweating and the velocity of air flow.

In physics, there are several cases of transport phenomena caused by a difference in a physical variable over a length. For example, we have seen that in thermal conduction energy will flow from the hot end of a bar to the colder end because of the temperature gradient which has been set up (Fourier's law). For water to flow along a pipe, a pressure gradient has to exist (Poiseuille's law); for current to flow along an electrical conductor a potential gradient is required (Ohm's law), the rate of diffusion depending on the concentration gradient (Fick's law); and for water to permeate soil an osmotic potential gradient is needed (Darcy's law). In the case of evaporation, the mechanism operates through a difference in vapour pressure of water.

Evaporation incorporates the passage of water vapour through the body's boundary layer. The rate of evaporation from the body can be expressed as:

$$\frac{dQ}{dt} = h(p_s - p_0) \tag{2.25}$$

where dQ/dt is the rate of energy transfer per unit area, h is the evaporative energy transfer coefficient ($\text{W m}^{-2} \text{kPa}^{-1}$), p_s is the water vapour pressure adjacent to the skin and p_0 is the water vapour pressure in the surrounding air.

The vapour pressure at the skin's surface depends on the environmental humidity and the sweating rate. The implications of the wet–cold interface for humans in exposed situations is obvious, and if the wind velocity increases, the rate of evaporation will increase. Experiments suggest that h depends on the wind velocity (v) in the form:

$$h = 124\sqrt{v}$$

This can be seen on a very warm humid day, when the temperature could be 35°C, i.e. the temperature is higher than the body's surface temperature. Not only can this produce an inward radiant and convective energy flux, but also it can increase sweating.

For non-evaporative energy loss (such as in conduction and convection) the driving mechanism is the temperature gradient between a body and its surroundings, but for evaporation it is the *vapour pressure gradient*. In regions of very high humidity, evaporative cooling becomes less efficient because this gradient changes. Evaporative energy loss occurs during perspiration and respiration. For a thermally comfortable and resting person, evaporative water loss can be about $30 \, \text{g h}^{-1}$. In comparison, exercising hard can produce a rate of sweating of 2 litres h^{-1}. Under normal conditions, energy loss by perspiration is the dominant factor.

2.4.5   Energy budget equation

The various energy transfer modes may be *collected* into a general energy budget equation that embodies an energy audit for the body. For a person under steady-state conditions, with energy being dissipated through radiation (R), convection (Conv), conduction (Cond) and evaporation (E), then:

$$\text{energy input} = \text{energy output}$$
$$M + R_{\text{ext}} \quad = R + \text{Conv} + \text{Cond} + E \text{ from the body} \tag{2.26}$$

where M is the metabolic energy production and R_{ex} is the radiation input from the surroundings.

If the input is not balanced by the output, then the body temperature can rise as thermal energy is stored, or fall as energy is lost. Thus, the general equation becomes:

$$M + R_{\text{ext}} = \pm R \pm \text{Conv} \pm \text{Cond} + E + S \tag{2.27}$$

where S is the rate at which energy is stored.

For Britain, where the environmental temperature is usually less than the body temperature, the radiation, convection and conduction components of energy loss are taken as being negative, implying that the energy flow is outwards. In the situation where the environmental temperature is greater than the body temperature, the positive sign would be used. Evaporative energy loss is always taken as being negative.

The relative proportion of each of the transfer modes depends on the activity and temperature difference between the person and the ambient conditions. For a thermally comfortable seated nude at 25°C, 67% of the energy loss is due to evaporation. As the ambient temperature increases the rate of energy loss due to convection increases, and at 40°C the energy loss is mostly evaporative with radiation and convection making small contributions in comparison.

Worked example 2.5   Surface area-to-volume ratio.

The amount of energy converted for bodily functions depends mostly on the volume of body tissues. Most of the heat loss from a body depends on the surface area of that body. By letting a cube of 1 m length represent a child and a cube of 2 m to represent an adult, show that the rate of loss of heat per unit volume of tissue is greater in a child than in an adult.

Solution

The volume of the child would be $1 \times 1 \times 1 = 1\,\text{m}^3$ and its surface area $= 6\,\text{m}^2$, while the volume of the adult $= 8\,\text{m}^3$ and surface area $= 24\,\text{m}^2$.

Therefore, for the child the surface area to volume ratio $= 6$, while for the adult it is 3.

This implies that babies lose energy faster than adults and, therefore, have to have higher metabolic rates especially for growth and development and to offset the energy they easily lose.

Case study 2.2 Human and animal thermoregulation.

It has been shown that, in temperature terms, there are two classes of animal, and how the human body maintains its temperature is an example of a servo-mechanism, involving a biological feedback process. This results in *homeostatis*, which implies the attainment and maintenance of a temperature equilibrium. When it is cold, the body attempts to maintain its energy balance by increasing its metabolic energy production. When it is warm, perspiration increases, energy from the body bringing about a change of state and the water in the sweat vaporizing, resulting in a cooling of the body.

Temperature sensors exist in the skin, the spinal chord and parts of the brain, especially the part called the *hypothalamus*. Peripheral receptors in the skin trigger nerve impulses under varying conditions. The 'data' from the receptors affect the body temperature control, which is coordinated in the hypothalamus (Figure 2.12). Any variations in body temperature can bring about physiological and behavioural responses. The set-point temperature is the 'optimum' temperature registered by the hypothalamus. If this is large, then energy will be lost through sweating and

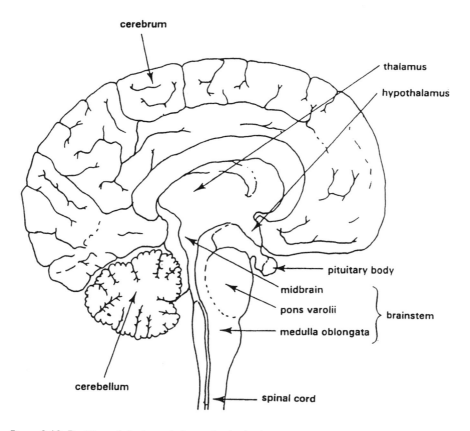

Figure 2.12 Position of the hypothalamus in the brain.

peripheral dilation of the capillary blood vessels in the skin. Below this set-point temperature the loss of energy is reduced by pilo-erection and peripheral vasocon-striction, i.e. the narrowing of the blood vessels of the skin.

Thus, at low temperatures, thermoregulation can break down when the rate at which energy is produced does not equal the rate at which energy is being lost. As a result, the core temperature decreases. As the average body temperature falls, the rate of metabolism (or energy production) also falls. Under these physiological con-ditions, a person can die, as shown below.

In analogous ways reference will be made to how an energy budget can regulate temper-atures in houses (see Chapter 3) and how an energy balance for the Earth can be achieved (see Chapter 7).

Finally, the surface area (A)-to-volume (V) ratio is important. The quantity of energy dissipated in metabolic processes depends on the volume, but the rate at which this energy is dissipated depends on the surface area. Babies have a high A/V ratio compared with adults.

2.5   Survival in the cold

As we grow older our metabolic rate decreases. The 'set point' seems to be less efficient in adjusting to external temperature changes. This means that for some people energy cannot be produced as fast as it is dissipated. Thus, in cold weather, without adequate insulation and heating *hypothermia* can set in. For some people the home can become a 'cold zone', especially for the elderly whose thermoregulatory controls are deteriorating. In contrast, in very hot summers babies and the elderly can become vulnerable to *heat stress*. Let us now examine thermal comfort and how humans may survive in what can be hostile thermal environments.

2.5.1   Thermal comfort and insulation

Thermal comfort in the home and in the work place depends on many interacting factors. These include the difference between the internal temperature (which will be affected by the building's insulation), the external temperature, how the space is being used and heated (including the possible impact of air-conditioning), the heat capacity of the ele-ments that comprise the structure of the building, the type of insulation, the prevailing weather conditions (e.g. rain, wind), the velocity of the air within the building (especially if there are draughts), and the degree of ventilation. All of these will be discussed in Chapter 3. The humidity of the room, the metabolic rate of the person and the clothing worn as thermal insulation should also be taken into account. Though we all respond in different ways to these, an individual can only feel comfortable when they are in equilib-rium.

Humans have the capacity to survive in inhospitable regions through the thermal insu-lation provided by suitable clothing. In contrast, some animals have evolved various mechanisms of insulation such as the use of fur. Clothes are worn for style and for the function of either keeping ourselves warm or cool depending on environmental con-

Figure 2.13 Types of clothing with insulations of 0.5, 1, 2, 3 and 6 clo respectively.

ditions. The right clothing in specific harsh conditions can save lives. The clothing for outdoor pursuits, especially for hill-walking and mountaineering, has steadily become more sophisticated.

There is currently a broad provision of such clothing and materials developed include breathable membranes, hydrophilics and waterproof fabrics. These materials come in two types: laminated (or membrane), such as Goretex, and coated, such as Cyclone. These can be produced either as microporous material, comprised of microscopic openings, which obstruct the possible penetration of water from outside, but which are sufficiently large to allow water vapour to pass through, or hydrophilic materials, which function in a similar way. If a fabric is not 'breathable', then water vapour from the body will accumulate as condensation on the inside of the fabric. In contrast to materials like Goretex, jeans (denim) are water-absorbent and when wet facilitate energy loss, especially in the wind.

To standardize the insulating properties of clothing, different units have been developed. The most common of these is the *clo*. This is the quantity of insulation provided by indoor clothing for a sedentary person in thermally comfortable conditions. If the skin temperature is 33°C and the ambient temperature is 21°C, 1 clo is 0.155°C m² W⁻¹. Figure 2.13 shows a variety of clothing with their corresponding clo values.

Nowhere can the importance of the clothing for a particular terrain be more appropriate than is the case for the polar explorer. S(he) will wear thermal layers, breathable water-proofs and wind-proof clothing, and a hooded parka to reduce the significant loss of energy through the head.

2.5.2 Boundary layer

The energy flow through a material is not quite as simple as it would appear. To explain this, consider the situation of a room at 25°C and an outside temperature of 5°C. Thermal energy is being lost through a window of area, say, 3 m² and thickness 0.5 cm. Assuming that there is no other loss of energy, and with a thermal conductivity (k) for glass of 1 W m⁻¹ K⁻¹, the rate of energy flow would be 12 kW. This is higher than would be expected. We have neglected the effect of the thin layer of stationary air that exists on either side of the glass. It is stationary because of its viscosity at the air–window interface.

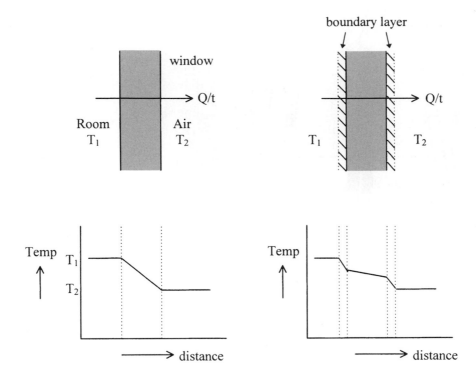

Figure 2.14 (a) Energy flow through a window without a boundary layer. (b) Energy flow through a window with boundary layers.

This thin layer is called a *boundary layer*, and it has a marked effect on the rate of energy outflow.

Assume that on both sides of the glass the boundary layer is 0.15 cm thick. This is a composite situation described earlier under thermal conduction (equation 2.12), and if one calculates the rate of energy flow, one finds it considerably reduced. Figure 2.14 illustrates the effect of the boundary layers on the temperature profile across the window.

This example serves to illustrate the widespread application of the boundary layer that surrounds people and exists in various types of clothing. Because of viscosity, the air in contact with the body is stationary, and air being a poor conductor (i.e. it has a low thermal conductivity of $25\,\mathrm{mW\,m^{-1\,\circ}C^{-1}}$), it is therefore a good insulator. A human boundary layer of a few millimetres can have a marked insulatory effect, and it acts like the composite layer discussed in section 2.4.1. Much of boundary layer theory was developed by the German engineer Ludwig Prandtl (1875–1953), and the relevance of this theme will be examined in Chapter 3.

2.5.3 Wind chill

Wind, like water, is a fluid, and the wind flows (see Chapter 5) can either be laminar (i.e. streamline flow) or turbulent (i.e. with eddies and vortices). Due to its geographical location, the British Isles experience a wide variation in wind direction and wind strengths.

Table 2.1 Wind-chill chart.

Air temperature (°F)	Wind speed (mph)							
	5	10	15	20	25	30	35	40
40	37	28	23	19	16	13	12	11
30	27	16	9	4	1	−2	−4	−5
20	16	3	−5	−10	−15	−18	−20	−21
10	6	−9	−18	−24	−29	−33	−35	−37
0	−5	−22	−31	−39	−44	−49	−52	−53
−10	−15	−34	−45	−53	−59	−64	−67	−69

This means that the flow of the wind can change from orderly to turbulent flow. We are now in a position to see how wind might affect energy loss through convection. For both natural and forced convection the convective airflow patterns can either be laminar, especially at lower speeds, or turbulent at higher speeds. The wind can, therefore, have a major impact on the human body.

This discussion can be extended to the impact of wind, the increase in energy outflow and the depression of the body's temperature. While walking in the hills bareheaded and with the wind blowing you may have a slight headache. This is the result of the wind facilitating the flow of thermal energy out from your body through the head. What is in operation here is the *wind-chill factor*. This is the factor that increases the rate of thermal energy flow from an object, and which results in the discomfort arising from various combinations of wind speed and air temperature. Table 2.1 shows how wind and air temperature combine to give a resulting wind-chill temperature.

For example, the figures suggest that a 15 mph wind on air at 20°F has a similar effect to a wind of 5 mph when the temperature is 0°F.

The wind-chill temperature is the temperature that prevails from a particular combination of wind speed and air temperature. As wind speed increases, wind-chill temperature is depressed. As a result of investigations carried out in Antarctica in the early 1940s, the American meteorologist and polar explorer Paul Siple formulated an expression for the wind-chill factor:

$$Q = (10 \sqrt{v} + 10.45 - v)(33 - T_0) \tag{2.28}$$

where Q is the wind-chill or the rate of energy loss (kcal cm^{-2} min^{-1}), v is the wind speed (m s^{-1}) and 33 is the is skin temperature in (°C) and T_0 the ambient air temperature.

Obviously, the walker has to appreciate the impact of wind-chill and be prepared accordingly, especially in terms of clothing. Under windy conditions, therefore, headgear should be worn. What then is the effect of the wind on the natural convection boundary layer? If a person is inadequately insulated, the wind effectively reduces the width of the layer, implying that the insulating properties of the layer are reduced. As a result, the rate of energy loss is increased. This and the wind-chill temperature can also facilitate the onset of *hypothermia*.

2.5.4   Hypothermia

This is the lowering of the normal body temperature and is an example of a switching mechanism from negative to positive biological–physical feedback. Under normal conditions, if a person feels cold the body readjusts and in the temperature range 37–35°C as the body temperature falls, the body works hard to produce the extra energy to compensate. This is negative feedback. However, below 35°C the body cannot generate sufficient energy as fast as it is being lost. The body's temperature, therefore, continues to fall, and the more the body tries to supplement the body's temperature, the faster it falls. In this perilous situation, in which death can result, there is positive feedback.

Hypothermia has several stages:

Stage 1 – mild hypothermia: core temperature starts to fall from 37 to 36°C
Stage 2 – moderate hypothermia: temperature falls from 35 to 32°C; thermoregulation begins to breakdown
Stage 3 – severe hypothermia: temperature falls from 31 to 30°C
Stage 4 – acute hypothermia: 29°C, i.e. death.

In addition to wind-swept areas, the sea is another context in which people can become victims of hypothermia. Immersion in cold, swirling water can facilitate the loss of energy by forced convection. Lean people are more susceptible.

Case study 2.3   Scott's Antarctic expedition (1911).

In June 1910, the naval captain Robert F. Scott departed for Antarctica. The expedition included both scientific research and a race to the South Pole, a race against the legendary Norwegian explorer Roald Amundsen. The scientific investigations included work in the areas of meteorology, glaciology, geology cartography and the biology of the Emperor penguins.

Whereas Amundsen had decided to use dogs, Scott used a mixture of motor sledges, ponies and dogs. The ponies and the motors proved unsuitable. Scott and his four-man team reached the South Pole (Figure 2.15) only to find that Amundsen had arrived there 34 days before.

Being extremely disappointed, the tribulations of the return journey have now entered the annals of heroic iconography. Scott noted the ensuing catastrophe in his diary. Evans was the first to die, then Oates. There was a deterioration in their physical condition. Towards the end as they weakened, the weather worsened. It was far worse than Scott had anticipated. Gales, blizzards, declining rations, frostbite, the effect of wind passing through their clothing, shortage of oil and temperatures as low as −43°C all contributed to a general level of hypothermia. Scott was the last to die. He, Wilson and Bowers were found 8 months later only 11 miles from one of their bases.

Figure 2.15 Scott and his team at the South Pole.

2.6   Survival in hot climates

2.6.1   *Effect of heat on the human body*

So far, we have dealt with cold environments in which the temperature of the body is greater than the temperature of the environment. Energy will therefore be dissipated outwards. Now we will consider the reverse situation, where $T_0 > T_B$, in which the body can be heated by both radiation and convection. If energy now transfers from the environment into the human body, without some dissipating mechanism the body's temperature will increase, perhaps to the point of *heat stress* and beyond to *heat stroke*, and ultimately death. Fortunately, the balance can be brought to an equilibrium steady-state, by perspiration in humans and panting in some animals. Energy from the body is used to vaporize sweat and a cooling effect results. Working in very hot climates, such as in deserts, can result in a water loss of 10–$12\,\mathrm{kg\,day^{-1}}$, but no cooling results if the sweat drips off the body rather than being vaporized. In hot climates evaporative cooling becomes the dominant mechanism of energy transfer and thermoregulation.

Heat stress is the medical condition that results from an increasing body temperature. It can occur in outdoor and occupational settings (e.g. miners and steelworkers), and also while walking in the mountains or sunbathing by the sea. *Dehydration* can also play a significant role in heat stress. This is the process that results in the loss of body fluid through sweating, and can become acute if the water lost is not replaced. Like

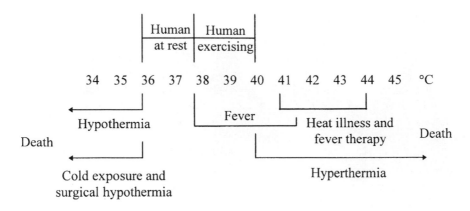

Figure 2.16 Spectrum of hypothermia and hyperthermia.

hypothermia, heat stress or *hyperthermia* has stages, all of which are accompanied by continuing dehydration:

Stage 1 – sweating and vasodilation: the effect of muscular work is to destabilize the energy balance with the result that the body temperature increases with vaso-dilation and sweating.

Stage 2 – heat cramp: results from continuing dehydration and may affect certain muscles, particularly in the legs and stomach. There may also be some dizziness.

Stage 3 – heat exhaustion: occurs when water and salt are not consumed to replenish those lost in sweat. Both dehydration and a continuing rise in the core body temperature may be in evidence.

Stage 4 – heat stroke: this is very serious and can occur when the body temperature is in excess of 41°C. The thermoregulatory system, especially the process of vasodila-tion, starts to collapse, with the result that the body cannot effectively dissipate energy and, therefore, the body temperature continues to rise. During this time the cardiovascular (i.e. circulatory) system is put under increasing stress and the flow of blood to the brain can be reduced. Unconsciousness may then result and death in the extreme case.

We are now in a position to summarize (Figure 2.16) the spectrum of medical conditions that can result from the variation in the core body temperature.

We can now perceive how animals too have evolved a wide variety of survival strat-egies in adapting to extreme climatic conditions. Polar bears have a layer of blubber covered by thick fur, and small ears so that the energy loss can be reduced, unlike the ele-phant whose large ears easily dissipate excess thermal energy. Penguins also have blubber covered by close-packed feathers and a means of recycling their respired air. The blubber not only has marked insulating properties, but also it acts as a source of energy. They also group together in large numbers to conserve body heat. There are some mammals (such as marmots and grizzly bears, but not polar bears) whose metabolic rate fluctuates seasonally. In the cold winter months some engage in winter sleep. This hibernation is characterized by a low metabolic rate, breathing and consumption of oxygen. For some hibernators, the

Case study 2.4   The camel and the Nefud Desert.

Colonel T. E. Lawrence (also known as Lawrence of Arabia) tells us in *The Seven Pillars of Wisdom* of how he conceived and executed a cunning plan. It involved leading a group of Bedouin on camels across the Nefud Desert and striking at the Turkish garrison at Aqaba on the Red Sea – from the rear. The Turkish commanders could not believe that anybody could come through the desert. Thus the appearance of the Bedouin from the desert caught the Turks unawares and led to the swift capture of Aqaba by Lawrence.

The Nefud is called locally the 'Sun's anvil', and the Bedouin, who were generally nomadic herdsmen of sheep and goats in the Arabian peninsula, could withstand daytime temperatures of between 40° and 45°C, and between 15 and 20°C at night. Even in deserts, freezing can sometimes occur. Their dependence on camels is wholly justified.

Camels are remarkable mammals, with a unique water economy and thermoregulation. It might be expected that the camel could suffer from dehydration and hyperthermia, but like other desert animals, it can respond to heat stress behaviourally and physiologically. To conserve its water supply it has a wide variation of body temperature. The camel's thermoregulation is such that during the day, as its temperature increases, the difference between it and the environmental temperature decreases. This implies that not as much sweat is needed to cool the animal as its temperature rises.

The one-humped dromedary does not store water in its hump, and neither does it contain water in the rumen 'water-sacs'. The hump stores food as fat, and the anterior part of the stomach, the rumen, has sacs containing a green fluid. When the fat in the hump is oxidized 'metabolic' water is generated. Because the fat is focused in one place, radiative cooling is enhanced from the rest of the surface area of the body. In addition, it loses water slowly, and the rough surface hair protects it against solar insolation and the inward energy transfer through conduction.

The camel can take in a considerable quantity of water, as much as 115 litres, in a short time, and because it possesses other features that enhance water retention, it can survive for longer periods than other animals in the desert. It can store up to 30% of its body weight as water, and it thermoregulates through fluctuations in its body temperature. Sweating begins at a higher temperature than in humans, of about 40°C, with the consequence that less water is lost in evaporative cooling. At night, any increase in metabolism can occur at a lower temperature. This mechanism illustrates the camel's ability to store thermal energy during the day and dissipate it at night. In addition to these physiological responses the camel also reveals behavioural adaptation through such means as orientating itself so that it presents a minimum of its surface area to solar radiation. Ultimately, it is a ruminant and both its metabolic rate and its rate of water exchange with its surroundings (such as through urination) are low. Evaporative cooling can also be facilitated by desert breezes.

body temperature can fall to just above the environmental temperature. Hibernation is a survival mechanism in harsh conditions, especially if there is insufficient food available, and hibernators are sustained by their own food reserves, especially fat. Even in deserts plants and animals can lie dormant for considerable periods, for example legend has it that once in every 7 years it rains over the Atacama Desert of northern Chile and the desert will briefly burst into bloom. Desert hibernators, too, can survive prolonged droughts.

2.7 Taking risks, weather and survival

The British Isles, which lie between 50° and 60°N latitude, is fortunate in having a temperate climate. Nevertheless, British winters can occasionally be very severe (as in 1947 and 1963) and the summers very warm (as in 1976). In addition to the fluctuations in environmental temperature which can affect the body, three other points should be noted.

First, in walking through mountainous terrain the temperature falls with increasing altitude (see Appendix 6). Second, as the air becomes thinner with altitude the pressure decreases exponentially (see Appendix 5). Third, wind speed can increase with altitude. The implications of changes in these variables can be acute. As temperature falls hypothermia can set in, and as pressure falls so does the pressure of the inspired oxygen, which in turn reduces the quantity of oxygen cycled by the blood. The impact of the weather can be dramatic.

A detailed study of hypothermia mortality (Pugh 1964) concluded that inappropriate clothing was a major factor. Fatalities, particularly in mountains, are often due to ignorance of the physical principles involved and precautions not being taken. There can be a failure to adjust to changing weather conditions, carrying inadequate equipment, defective planning and preparation, and a lack of appreciation of unknown terrain. There is also the psychological impact of over-estimating what one is capable of achieving and underestimating the extremes and rapidity with which weather conditions can change.

2.8 Summary

In this chapter, the physical principles underlying human survival in the local environment have been discussed. We have seen how cold and hot climates can affect a person's energy exchanges and their ability to survive. In subsequent chapters, we will see how the physical principles that explain survival link with a wide variety of environmental phenomena. We will begin by considering the artificial environments that human beings have designed for themselves, namely the home (the 'built environment') and then the city (the 'urban environment').

References

Energy transfer from human beings and animals

Blaxter, K., *Energy Metabolism in Animals and Man*. Cambridge: Cambridge University Press, 1989.
Case, R. M. and Waterhouse, J. M., *Human Physiology: Age, Stress and the Environment*. Oxford: Oxford University Press, 1994.
Clark, R. P., *Man and his Thermal Environment*. London: Edward Arnold, 1985.
Clark, R. P. and Edholm, O. G., *Man and his Thermal Environment*. London: Edward Arnold, 1985.

Hardy, R. N., *Temperature and Animal Life*, 2nd edn. London: Edward Arnold, 1979.

Haymes, E. M. and Wells, C. L., *Environment and Human Performance*. Champaign: Human Kinetics, 1986.

Ingram, D. L. and Mount, L. E., *Man and Animals in Hot Environments*. New York: Springer-Verlag, 1975.

Kleiber, M., *The Fire of Life: An Introduction to Animal Energetics*. New York: Kreiger, 1975.

Mount, L. E., *Adaptation to Thermal Environments: Man and his Productive Animals*. London: Edward Arnold, 1979.

Parsons, K. C., *Human Thermal Environments: The Principles and the Practice*. London: Taylor and Francis, 1993.

Stanier, M. W., Mount, L. E. and Bligh, J., *Energy Balance and Temperature Regulation*. Cambridge: Cambridge University Press, 1984.

Wrigglesworth, J. M., *Energy and Life*. London: Taylor and Francis, 1997.

Survival

Barton, R., *Outward Bound Survival Handbook*. London: Ward Lock, 1997.

Langmuir, E., *Mountaincraft and Leadership*, 3rd edn. Edinburgh: Scottish Sports Council, 1995.

Stroud, M., *Survival of the Fittest: Understanding Health and Peak Physical Performance*. London: Cape, 1998.

Walker, K., *Safety on the Hills*. Skipton: Dalesman, 1995.

Journal articles

Pugh, L. G. C. E., 'Deaths from exposure on Four Inns walking competition', *Lancet*, 1 1964, 1210–1212.

Readings from Scientific American: Vertebrate Structures and Functions. Freeman, San Francisco, 1974. The relevant section is Part 5: Temperature Adaptation, including a section on the human thermostat.

Web sites

Deserts
http://www.general.uwa.edu.au/u/uwazool/pcwlec1.htm
Environmental Physiology
http://www.general.uwa.edu.au/u/uwazool/pcwlec4.htm
This includes references to temperature regulation and energy exchanges of animals
Polar exploration
http://www.pbs.org/wgbhnova/shackleton
This refers to Shackleton's Antarctic odyssey of 1914
http://wwwgeocities.com/RainForest/Canopy/8947/scott/htm
This is the R. F. Scott web-site
http://mnc.net/norway/roald.html
This is the R. Amundsen web-site
http://www.south-pole.com/homepage.html
Refers to South Pole expeditions
http://www.terraquest.com/antarctica/index.html
Refers to Antarctic science and history
http://www.nerc-bas.ac.uk

The British Antarctic Survey web site
http://www.spri.cam.ac.uk
The Scott Polar Research Institute is the oldest polar research institute in the world and
covers both the Antarctic and Arctic
http://205.174.118.254/nspt/home.htm
This is a Newsletter prepared by personnel at the Amundsen-Scott South Pole Station.

Discussion questions

1 What is thermoregulation and explain the role the hypothalamus plays in maintaining human
   thermoregulation?
2 Distinguish between homeothermy and poikilothermy.
3 Explain the term 'homeostatis'.
4 Explain how the principle of the conservation of energy governs the maintenance of the core
   body temperature.
5 Why are many thin layers better as insulation than one thick pullover when walking in the
   hills?
6 What are the factors that influence how much energy is exchanged between a person and the
   environment by radiation?
7 Explain what is meant by a phase change, and how does it apply to evaporative energy loss?
8 Distinguish between vasodilation and vasoconstriction in thermoregulation.
9 What are the physical principles underlying: (a) animal hibernation, (b) how Eskimos survive
   in igloos and (c) the inadvisability of wearing jeans in rainy and wind-swept terrain.
10 Energy outflow from buildings can be reduced by (a) lagging (b) double glazing and (c) cavity-
   insulation. What relevance do the physical principles that underline the above examples have
   for clothing for humans and fur (or other external surface) for animals?
11 To what extent is a baby in a hospital incubator an example of radiative exchange?
12 How do different types of animal dissipate surplus heat, for example, hippopotami in rivers, and
   insects and birds in flight? Is the surface area-to-volume ratio a relevant factor?
13 You are planning a week's camping and hiking holiday in the Scottish Highlands in April.
   What equipment would you take, especially clothing, and outline the physical principles underpinning the list. Consider the problems that you could encounter and how you might reduce
   them. In addition to enquiring about the prevailing weather conditions, what other precautions
   would you take?
14 Conduct a daily energy audit for the following scenarios: (a) a sedentary student, (b) a hill-
   walker, (c) a cow, (d) a foot-soldier in Harold's Anglo-Saxon army during the critical weeks of
   autumn 1066, following the Viking and Norman invasions of England, and (e) a member of
   Scott's fateful expedition to the South Pole in 1912. To what extent can this expedition be
   considered a failure. To what did his Norwegian rival, Roald Amundsen, owe his success?
15 With consideration of the physical principles involved, what action would you take, and not
   take, to relieve somebody suffering from hypothermia?
16 (a) Outline the processes by which a healthy person loses thermal energy.
   (b) By reference to the processes outlined in (a), explain how the body responds to regulate
       loss of thermal energy.
   (c) Hypothermia occurs when the mechanisms controlling thermal energy losses fail. Distin-
       guish clearly between the different stages of hypothermia.
                    (UCLES: 'A' Level Physics: Radioactivity and Health Physics: June 1993)

Quantitative questions

1   During a warm summer's day a walker loses 1.5 kg of perspiration by evaporation. Given that the latent heat of vaporization is 2.25 MJ kg^{-1}, calculate how much thermal energy is required to achieve this.

2   (a)   The basal metabolic rate (BMR) varies from individual to individual but follows a general pattern with respect to gender and age.
    (i)   Explain the term basal metabolic rate.
    (ii)   Explain why metabolic rate increases during physical activity.
    (b)   An athlete expends 1.62MJ in total during a run lasting 30 minutes. The average efficiency of the athlete's muscles is 30%.
    (i)   Show that the average rate of production of thermal energy by the athlete is 630W.
    (ii)   During the athlete's run, 40% of the thermal energy produced is transferred from the body through perspiration. Calculate the mass of sweat which evaporates as a result of the run. (2.3×10^3J is required to evaporate 1g of sweat).
    (iii)   State and explain other mechanisms by which excess thermal energy may be removed from the body.
    (c)   Suggest why an overweight person feels hot after only moderate exercise.

(Cambridge University: 'A' Level Physics: Medical Physics: 1995)

3   (a)   Explain briefly why body temperature rises during vigorous exercise, even though the heat lost from the body due to respiration increases.
    (b)   (i)   A person inhales in one breath 5×10^{-4}m^3 of dry air at atmospheric pressure and 20°C. The air is then warmed to the body core temperature of 37°C in the lungs. If the person takes 12 breaths per minute, calculate the heat transferred per minute to the air from the body. Assume that there are no pressure changes in the inhaled air during respiration.
    (ii)   During each breath 2.2×10^{-5}kg of water vapour is formed in the lungs and is then expelled with the exhaled air. If the specific latent heat of evaporation of water at 37°C is 2.3×10^6J/kg, calculate the heat lost per minute due to this process.

Comment on your result.
Density of dry air at 20°C and atmospheric pressure = 1.2kg/m^3.
Specific heat capacity of dry air at atmospheric pressure = 1×10^3J kg^{-1}K^{-1}

(Northern Board: 1991)

4   (a)   What three types of food provide energy to a person?
    (b)   The rate of energy supply needed by a person resting but awake is called the basal metabolic rate. (A typical value for a 20 year old woman is 1.1W/Kg). State three different needs which this rate of energy supply meets.
    (c)   A woman whose mass is 55kg has a metabolic rate of 9W/kg when she is running up a hill at an angle of 5° to the horizontal with a constant speed of 6m/s.
    (i)   At what rate is she gaining potential energy?
    (ii)   At what rate is she using energy?
    (iii)   Why are the answers to (i) and (ii) different?
    (iv)   What is the effect on the woman of the fact that the answers to (i) and (ii) are different? What should she do to limit any adverse effect on her?
    (v)   Suggest a reason why her kinetic energy is not constant when she is running up the hill at a constant speed.

(Cambridge University: 'A' Level Physics: Medical Physics: 1991)

5   The basal metabolic rate is usually expressed in terms of energy expended per unit time per unit surface area of the body. The variation of BMR with age and sex is shown in Figure 2.17:
    (a)   (i)   Explain why the BMR is higher at an early age, dropping rapidly after the age of 7.
    (ii)   A 30 year old woman has a surface area of 1.92m^2. Calculate her basal rate of energy expenditure in watts.

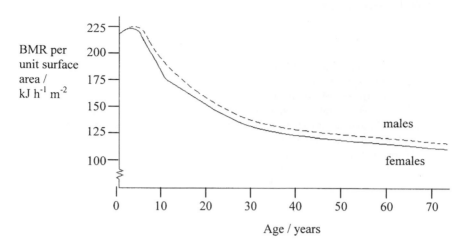

Figure 2.17 Basal metabolic rate as a function of age.

(b) During a game of squash lasting half an hour, the average power consumption of the woman is 1080W. The muscles of her body develop power with an efficiency of 20%. Calculate the rate of production of heat in the woman's body during the game.

(Cambridge University: 'A' Level Physics: Radioactivity and Health Physics: 1992)

6　(a) Differentiate and test for a maximum and (b) integrate Planck's expression, given as equation 2.16 to generate expressions for Wien's law and the Stefan–Boltzmann law respectively. In case (b), take the limits of λ from 0 to infinity.

7　The energy falling on a person sunbathing in Bognor is $1000\,\mathrm{W\,m^{-2}}$.

(a) Assuming that no energy is dissipated or reflected from the body, calculate the rate of its temperature increase, given that the specific heat capacity of the body is $3700\,\mathrm{J\,kg^{-1}\,K^{-1}}$ and its mass per unit area is $39\,\mathrm{kg\,m^{-2}}$.

(b) Determine the temperature of the body if all its energy was dissipated through the process of radiation. Comment on your answer. (Assume Stefan's constant and that the body acts as a perfect black body.)

(c) Your answer to (b) will be too high. Calculate the perspiration rate required to provide a temperature of 37°C. (Assume that the latent of vaporization of water at 37°C is $2.45 \times 10^6\,\mathrm{J\,kg^{-1}}$.)

(d) An easterly wind starts to blow across the body with a speed of $5\,\mathrm{m\,s^{-1}}$. What mechanism of energy transfer is now in operation? Determine the fraction of the total incident radiation released under these circumstances when there is a difference in temperature of 15°C between the body and the surroundings. Assume a radiative area of $1\,\mathrm{m^2}$, and that the energy transfer convection coefficient $K = 4 \times \sqrt{(v/L)}$ where v is the velocity of air-flow and L is the height of a person (say 1.70m).

8　Give an account of the mechanism of temperature regulation in mammals.

(BSc Biological Sciences: Animal Physiology and Environment, May 1995, Birkbeck College, London University)

9　Dulong and Petit's model for the natural convection of a cooling body suggests that the rate of temperature change is given by the following relationship:

$$dT/dt = k.(T - T_0)^{5/4}$$

Derive an expression for the temperature of the body at any time t.

10. Holmes looked out of the window reflectively as he saw Moriaty disappearing through Russell Square. 'The body is still warm', said Watson. 'Well', Holmes replied, 'we will follow our time-honoured procedure of applying Newton's law of cooling to calculate the time of death. Take the temperature now, at 10 a.m., and again at noon'. 'Remind me Holmes of the formulae.' Holmes wrote the following in the dust on the window:

$$dQ/dt = -kA\,(T - T_0)$$

$$dT/dt = -\frac{kA}{mc}\,(T - T_0)$$

where kA/mc, since all these variables are constant, constitutes another constant, β.

'Is there anything else I should assume, Holmes?' 'Yes, Watson, the core body temperature is 37°C and the ambient temperature of this room is 29°C. I suggest that you calculate β, and hence the time of death.' Watson duly found the temperatures to be 33°C at 10 a.m. and 30°C at noon. Assume that $k = 8\,\mathrm{W\,m^{-2}\,K^{-1}}$, $m = 70\,\mathrm{kg}$, $A = 1.8\,\mathrm{m^2}$, and that $c = 3700\,\mathrm{J\,kg^{-1}K^{-1}}$. Determine the time of death and sketch a graph indicating how the body's temperature varies with time since the moment of death.

Chapter 3

The built environment

3.1 Introduction

From the earliest times, people have adapted their habitat to provide shelter from the weather. In little more than 5000 years humanity has developed from living in shelters provided by the natural landscape to the gigantic tower block structures of today; yet for the individual the cave or flat, cottage or castle in which they lived have one common theme, for that individual it was 'home'. Though the interpretation of 'home' may be complex and wide ranging, for every individual the importance of home is an essential part of their lives; next to the procreation of children the most dramatic event of most people's lives is the establishment of their own home. Hence, it is not surprising that the construction of buildings has proven to be of major significance in the development of human evolution and civilization. Indeed, the course of civilization is often traced through the architecture of its buildings.

Scientific progress has therefore been reflected in the design of buildings. Most inhabitants of ancient Egypt lived in huts built of mud bricks constructed from the rich silt of the Nile, while in Europe and the Far East people used the plentiful supplies of timber to build their houses. Even today the majority of the world's population lives in homes built from material available from their local environment (Figure 3.1). Yet as science developed and the understanding of the properties of materials grew, we have begun to fashion our own materials (e.g. concrete and glass) which can be used to construct stronger, warmer and more pleasant homes. Thus, from the *natural environment* we have gradually fashioned across the world a new environment in which we live, the *built environment*.

The built environment is formed by the buildings and other objects constructed in the natural environment. Modern buildings have various purposes but all provide an internal environment different from the external environment. The climate of the natural environment, the shelter of the built environment, and human activities are linked by physical processes and an outline of the interdependent relationships is given in Figure 3.2.

Inevitably, there has to be a balance between the technical and aesthetic issues. The technical choices for buildings of the future are relatively easy compared with other decisions that have to be made about the built environment. These decisions include agreement about what areas of land should be allowed to contain housing and what external features look agreeable to everyone. In those countries that have an established stock of buildings there is the additional challenge of refurbishing existing buildings while maintaining their appearance.

In addition to the scientific principles controlling the built environment, the design of

(a)

Figure 3.1 Examples of homes around the world: (a) English 'standard' house under construction in South West England and (b) African hut in Botswana.

(b)

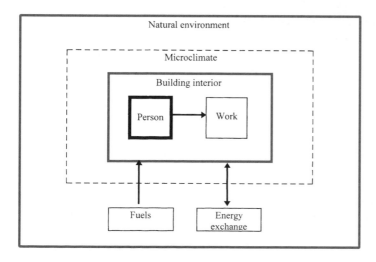

Figure 3.2 Schematic diagram of the environmental relationships between people, buildings and the local environment.

buildings includes many human choices about how we use the environment and the style in which people live. For example, a building with small areas of glazing, or even no windows at all, uses less energy than a building with large windows. Yet most people will prefer to live in a home with a view and so use more fuel. Fortunately, there are techno-logical solutions to building design that allow reasonable compromises. All require choices between how technology is applied and how the appearance of the built environment is planned, but the decisions should always be informed by good applications of physics.

This chapter looks at how the thermal comfort, air quality and moisture behaviour inside buildings may be controlled and it describes the physics required to understand the technology being used in modern building design.

3.2   Thermal regulation in buildings

3.2.1   Thermal insulation

In Chapter 2 it was pointed out that appropriate insulation can save lives. Reducing thermal transfer and hence energy loss from buildings is the main aim of thermal insula-tion. Adequate thermal insulation should, therefore, be a feature of any good building design. However, only in the past 30 years, as fuel prices have increased, has insulation become an integral part of the architect's portfolio. Much of the housing stock in the UK (and the rest of Europe) is more than 70 years old and, in general, was not designed with good insulation properties. Large windows, draughts and poor loft insulation lead to as much as 30% of the heat generated within the house being lost. However, insulation can be added to existing buildings. The relatively small cost of extra insulating materials is quickly paid for by the reduction in the size of the heating or cooling plant required and by the annual savings in the amount of fuel needed. The increases in energy efficiency and reduction in fossil fuel consumption necessary in the twenty-first century ensure that thermal insulation of the existing housing stock must be an integral part of any national energy policy.

Thermal insulation is the major factor in reducing the transfer of thermal energy between the internal and external environment of a building. In a cool or temperate climate it is easy to understand how insulation helps reduce the loss of heat energy from a building. Insulation also has an equally important role in reducing heat gains in a building. A well-insulated roof, for example, will help prevent a house overheating in summer by limiting the amount of solar radiation passing through the roof fabric. Hence, even in British cities there are some office blocks that use more energy per day for cooling the building in summer than they do for heating the same building in winter (e.g. Canary Wharf in London) (Figure 3.3).

When choosing insulating materials the physical properties of the building materials also need to be considered. An aerated concrete block, for example, must be capable of supporting a great weight. The properties listed below are relevant to many situations, although different balances of these properties may be acceptable for different purposes:

- Thermal insulation suitable for the purpose (Figure 3.4)
- Strength or rigidity suitable for the purpose
- Moisture resistance
- Fire resistance

Figure 3.3 The Canary Wharf group.

- Resistance to pests and fungi
- Compatibility with adjacent materials
- Harmless to humans and the environment.

The practical materials used for thermal insulation in construction appear in many forms and can be grouped (Table 3.1).

The physical mechanisms important in the insulation of buildings include thermal

Table 3.1 Materials used in building insulation.

Type of insulating materials used for building	Example
Rigid pre-formed materials	aerated concrete blocks, such as used for wall construction
Flexible materials	fibreglass quilts, such as laid in roofs, wall cavities and floor spaces
Loose fill materials	expanded polystyrene granules, such as used in roofs and wall cavities
Materials formed on site	foamed polyurethane, such as pumped into wall cavities
Reflective materials	aluminium foil, such as attached to plaster board

Figure 3.4 Roof insulation in a house in western Germany.

conduction, convection, thermal radiation and condensation of water vapour. The Laws of Thermodynamics apply equally to the thermal performance of buildings as to the thermal performance of the human body (see Chapter 2).

3.2.2 Thermal conduction effects

The theory of thermal conduction, as discussed in Section 2.5.1 for clothing, applies equally to the materials used in buildings. The thermal conductivities of practical building materials are often termed k *values* and Table 3.2 gives some typical examples. It shows that, in approximate figures, aluminium ($k = 160$) conducts heat 160 times faster than brick (≈ 1.0) or glass (≈ 1.0), assuming that one uses the same thickness of each material. Similarly, glassfibre (≈ 0.04) conducts heat at only 4% of the rate of glass.

The 'active ingredient' in glassfibre is air, which at first may seem a strange construction material. However, the low conductivity of air is the main mechanism of many insulating materials for buildings. The air must be kept stationary to avoid convection effects and the purpose of the glass is to keep the air still. Fortunately, as shown in Chapter 2, a *boundary layer* of stationary air is attached to open surfaces, such as window panes, and provides important thermal insulation.

It is also important to remember that the thermal conductivity of building materials varies with moisture content because the presence of water increases conduction. All the batches of brick used to construct a house may be made with the same k but after several years those on a north-facing wall will have been exposed to different amounts of sun and rain than those on a south-facing wall and they will have different moisture contents and hence different k-values.

Table 3.2 Thermal conductivities of typical construction materials.

Material	k $(W m^{-1} K^{-1})$
Aluminium alloy, typical	160.00
Brickwork, exposed ($1700 \, kg \, m^{-3}$)	0.84
Concrete, lightweight ($1200 \, kg \, m^{-3}$)	0.38
Glass sheet	1.022
Glass wool, mat or fibre	0.04
Polystyrene, expanded (EPS)	0.035
Steel, carbon	50.00
Timber, softwood	0.13

Variations in density also have significant effects on k values of brickwork, concrete and stone, and manufacturers of insulating materials publish certified values of thermal conductivity for their products. These values are then used for standard calculations of insulation, including calculations which prove that the insulation of a building meets the minimum standards required by legislation, such as the Building Regulations.

The *thermal resistance* or R-value $m^2 K W^{-1}$ is a coefficient used to describe the opposition to thermal transfer offered by a particular component of a building structure. For a particular thickness of material (L) with a thermal conductivity of k, the thermal resistance is given by

$$R = L/k.$$

Worked example 3.1   Thermal conduction through brickwork.

Consider the wall of a house built of a certain type of brickwork having $k = 0.84 \, W \, m^{-1} K^{-1}$ and 225 mm thick. The temperature on the inside of the building is 22°C while on the outside it is 2°C. Ignoring other effects, calculate the rate of flow of thermal energy through $1 \, m^2$ of the wall.

Solution
Using the equation (2.11)

$$dQ/dt = -kAd\theta/dL,$$

and knowing $k = 0.84 \, W \, m^{-1} K^{-1}$, $A = 1 \, m^2$, $d\theta = 22 - 2 = 20°C$ and $dL = 0.225 \, m$, then the thermal energy loss rate $dQ/dt = -74.67 \, W$, the negative sign showing that there is a net flow of heat from the inside to the outside of the building, that is in the direction of the falling temperature.

Note: in practical locations this brickwork has a boundary layer of air on each side and these layers provide additional insulation (see Section 2.5.2). These effects are incorporated into building design through the definition of a U-value, described in Section 3.2.5.

Table 3.3 Examples of standard thermal resistance values for surfaces.

Type of resistance	Construction element	Direction of heat flow	Surface emissivity	Standard resistance values $(m^2 K W^{-1})$
Inside surfaces	walls	horizontal	high	0.123
			low	0.304
	roofs – pitched or flat	upward	high	0.106
			low	0.281
	ceilings, floors	downward	high	0.150
			low	0.562
Outside surfaces (normal exposure)	walls	horizontal	high	0.055
			low	0.067
		upward	high	0.045
			low	0.053

Standard thermal resistances for surfaces (Table 3.3) also include allowances for the effects of boundary layers, and for the radiation effects described below. Technical sources for these standard values include the Chartered Institution of Building Services Engineers (CIBSE) and Building Research Establishment (BRE).

3.2.3   Convection effects

Thermal energy is transferred away from the outside surface of a building by convection (Section 2.4.2) or radiation effects (Section 2.4.3). Buildings are sufficiently large to always have air moving over their surfaces by natural convection currents. This convection transfers energy away from the building and this energy transfer is increased when the building is exposed to wind. Convection effects also exist within the cavities of walls and the air spaces within roofs.

Newton's law of cooling (Section 2.4.2) also applies to the built environment although the theoretical behaviour of convection varies greatly over a building and the convection coefficients for a building are difficult to derive. Instead, building scientists work with composite coefficients for a building or factors which are published for the various types of surfaces found in and around buildings. These standard values are usually obtained from practical experiments with structures under controlled conditions of thermal energy flow.

3.2.4   Radiation effects

In the physics of building performance use is made of the *emissivity coefficient* (ϵ) found in expressions of the Stefan–Boltzmann law (Section 2.4.3). Emissivity is the fraction of energy radiated by a body compared with that radiated by a black body at the same temperature. Similarly, the *absorptivity*, or absorption factor, is the fraction of radiant energy absorbed by a body compared with that absorbed by a black body.

Emissivity and absorptivity depend on the wavelength of the radiation and the wavelength in turn depends on the temperature of the source of the radiation. The Sun is a

Table 3.4 Surface coefficients for building materials.

Surface	Emissivity (thermal radiation)	Absorptivity (solar radiation)
Aluminium	0.05	0.2
Brick (dark)	0.9	0.6
Paint – white	0.9	0.3
Paint – dark	0.9	0.9

high-temperature source of radiation, and building materials are low-temperature sources. Therefore, different sets of values for emissivity and absorptivity are needed for the same surface, depending on the wavelength of the radiation.

Table 3.4 gives typical values for common building surfaces for solar radiation absorbed by the surface and for thermal radiation emitted from the building. Hence, unless there is a shiny surface such as aluminium or reflective paint, most building materials radiate 0.9, or 90%, of the maximum. This explains why it does not matter that hot water radiator panels in rooms are commonly painted white when simple physics suggests that they should be black for most effective emission of radiant energy. Indeed such panels actually transfer most of their thermal energy by the convection of air over the metal, not by radiation.

3.2.5   U-values

Since thermal energy is transferred through an element of a building, such as a wall, by a number of mechanisms it is convenient to combine all these factors in a single measurement that describes the total thermal transfer behaviour of a particular construction. This measurement is called the overall thermal transmittance coefficient or the U-value.

The U-value is the rate at which energy flows through $1\,\text{m}^2$ of material with a temperature difference of $1\,\text{K}$. U is, therefore, measured in Watts per square metre per Kelvin ($\text{W}\,\text{m}^{-2}\,\text{K}^{-1}$). U is inversely proportional to the total thermal resistance, such that for any structure of thickness L:

$$U = 1/\Sigma R. \tag{3.1}$$

Thus, the higher the U the lower the resistance and the poorer the insulator. For example, a wall with $U = 0.4\,\text{W}\,\text{m}^{-2}\,\text{K}^{-1}$ loses heat at half the rate of a wall with $U = 0.8\,\text{W}\,\text{m}^{-2}\,\text{K}^{-1}$. Therefore, the cost of replacing energy lost through the first wall will be half the cost of that for the second wall.

Worked example 3.2   *U*-value of a cavity wall.

A cavity wall consists of brick $10\,\text{cm}$ thick, a $5\,\text{cm}$ air cavity spacing, $10\,\text{cm}$ of concrete and $2\,\text{cm}$ of plaster. Given that the thermal conductivity of brick is 0.8, concrete 0.2 and plaster $0.1\,\text{W}\,\text{m}^{-1}\,\text{K}^{-1}$ respectively and the thermal resistance of the internal surface is 0.12, external surface 0.06 and the cavity $0.19\,\text{m}^2\,\text{K}\,\text{W}^{-1}$ respectively, determine the U-value for the cavity wall (Figure 3.5).

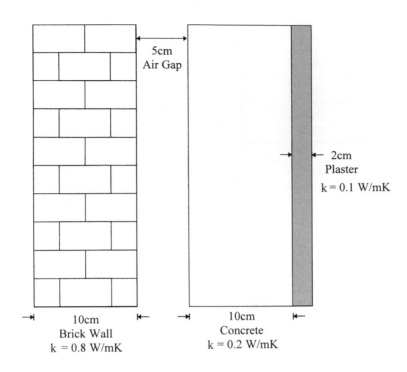

Figure 3.5 Schematic diagram of a typical cavity wall.

Solution

Calculate the thermal resistance of the brick, concrete and plaster ($R = L/k$), where $R_{brick} = 0.1/0.8 = 0.125$, $R_{concrete} = 0.1/0.2 = 0.5$ and $R_{plaster} = 0.02/0.1 = 0.2$. The total thermal resistance of the whole cavity wall system may be calculated by the summation of the thermal resistances of each element $= \Sigma R = 1.195$, so U-value for whole system $= 1/\Sigma R = 0.84\,\mathrm{W\,m^{-2}K^{-1}}$.

U-values for some common types of construction material are given in Table 3.5.

Table 3.5 Typical U-values for construction materials.

Element	Composition	U $(Wm^{-2}K^{-1})$
Traditional solid wall	solid brickwork	2.3
Early cavity wall (e.g. 1930s)	brickwork, cavity, brickwork	1.5
Modern cavity wall	brickwork, cavity with insulation, lightweight concrete block	0.6–0.3
Modern timber frame wall	brickwork, cavity, insulation, air gap, plasterboard	0.45
Traditional pitched roof	without insulation	2.0
Modern roof	with insulation	0.25
Window	with single-glazing	5.7
Modern window	with double-glazing	2.8

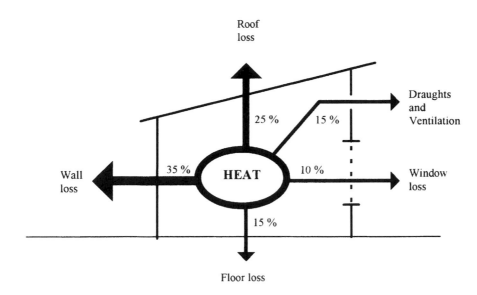

Figure 3.6 Typical percentage energy losses from a traditional building.

3.3   Energy use in buildings

3.3.1   *Efficiency*

Buildings in Europe and in countries with a similar climate suffer an overall loss of thermal energy during the year (Figure 3.6). The energy required to replace thermal losses from buildings represents a major portion of the country's total energy consumption. Building services such as heating and air-conditioning make up typically 40–50% of the national consumption of primary energy with about half of this used in domestic buildings.

The thermal energy required for buildings is commonly obtained from fuels such as coal, gas and oil, even if the energy is delivered in the form of electricity. Each type of fuel must be converted to thermal energy in an appropriate piece of equipment, such as a boiler, and the heat distributed to the place of use, such as by hot water pipes. The amount of heat finally available to the users of a building depends on the original energy content of the fuel and on the efficiency of the system in converting and distributing this energy.

As discussed in Chapters 2 and 5, the Second Law of Thermodynamics can be expressed by the idea of efficiency. For heating a building, the efficiency index or percentage is calculated by comparing the output power or energy with the input power or energy:

$$\frac{\text{Efficiency (\%)}}{100} = \frac{\text{output}}{\text{input}} = \frac{\text{useful energy}}{\text{delivered energy}}. \tag{3.2}$$

In general, the useful energy is the output energy from the heating system, which is used to balance the heat losses and heat gains. The delivered energy is the input energy needed for the boiler, or other device, and it is the energy which you have to pay for.

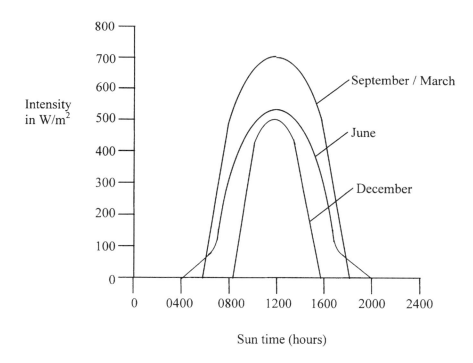

Figure 3.8 Solar intensities on vertical surfaces at latitude 51.7°N.

causes the largest solar gains for buildings in the British Isles. Figure 3.8 illustrates this effect by displaying the intensity of solar radiation $(\mathrm{W\,m^{-2}})$ on a vertical surface at different times of the year. The figures are for London or other locations at the same latitude, and they assume cloudless days.

Other energy gains are generated by activities and equipment in buildings that dissipate heat even though they are not intended for heating. For example, as discussed in Section 2.3.1, each person sitting in a room emits energy at a rate of approximately 100 W. Ten people together in a room therefore provide as much heat as a 1 kW (one bar) electric fire. Artificial lighting is another major source of energy gain and ordinary tungsten filament lamps waste about 90% of their electrical energy as heat. A 100 W light bulb is therefore really a 90 W heater!

Calculations of exact energy gains are relatively complicated but the use of standard figures gives results which are of an acceptable order of accuracy for practical purposes. Typical values are listed in Table 3.8.

Table 3.8 Typical energy gains in a building.

Type of source	Typical rate of thermal emission (W)
For a typical adult:	
Seated or at rest	100
Medium work (e.g. decorating)	140
Lighting:	
Fluorescent system (e.g. classroom)	$20\,\mathrm{W\,m^{-2}}$
Tungsten system (e.g. home)	$40\,\mathrm{W\,m^{-2}}$
Equipment:	
Desktop computer	150
Photocopier	800
Gas cooker	350 per burner
Colour TV	100

3.4   Air regulation in buildings

3.4.1   Ventilation requirements

Ventilation in buildings is the process of changing the air in a room or other internal space. This process should be continuous with new air taken from a clean source. Although oxygen is required for life, a build-up of carbon dioxide is more life-threatening than a shortage of oxygen, and a general build-up of odours will be objectionable long before there is danger to life.

The ventilation in a building, therefore, serves many purposes in addition to the requirements for basic life and comfort. In addition to the supply of oxygen and the removal of carbon dioxide, ventilation is used to control humidity and control air velocity. Ventilation also removes odours, heat, microorganisms, mites, moulds, fungi, particles such as smoke and dust, and organic vapours from sources such as cleaning solvents, furniture and building products. Water vapour needs to be removed close to its source; such as in a kitchen where combustion products from heating and cooking also need to be removed. More specialized purposes of ventilation include the removal of ozone gas from photocopiers and laser printers and the removal of methane gas and decay products which may build up from certain ground conditions such as land-fill. The 'old' air being replaced in the ventilation process has often been heated and, therefore, to conserve energy, the rate of ventilation may be limited, and thermal energy can be recovered from the extracted air and recycled.

Adequate ventilation of buildings and spaces is required by several sources of regulations in Britain, such as the Building Regulations, Workplace Regulations, Housing Acts, and the Health and Safety at Work legislation. Various authorities produce technical guidance for providing adequate ventilation and important sources include British Standards, the Building Research Establishment (BRE) and the Chartered Institution of Building Services Engineers (CIBSE).

The method of specifying the quantity and rate of air renewal varies and includes:

- Air supply to a space – such as 1.5 room air changes $\mathrm{h^{-1}}$
- Air supply to a person – such as 8 litres $\mathrm{s^{-1}}$ per person.

Table 3.9 Typical ventilation rates.

Element	Rate (air changes per hour (h^{-1}))
Commercial kitchens	20–40
Restaurants (with smoking)	10–15
Classrooms	3–4
Offices	2–6
Domestic rooms	1
Hospital ward	8 litres s^{-1} ($=30$ m^3 h^{-1}) fresh air per occupant

Some examples of ventilation rates are given in Table 3.9. It can be assumed that if air is actively extracted, such as by a fan, then it will flow in to replace the extracted air. So the supply rate will match the extraction rate, although the source of the supply may need to be considered in the ventilation design.

Worked example 3.4   Ventilation and air supply rates.

An extractor fan is required for a toilet. What is the air supply rate required in such a system if there are 3 air changes per hour and the room has a volume of $2 \times 3 \times 2.5$ m?

Solution

$\quad$ 3 air changes h^{-1} $= 3 \times 15 = 45$ m^3 h^{-1} $= 45\,000$ litres h^{-1}
$\quad$ $45\,000$ litres h^{-1} $= 45\,000/60 \times 60 = 15.5$ litres s^{-1}.

Therefore, the extractor fan needs to have an extraction rate of at least 15.5 litres s^{-1}.

3.4.2   Ventilation installations

The control of air in buildings, by ventilation and air-conditioning, is governed by the principles and applications of physics, such as:

- Convection and the *stack effect*
- Pressure differences and types of fluid flow.

The common systems used to provide ventilation can be considered as two broad types: natural and mechanical.

Natural ventilation

This is provided by two main mechanisms: air pressure differences and the stack effect. Air pressure differences are caused by wind direction and movement over and around a

Figure 3.9 A tall stairwell in a block of flats.

building. Examples of using air pressure differences are opening the windows on either side of a building for ventilation, and the stack effect, which is caused by the natural convection of warm air rising within a building. This effect is particularly noticeable when there is a considerable difference in height, such as in the stairwell of a tall building (Figure 3.9). This effect in a stairwell is a nuisance and a fire danger but, when appropriately designed, it can be used to drive a ventilation system. Some modern environmentally friendly buildings use the stack effect to avoid the need for mechanical ventilation.

Mechanical ventilation

There are some spaces that cannot be adequately ventilated by natural means. Mechanical ventilation, driven by fans, is then required and these systems depend on a supply of energy, usually electricity.

A simple extractor or input mechanical system provides movement of air but only a limited degree of control. A *plenum system*, using ductwork, gives better control of air and usually has the ability to heat the air if necessary.

Air-conditioning

An air-conditioning installation has the aim of producing and maintaining a predesigned internal air environment, despite the variations in external air conditions. In addition to mechanical ventilation described above, the equipment has to heat and cool the air, to humidify and dehumidify the air, and to respond automatically to changes in the external air. Thus, such equipment is a major user of energy in any building. Typical features of air-conditioning equipment are summarized in Table 3.10 and in Figure 3.10.

Table 3.10 Features of typical air-conditioning installation.

Installation feature	Purpose and features
Fresh air intake	must be carefully situated
Recirculated air	can be mixed with fresh air
Air filter	removes particles and contaminants
Heater	elements or coils over which air passes
Preheater	to provide initial heat energy if needed
Chiller	lowers temperature and therefore removes moisture
Humidifier	adds moisture to the air if needed
Reheater	adjusts final temperature of air if necessary
Fan	provides air movement for intake and distribution
Ductwork	guides distribution of air
Diffusers	controls distribution of air into room

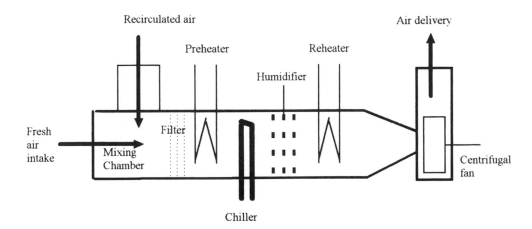

Figure 3.10 Schematic diagram of a modern air-conditioning plant.

A medical condition often linked to air-conditioning is *Legionnaires' disease*, first identified in 1976 when 29 members of the American Legion died after staying in a hotel in Pennsylvania. Legionnaires' disease is a form of pneumonia, sometimes fatal, caused by bacteria that may live in the water and air-conditioning systems of buildings. These bacteria grow best in temperatures of 20–50°C. Hence, all hotels and large institutions using air-conditioning systems should keep their water tanks and air-conditioning plants below the minimum temperature required for bacterial growth. Since the 1980s it is a legal requirement that all public facilities and offices are regularly checked for the presence of such bacteria and that the water supplies and air-conditioning plants are regularly maintained.

3.5   Heat pumps

The maintenance of comfortable living conditions within buildings needs a source of thermal energy. The source of thermal energy chosen will depend on the technical convenience, energy costs and environmental consequences of the system selected. A *heat pump* is an interesting application of physics that allows us to recover energy from systems or allows one to make use of low temperature sources such as heat naturally present in the air, in the ground, or from bodies of water. A heat pump is a device that extracts thermal energy from a low temperature source and upgrades it to a higher temperature.

A heat pump at first sight appears to defy the laws of physics. Typically it gives out thermal energy at a rate of 3 kW while using electrical energy at a rate of only 1 kW. The 'missing' energy is taken from sources of heat which are overlooked. Every object in the environment is a potential source of thermal energy, even ground that has been frozen solid. Although the quantity and the temperature of this 'low grade' heat is usually too low for it to be useful, with some sources it can be profitable to 'pump' the heat to a higher temperature.

Most heat pumps employ a standard refrigeration cycle and a compression system as shown in Figure 3.11. The extraction coils (evaporator) of a heat pump should ideally be placed in a source of thermal energy which remains at a constant temperature. Air is the most commonly used thermal source for heat pumps but has the disadvantage of low thermal capacity. The exhaust air of a ventilation system is also a useful source of heat at constant temperature, so heat pumps are a useful device in the recovery of thermal energy from ventilation systems.

3.5.1   Heat pump efficiency

The main feature of the heat pump is that it produces more usable energy than it consumes. This efficiency is measured by a *coefficient of performance* for heating. The Coefficient of Performance (COP_H) of a heat pump is the ratio of thermal energy output to the energy, such as electrical energy, needed to operate the pump.

$$COP_H = \frac{\text{Heat energy output}}{\text{Pump energy output}} = \frac{T_1}{T_1 - T_2} \tag{3.5}$$

where T_1 = absolute temperature of heat output and T_2 = absolute temperature of heat source.

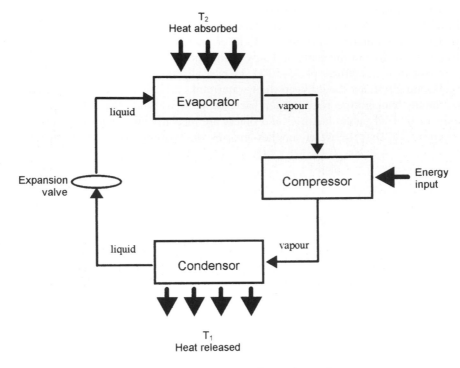

Figure 3.11 A heat pump using a compression–refrigeration cycle.

The above expression shows that the COP_H is always greater than unity, so that a heat pump always gives out more energy than is directly supplied to it. It therefore uses electricity more efficiently than any other system. The temperature equation also shows that the COP_H decreases as the temperature of the heat source decreases, which unfortunately means lower efficiency in colder weather. A heat pump does not actually give energy for nothing, as might be suggested by the coefficient of performance. It is important to realize that the COP_H only takes account of the input energy *from the user* and ignores the energy taken *from the source*.

The COP_H values achieved by heat pumps under practical conditions are between 2 and 3. These values take account of the extra energy needed to run circulatory fans and pumps, to defrost the heat extraction coils, and to supply supplementary heating, if necessary. The ground can be a useful source of low-grade heat in which to bury the extraction coils of a heat pump. Ground temperatures remain relatively constant but large areas may need to be used, otherwise the heat pump will cause the ground to freeze. Water is a good source of thermal energy for heat pumps provided that the supply does not freeze under operating conditions. Rivers, lakes, the sea and supplies of waste water can all be used.

3.6   Condensation

Condensation in buildings is a form of dampness caused by water vapour in the air. Among the effects of condensation in the domestic house are misting of windows, beads of water on non-absorbent surfaces, dampness of absorbent materials and mould growth

(a)

(b)

Figure 3.12 (a) Condensation on a window and (b) mould in a house.

(Figure 3.12). Unwanted condensation is a problem in any building since it causes unhealthy living conditions and damage to materials, structures and/or decoration. Condensation as a problem in building design is a relatively recent problem but one that has been increasing. It is affected by the design of modern buildings and by the way in which buildings are heated, ventilated and occupied.

To understand the problems caused by condensation, how they arise and how they may be overcome it is necessary to consider the role of water vapour in the air. The maximum proportion of water vapour in air is about 5% by weight, yet this relatively small amount of moisture produces considerable effects for our life on Earth. Water vapour is an essential component of the Earth's atmosphere and biosphere, playing a pivotal role in the Earth's temperature balance, the formation of clouds and in cycling the fresh water that all life needs to survive (see Chapter 8). Human survival, comfort, weather conditions, water supplies and the growth of plants are environmental topics discussed in this book and which are all affected by humidity. The amount of moisture in the air also influences the durability of materials and drying of materials, and the operation of industrial processes.

3.6.1   Water vapour

A vapour can be defined as a substance in the gaseous state that may be liquefied by compression. Water vapour, for example, is formed naturally in the space above liquid water left open to the air. This process of *evaporation* occurs because some liquid molecules gain enough energy, from chance collisions with other molecules, to escape from the liquid surface and become gas molecules. The latent heat required for this change of state is taken from the liquid, which therefore cools.

The amount of moisture in a sample of air can be defined by the following different parameters:

- Vapour pressure
- Moisture content
- Dew point
- Relative humidity.

Vapour pressure

Water vapour is invisible. Steam and mist, which can be seen, are suspended droplets of water liquid, not water vapour. The molecules of water vapour rapidly occupy any given space and exert a *vapour pressure* on any surface in which they are in contact. The molecules of water vapour in the air exert a pressure that increases as the amount of water vapour increases. *Vapour pressure* is then the partial pressure exerted by the molecules of a vapour and in the SI system is measured in Pascals (Pa).

This pressure of moisture in the air behaves independently of the other gases in the same sample of air. The pressure of this water vapour can, therefore, be considered independently of the other gases in the air. This is a manifestation of *Dalton's law of partial pressures*, which states that the pressure of a mixture of gases is the sum of the partial pressures of the individual gases present.

If the air space above any liquid is enclosed, then the evaporated vapour molecules collect in the space and the *vapour pressure* increases. Some molecules are continually returning to the liquid state and eventually the number of molecules evaporating is equal to the number of molecules condensing. This is dynamic equilibrium and the air in the space is then said to be *saturated*.

The vapour pressure of any substance increases as the amount of vapour evaporating from a surface increases and at saturation the vapour pressure reaches a steady value called

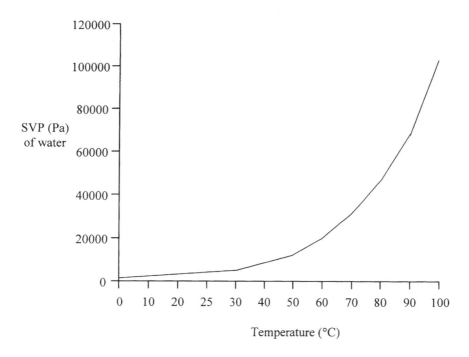

Figure 3.13 Saturated vapour pressure (SVP) of water vapour as function of temperature.

the *saturated vapour pressure*. The saturated vapour pressure of water (SVP) is, therefore, defined as the vapour pressure of the water vapour in an air sample that contains the maximum amount of vapour possible at that temperature.

Saturated vapour pressure is found to increase with increase in temperature. Figure 3.13 shows the relationship between the saturated vapour pressure of water vapour at different temperatures. An increase in vapour pressure indicates an increased moisture content. The saturation of a fixed sample of air is delayed if the temperature of that air is raised. This property gives rise to the important general principle: *warm air can hold more moisture than cold air.*

If an unsaturated sample of moist air is cooled then, at a certain temperature, the sample will become saturated. If the sample is further cooled below this *dew point temperature* then some of the water vapour must condense to liquid. This condensation may occur on surfaces, or inside materials, or around dust particles in the form of particles of cloud or fog as discussed in Section 8.6. The *dew point* is, therefore, defined as the temperature at which a fixed sample of air becomes saturated.

A sample of air with a fixed moisture content has a constant dew point, even if the air temperature changes. The dew point is particularly relevant in the study of condensation in buildings. Dew points can be measured directly by instruments that detect the first signs of misting on a silvered surface, similar to the observation of the first signs of misting on a bathroom mirror when you have a shower or bath, while also measuring the temperature of the surface. However, it is often more convenient to measure other qualities of moist air, such as the temperatures of a dry thermometer bulb (dry bulb) and a damp thermometer bulb (wet bulb). There is a constant relationship between these pairs of temperature and equivalent dew points, so that the dew point can be found from standard graphs and tables. Measurement of humidity is also an essential part of weather monitoring and is discussed in more detail in Section 8.2.4.

3.6.2 Humidity

Humidity is a measure of the moisture content of the atmosphere. Most of the moisture in the atmosphere is a result of evaporation from the surfaces of the sea, land and vegetation. At any particular place natural humidity is dependent on local weather conditions but inside a building humidity is further affected by the thermal properties and the use of the building.

The *relative humidity* (RH) of a sample of air compares the actual amount of moisture in the air with the maximum amount of moisture the air can contain at that temperature. The definition of relative humidity is

$$RH = \frac{\text{vapour pressure of a sample}}{\text{SVP of the sample at same temperature}} \times 100\% \qquad (3.6)$$

It is also common practice to describe humidity in terms of percentage saturation, which is defined by:

$$\% \text{ saturation } = \frac{\text{mass of water vapour in an air sample}}{\substack{\text{mass of water vapour required to} \\ \text{saturate the sample at same temperature}}} \times 100\% \qquad (3.7)$$

Percentage saturation and relative humidity are identical only when air is perfectly dry (0%) or fully saturated (100%). However, for temperatures in the range 0–25°C, the difference between relative humidity and percentage saturation is small.

A convenient alternative formula for relative humidity is:

$$RH = \frac{\text{SVP at dew point}}{\text{SVP at room temperature}} \qquad (3.8)$$

A relative humidity of 100% represents fully saturated air, such as occurs in a mist or fog. A relative humidity of 0% represents perfectly dry air; such a condition may be approached in some desert conditions or permanently frozen arctic environments.

The SVP or saturated moisture content varies with temperature; therefore, relative humidity changes as the temperature of the air changes. Despite this dependence on temperature, relative humidities are a good measurement of how humidity affects human comfort and drying processes. Since warm air can hold more moisture than cool air, raising the temperature increases the SVP or the saturation moisture content and so RH decreases. This property explains why heating the air lowers the RH and cooling the air increases the RH of any sample of air.

The actual quantity of water vapour present in the air is defined by the *moisture content*:

$$\text{Moisture content} = \frac{\text{mass of water vapour in an air sample}}{\text{mass of that air sample when dry}} \qquad (3.9)$$

Moisture content is not usually measured directly but it can be obtained from other types of measurement, most commonly from the humidity. Moisture content is needed, for example, in determining what quantity of water in an air-conditioning plant needs to be added or to be extracted from a sample of air.

Worked example 3.5   Water vapour pressure of air.

A sample of air has a relative humidity of 40%. Given that the SVP of water vapour = 2340 Pa, calculate the vapour pressure of the air at 20°C.

Solution
Relative humidity (RH) = 40%, vapour pressure (VP) = x, SVP = 2340 Pa.
Using

$$\text{RH} = \frac{x(\text{VP})}{\text{SVP}} \times 100$$

$$x = \text{VP} = \frac{2340 \times 40}{100} = 936 \, \text{Pa}.$$

An alternative way of expressing humidity, which is widely used when discussing evaporation processes, is in terms of the *saturation deficit* (*D*) of the air. Saturation deficit is a measure of the amount of water vapour required to be added to the air in order for the air to become saturated with water vapour. In terms of vapour pressure, the saturation deficit is the difference between the actual vapour pressure of the air and the saturated vapour pressure at the same temperature. In Chapter 10, the saturation deficit is expressed in terms of the concentration of water vapour ($g\,m^{-3}$) rather than the vapour pressure, where the concentration of water vapour and the vapour pressure are related via the gas laws, such that:

$$\text{concentration} = 2.165 \times \text{vapour pressure/air temperature}, \tag{3.10}$$

where the concentration is in $g\,m^{-3}$, vapour pressure is in kPa and air temperature is in Kelvin.

Worked example 3.6   Saturation deficit of the air.

Take the case above of air at 40% relative humidity and at 20°C. The saturated vapour pressure of the air is 2.34 kPa. Determine the saturation deficit.

Solution
The concentration at saturation (equation 3.10) is $18\,g\,m^{-3}$. The *actual* concentration is $40\% \times 18 = 7.2\,g\,m^{-3}$, so the saturation deficit is:

$$18 - 7.2 = 10.8\,g\,m^{-3}.$$

This is a useful way of expressing humidity when considering evaporation because $D = 10.8\,\mathrm{g\,m^{-3}}$ represents the magnitude of the difference in vapour concentration between the air immediately adjacent to the wet surface from which water is evaporating and that in the overlying air (assuming that the surface and air temperatures are similar). It is this difference in vapour concentration that is one of the factors controlling the rate at which water evaporates, as will be discussed in Chapter 10.

3.6.3   Condensation in buildings

Condensation in buildings occurs whenever warm moist air meets surfaces whose temperatures are at or are below the dew point of the moist air. It is convenient to classify the effects of condensation into two main types: *surface*, which occurs on the surfaces of the walls, windows, ceilings and floors; and *interstitial*, which occurs within the structure of a building.

Surface condensation appears as a film of moisture or as beads of water on a surface and is most obvious on the harder, more impervious surfaces, such as a pane of window glass or a cold water pipe where visible misting or droplets of water are seen. The initial surface condensation, occurring on the surfaces which are at the dew point temperature, also provides a 'map' of areas of a room which have the least effective thermal insulation.

Interstitial condensation occurs within the materials of a building, such as in walls and roofs. Although unseen, the dampness caused by this interstitial condensation can damage important structural materials, such as steel work. It also causes thermal insulating materials, such as expanded plastics, to be less effective as their thermal conductivity increases when they are wet. Water vapour can reach the interior of structures because building materials such as plaster, timber and brick are relatively permeable to water vapour.

The passage of water vapour through a typical wall or roof is driven by the difference in vapour pressure between the inside and outside. In most northern European countries, such as Britain, the air inside the building is usually at a higher temperature than outside. Therefore, the vapour pressure is higher inside than outside and the water vapour will then move from inside to outside. Knowing this direction of flow allows designers to specify vapour-resisting layers, such as polythene, where analysis shows there is a risk of low dew points within the structure. Figure 3.14 shows a wall with a region where water vapour might condense if vapour is allowed to penetrate the wall.

Remedies for condensation

The various preventive measures for condensation can be summarized under three main mechanisms:

- *Ventilation*: helps to physically remove the moist air.
- *Heating*: allows the air to hold more moisture as vapour and helps to keep surface temperatures above the dew point temperature, and
- *Insulation*: helps to keep surface temperatures above the dew point temperature.

A combination of these three remedies is usually necessary for controlling condensation. In domestic homes, for example, it is helpful to have enhanced ventilation (e.g. extractor fans) in moist places, such as kitchens; while in rooms like bedrooms, a smaller but a constant

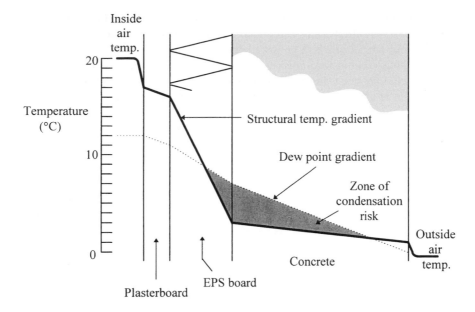

Figure 3.14 Sample profile of temperatures within a structure suggesting where condensation may occur.

source of ventilation is required (usually through a window). Background levels of heating in all areas, even those not permanently occupied, also helps to prevent condensation by allowing the air to safely hold more water vapour and by keeping surfaces at temperatures above the dew point. Dehumidifiers and 'anti-condensation' paint can be useful for controlling temporary condensation but are not usually a permanent remedy for condensation.

To prevent possible interstitial condensation in the structure of buildings, such as within walls or roofs, it is also possible to install a layer of material which has a high resistance to the passage of water vapour (Figure 3.15). The material used for such a 'vapour

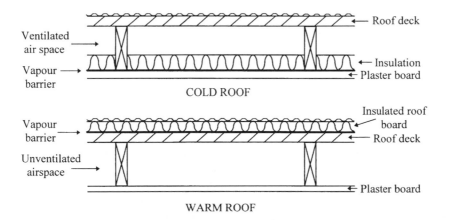

Figure 3.15 Roof construction showing use of ventilation and vapour barriers to prevent the risk of condensation.

barrier', or 'vapour check', is typically a layer of polythene or aluminium foil and it is usually attached to another component, such as plaster board or insulation board. To be effective, a vapour barrier needs to be installed on the warm side of the insulation layer to prevent the vapour coming into any cooler parts of the structure where the vapour will become saturated. Like the walls of a balloon, a vapour barrier is only effective if it is free from defects such as gaps, holes or tears. The processes of construction often makes holes and gaps so the success of a vapour barrier depends on the correct design and careful installation.

Christopher Wren (1632–1723) was born in Wiltshire, England. In addition to being an outstanding master builder architect, he was both a mathematician and a scientist, being interested in anatomy and physiology. In 1657 he was appointed Professor of Astronomy at Gresham College, London, at the age of 25. He was a founding member of the Royal Society (1660), and subsequently was its President. In 1661 he was appointed Savilian Professor of Astronomy in Oxford, and while there he designed several college buildings including the Sheldonian Theatre. Following the Great Fire of London (1666) he was appointed Surveyor-General (1669) and was commissioned to rebuild 55 out of 87 churches, including the stupendous St Paul's Cathedral (1675–1710) [see right].

3.7   Buildings of the future

Buildings of the future need to offer better comfort and convenience than the buildings of today and at the same time they must use the resources of the Earth in a more responsible manner. Homes are a major factor in the perceived standard of living and most people expect them to improve with time. They also expect the quality of other buildings that they use, such as offices, shops and schools to steadily improve. The checklist for a future house below contains features that are attainable now; we do not need to wait until the year 3001!

3.7.1   Checklist for a future house

- Building positioned and orientated to make best use of site and local climate
- Agreeable visual features acceptable to others
- Building materials produced from recycled or sustainable sources, such as renewable forests
- Building materials requiring low energy input during manufacture
- Safe environment with low danger from toxic materials and pollution
- Minimum energy losses through fabric by using maximum thermal insulation
- Minimum energy losses through air infiltration loss by control of openings, such as door lobbies

- Use of passive features, such as the thermal capacity of materials and natural convection currents
- Use of natural energy sources, such as wind power, photovoltaic cells and solar water heating (Chapter 5)
- Local conversion of fuel to thermal energy, such as combined heat and power generation
- More efficient conversion of fuels, such as by installation of condensing boilers
- More efficient use of energy by appliances, such as energy-efficient fridge-freezers
- More efficient use of energy for lighting, such as compact fluorescent lamps (CFL)
- Water-saving devices such as showers and water-saving toilets
- Reclamation of energy, such as extraction of thermal energy from exhausted air and from warm waste water
- Intelligent controls for heating and lighting that sense when energy is not required
- Low running costs for energy use and for maintenance.

Case study 3.1   Energy-efficient buildings.

An energy efficient office
Offices for the Building Research Establishment (BRE) at Garston in Watford (Figure 3.16) have the following features:

- Maximum use of recycled and waste materials
- Use of natural ventilation using the stack effect
- Use of daylight with glass areas optimized to give low-energy losses and allow solar heating
- Use of the building's mass to moderate temperature. Concrete floor slabs can also be cooled by recycled well-water

Figure 3.16 Building Research Establishment at Garston, Watford.

References

Burberry, P., *Environment and Services*, 8th edn. Harlow: Longman, 1997.

Havrella, R. A., *Heating, Ventilating and Air Conditioning Fundamentals*. Englewood Cliffs: Prentice-Hall, New Jersey 1995.

Littler, J. and Thomas, R., *Design with Energy: The Conservation and Use of Energy in Buildings*. Cambridge: Cambridge University Press, 1984.

McMullan, R., *Environmental Science in Building*, 4th edn. London: Macmillan Press, 1998.

O'Reilly, J. T., Hagan, P., Gots, R. and Hedge, A., *Keeping Buildings Healthy: How to Monitor and Prevent Indoor Environmental Problems*. Chichester: Wiley, 1998.

Oliver, A., *Dampness in Buildings*, 2nd revd edn by J. Douglas and J. S. Stirling. Oxford: Blackwell, 1997.

Seeley, I. H., *Building Technology*, 5th edn. Basingstoke: Macmillan, 1995.

Smith, B. J., Peters, R. J. and Owen, S., *Acoustics and Noise Control*, 2nd edn. Harlow: Addison Wesley Longman, 1996.

Vale, B. and Vale, R., *Green Architecture: Design for a Sustainable Future*. London: Thames and Hudson, 1998.

Other sources

BRE Digests, BRE Information Papers, and other information from the Building Research Establishment [http://www.bre.co.uk].

CIBSE Guides, from the Chartered Institution of Building Services Engineers, London [http://www.cibse.org].

CIBSE Code for Interior Lighting, from the Chartered Institution of Building Services Engineers, London [http://www.cibse.org].

Energy Audits and Savings, from the Chartered Institution of Building Services Engineers [http://www.cibse.org].

Information on National Energy Saving from the Energy Saving Trust [http://www.est.org.uk].

Discussion questions

1   Choose three different insulating materials used in modern buildings. For each material, use the physical properties of the material and the principles of energy transfer to explain why the material acts as a good thermal insulator.

2   Use definitions, units and examples to distinguish between thermal conductivity (k) and thermal transmittance (U).

3   A factory roof has an external surface of aluminium alloy. Describe in terms of thermal transfer and other principles of physics the thermal behaviour of the roof during the course of a winter's day and a night under clear skies.

4   Select three energy-saving measures and discuss the difficulties of implementing these measures. In particular, give and explain your expectation of the take-up of these measures in: (a) 5 years' time and (b) in 10 years' time?

5   A customer likes to drink his tea white and hot. Should the waiter add the milk to the hot tea immediately at the time of serving or let the tea stand for a finite time before adding the milk to ensure the hottest yet whitest tea mixture?

Quantitative questions

1   Make a visual survey of two contrasting houses in your area, one older and one more recently built. Note the approximate era of construction, perimeter area, height, the wall type, roof type and glazing type used in each building. Use Table 3.5, or similar data, to establish approximate

U-values for the construction. Use the information to compare, in percentage terms, the rate at which each building loses energy through its external walls, roof and windows.

2   The outside wall of a building consists of two layers of brick each 10 cm thick with a thermal conductivity $0.54\,W\,m^{-1}K^{-1}$. The two layers are separated to provide an airspace 5 cm thick and having a thermal resistance $0.180\,m^2\,KW^{-1}$. Assuming that the thermal resistances of the inner and outermost surfaces are 0.123 and $0.055\,m^2\,KW^{-1}$ respectively, (i) calculate *U* for the wall and (ii) what would the *U*-value of the wall be if the cavity was filled with polyurethane foam of thermal conductivity $0.026\,W\,m^{-1}K^{-1}$? If a vertical sheet of aluminium foil had been placed in the cavity during construction, what effect would this have had on *U*-value of the wall?

3   A simple building has internal dimensions of $11 \times 4 \times 3\,m$ high. Of the wall area, 20% is glazed and the doors have a total area of $6\,m^2$. *U* $(W\,m^{-2}K^{-1})$ is: walls 1.6, windows 5.5, doors 2.5, roof 1.5 and floor 0.8. The inside air temperature is maintained at 18°C when the outside air temperature is −2°C. There are four air changes per hour and the volumetric specific heat capacity of air is $1300\,J\,m^{-3}K^{-1}$. The heat gains total 2200 W.
   (a)   Calculate the net rate of energy loss from this building.
   (b)   Calculate the surface area of the radiators required to maintain the internal temperature under the above conditions. The output of the radiators is $440\,W\,m^{-2}$ of radiating surface area.

4   A room has $7.5\,m^2$ area of single-glazed windows, which have $U = 5.6\,W\,m^{-2}K^{-1}$. It is proposed to double-glaze the windows and reduce *U* to $3.0\,W\,m^{-2}K^{-1}$. During a 33-week heating season, the average temperature difference across the windows is 7 K.
   (a)   For both types of glazing, calculate the total energy lost during the heating season.
   (b)   Obtain current figures for the cost of electrical energy and the approximate cost of double-glazing such windows. Estimate the number of years required for the annual fuel saving to pay for the cost of the double-glazing.

5   *U*-values of four construction components are given below:

Component	U $(Wm^{-2}K^{-1})$
Single-glazed window	5.6
Double-glazed window	3.2
Non-insulated roof	1.9
Well-insulated roof	0.4

A house has single-glazed windows of $24\,m^2$ and an non-insulated roof of area $60\,m^2$. The occupier heats the house for 3000 hours per year to a temperature which, on average, is 14 K above that of the air outside. Calculate the energy lost per year through: (a) the single-glazed windows and (b) the non-insulated roof, expressing your answer in kWh.

   If electricity costs 5.5 pence per unit, calculate the annual savings the occupier could make by (c) installing double-glazing and (d) insulating the roof. If double-glazing costs £3000 and roof insulation costs £100, which, if either, of the two energy-saving steps would you advise the occupier to take?

6   The average rates of energy loss for a particular house are 1580 W total fabric loss and 870 W ventilation loss. The seasonal heat gains of the house total 27 500 MJ. The fuel used has a calorific value of $32\,MJ\,kg^{-1}$ and the heating system has an overall efficiency of 75%.
   (a)   Calculate the input heat required during a heating season of 33 weeks.
   (b)   Calculate the mass of fuel required to supply one season's heating.

7   A sample of air at 1°C has a vapour pressure of 540 Pa. Calculate the RH of this air, given that the SVP of water is 1230 Pa at 10°C.

What are the key infrastructural requirements for a modern city?

- Sufficient food supplies for its population
- Sufficient water supply for domestic and industrial use
- Sufficient energy supplies for heating and lighting
- Transport links to/from the city
- Adequate sewage systems to allow waste disposal and
- Quality of housing to attract citizens.

A computer programme/game (*Sim City*) has been designed that enables the user to simulate the building and control of a city. In *Sim City* the player is invited to design a city, with all its necessary infrastructure. In planning residential, industrial and commercial areas together with the roads and power supplies needed to encourage the development of the city the 'mayor' has to decide the level of taxation required to improve and expand the city. City planners must try to strike the right balance between industry and residential areas, and the level of taxation such that migration into the town remains positive and that the 'residents' are satisfied. Should the mayor fail to achieve this s(he) will soon be voted out of office! Eventually the player can build a complete city with train systems, an airport/seaport, hospitals and even sports stadia. *Sim City*, although a game, allows the player to gain an appreciation of the difficulties of urban design with the attendant problems of pollution levels, population densities, crime and the funding needed to provide all the facilities of a modern city.

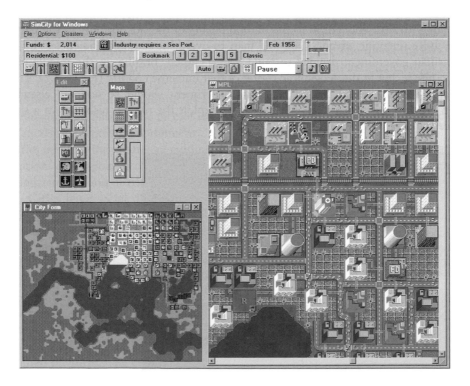

Figure 4.3 Sim City – a computer game in which the player designs a city.

4.2   Energy in the city

Modern civilization is a large energy consumer. Table 4.1 shows the average energy consumption of Britain between 1970 and 1989. With the increase in sole-occupancy households and the increasing number of domestic appliances energy consumption increased by some 10% and continued to increase throughout the 1990s until it will exceed 2000 PJ ($1 \, PJ = 10^{15} J$) in 2000. However, the development of more energy efficient devices and energy-saving building design (see Chapter 3) have led to an actual fall in energy per dwelling of some 10%.

Nevertheless, the actual consumption of any town or city is enormous. How is this energy demand to be met? How will it be supplied from power plants often many hundreds of miles from the city but closer to fuel sources? The energy that modern civilization utilizes is, to a large extent, electrical, and before discussing the transmission of electrical energy the major contribution of Michael Faraday should be cited.

Michael Faraday (1791–1867) was born in London, the son of a blacksmith and one of 10 children. He started work at 12 years of age as an apprentice to a bookbinder, and as a result he developed an interest in the books on science that passed through his hands. When he was 20 he attended lectures at the Royal Institution given by Humphrey Davy. He became increasingly interested in science and began conducting his own experiments. In 1813, he started work as a laboratory assistant at the Royal Institution, and in the years that followed developed further Davy's researches. Faraday made great discoveries in both physics and chemistry, particularly in the areas of electromagnetic induction, electrolysis, electrochemistry, electrical conduction, polarization and the Faraday effect, which is used in optic fibre communications. The idea of a 'field' was developed by him, and he is acknowledged as the father of electrical engineering and the modern electrical industry. He is possibly the greatest British experimental physicist of all time.

Table 4.1 Average consumption per dwelling in Britain.

Year	Number of households ('000)	Energy delivered (PJ)	Average external temperature (°C)	Average annual consumption per dwelling (GJ)
1970	17 759	1502	5.8	84.6
1989	21 340	1638	6.9	76.8

4.2.1   Electromagnetic induction

In 1819, the Danish physicist Hans Oersted discovered that a current-carrying conductor creates a magnetic field. For 7 years Faraday searched in vain for the opposite effect, i.e. to generate an electric current from a magnetic field. Eventually, his discovery that the relative motion between a magnetic field and a coil could generate an electric current by electromagnetic induction would have profound implications for the electrical energy generated by wind energy (see Section 5.6) and hydroelectric power (see Section 5.7).

Continually inserting and removing a magnet into and out from a coil induces an alternating voltage as the lines of magnetic flux, that comprise the magnetic field, cut the turns of the coil. As an analogy, suppose the magnet is rotating inside a fixed coil, as in a bicycle dynamo (Figure 4.4).

The magnet's motion can be depicted as a radius vector rotating anti-clockwise. The position of the base of the perpendicular on the diameter XOY changes with time as the sinusoidal trace shown in Figure 4.5.

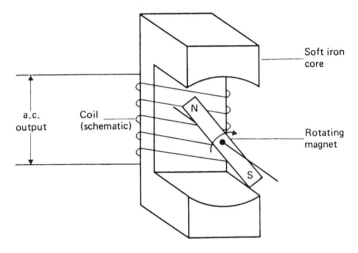

Figure 4.4 Schematic diagram of a bicycle dynamo.

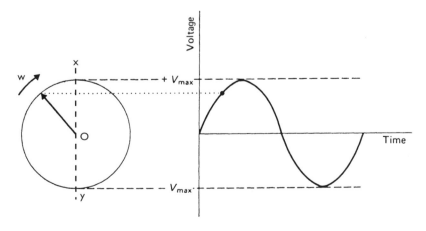

Figure 4.5 Vectorial representation of a rotating magnet.

This curve will have the expression $V = V_o \sin \omega t$, where V is the voltage at any time t, V_o is the maximum voltage, ω is the angular frequency $(= 2\pi f)$ and f is the frequency (Hz). The electrical energy is generated in power stations and transmitted for industrial, commercial and domestic use through the National Grid. Alternating (AC) voltages are preferred to direct current ones because AC voltages can be transformed (i.e. reduced or increased) and because the matching of high AC voltages and low currents produces low power losses, even over considerable distances. The power loss is calculated by $P = I^2 R$.

4.2.2 Electrical power transmission

Low AC voltages are more practical (and safer) for local use since they are more easily insulated against breakdown than high voltages. However, it is far more efficient to transport electrical energy at high voltages. This can be demonstrated by calculating the power loss (see Worked example 4.1). *Transformers* allow these different voltage requirements to be met.

To facilitate AC transmission a *three-phase system* is used employing three conductors. Imagine that the radius vector that 'represented' one rotating magnet is now replaced by three of them (one for each magnet), with the coils aligned at 120° to each other (Figure 4.6) four connecting wires are needed, of which one would be the neutral. The other three would be transmitting the phases. If the current in the neutral conductor is zero, then it will be unnecessary and only three conductors will be required.

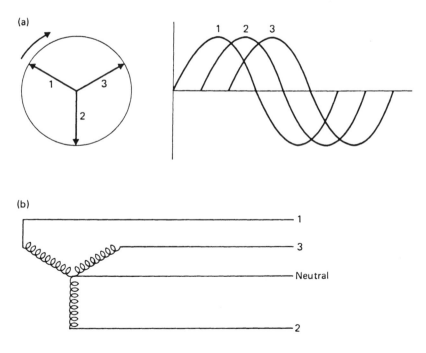

Figure 4.6 Three-phase system.

Worked example 4.1   Transmission along power lines.

Electrical power of 150 kW is provided to a college through power cables of resistance 0.5 ohm. If the potential difference at the college is 12 kV, determine (a) the potential drop cross the cables and (b) the percentage power loss.

Solution

(a) Since power = voltage × current, the current flowing $I = P/V = 150\,000/12\,000$
    $= 12.5\,A$.
    Applying Ohm's law, the potential drop across the cables:
    $V = $ current × resistance $= 12.5 \times 0.5 = 6.25\,V$.

(b) Power loss $= I^2R = (12.5)^2 \times 0.5 = 78.1\,W$.

The % power loss $= 78.1/150\,000 \times 100 = 0.05\%$. Note how small this is.

4.3   Transportation

Essential to the development of civilization has been the ability of people to travel and transport goods to and from towns and cities. The acquisition of food from the countryside, pure drinking water from neighbouring rivers and lakes, movement of armies for common defence, or simply the latest technological developments can only be achieved by the provision of an adequate transportation network. Those civilizations that have been largely cut off from the rest of the world (perhaps by geography, perhaps by choice) have ultimately stagnated and proven unable to adapt once their insularity has been broken by contact with the outside world. Indeed, the history of the world might be largely written as one of transportation and exploration with the development of the wheel and the boat being the most important of all of humanity's inventions. Ancient civilizations are often marked by the remains of their transport systems, canals, roads, aqueducts and today transportation plays a major role in everyday life. Today the design of any urban environment is dominated by the need for roads, indeed throughout most of the industrialized world the land area dedicated to roads is larger than that dedicated to natural parks. The energy used to power transportation is between 20 and 25% of the total energy needs of the country and on average 10% of any individual's disposable income will be spent on transportation (only food and housing having higher percentages). In the USA one-fifth of the gross national product is involved in the transport industry and one-sixth of the workforce!

By far the most popular form of transportation is road transport which in the USA accounts for 80% of the total (air transport is the next highest at 13%) and nearly 50% of the total oil consumption. Since the 1950s the growth in road transport has been dramatic (Table 4.2). The convenience, flexibility and personal freedom provided by the motor car means that there is little prospect of such growth slowing in the near future. Indeed, since the car is to many a considerable status symbol, the world market for cars is likely to increase as people in the 'developing' countries gain incomes that allow them to purchase

Table 4.2 Global growth in road transport since 1940.

Year	No. of motor vehicles (millions)
1940	25.6
1950	62.5
1960	129.6
1970	240.1
1980	409.6
1990	656.1

cars for the first time. Hence, an increasing proportion of the global energy supply will be required to support transportation while the problems for urban planners will multiply and the health implications from car exhaust emissions (see Section 4.9) will become more apparent.

4.3.1   Energy efficiency in transport

Two methodologies might be adopted to curb the increasing energy use of transport. The first is to plan cities such that the need for personalized transport is reduced – work, shopping and leisure needs all being catered for by an efficient public transport system. However, the provision of such an infrastructure requires large capital investment that few governments are willing to contemplate since they would in turn require significant increases in taxation, either indirect (through tolls and road taxes) or direct (e.g. through income tax).

Alternatively more efficient use must be made of the fuel used. While one cannot expect all cars to match the car that in 1992 covered 1200 km on 1 gallon of petrol, technology can greatly increase the efficiency with which we use fuel in present cars, planes and ships (Figure 4.7).

In determining how any system of transport may be made more efficient, it is necessary first to determine the power needed by the vehicle and, hence, the forces of resistance it must overcome to fulfil its purpose (Figure 4.8). The total force that must be provided for satisfactory performance by the vehicle (mass, m) can be written as:

$$F = F_a + F_h + F_r + F_{ad}$$
$$= ma + msg + C_r mv + C_{ad} v^2$$

where F_a ($=ma$) is the force needed to provide the acceleration (a) or change of speed; F_h ($=msg$) is the force needed to overcome a hill slope s; rolling friction F_r ($=C_r mv$) is the force necessary to overcome resistance that is characterized by a factor C_r (e.g. frictional losses of the road and within the axle); a force which is also dependent on the speed of the vehicle (v) and F_{ad} ($=C_{ad} v^2$) the force needed to overcome aerodynamic drag upon the vehicle, which is dependent on v^2, being small at low speeds but increasing rapidly as speed increases. The magnitude of the aerodynamic drag depends on a variety of factors such as the shape, size, smoothness of the outer surface and whether there is a tail wind or cross wind, all of which are conveniently folded into a single coefficient (C_{ad}). Indeed, for most vehicles travelling above 60 km h^{-1} aerodynamic drag is the main reason for the decrease in fuel economy. Today, car manufacturers are using the opportunities made

Figure 4.7 An energy efficient car in Munich, Germany.

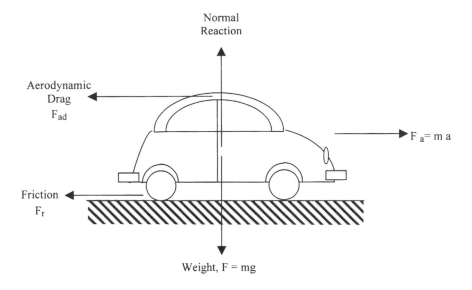

Figure 4.8 Forces on a car.

available by computers to design ever more aerodynamically-efficient cars; thus 'stream-lining' the car to minimize C_{ad}. Small changes to the design such as setting the headlights flush with the body and using flatter wheel covers have led to reductions in C_{ad} from an average of 0.5 in the 1970s to 0.35 in the 1990s. Such changes can lead to substantial gains in fuel economy, such that modern cars are 10–20% more efficient in their use of fuel than those on the market 10 years ago. Further developments (including the use of lighter materials) will lead to even more efficient cars and coincidentally will ensure that the cars of the twenty-first century (Figure 4.7) look as different to our children as the cars of the 1920s look to us today!

However, it is equally important for car owners to remember that the efficiency of a car is also dependent on the state of maintenance. Frequent tune-ups with attention to the ignition timing, spark plugs, air-filter, correct air–fuel mixture adjustment and proper infla-tion of the tyres (to reduce the resistance on the road) can easily provide a 10% fuel saving.

4.4   Water for the urban environment

Essential to the development of any human community has been the provision of fresh water supplies. Hence, many of the oldest towns and cities were built on or close to rivers. If the river ran dry or the size of population exceeded the local water supply, then the city was soon abandoned, e.g. Fatehpur Sikri (Figure 4.9) the old capital of Muslim India. Each

Figure 4.9 Fatehpur Sikri, the ancient capital of Muslim India abandoned after only 50 years owing to a lack of water.

person requires 1 litre of water a day to survive; we can lose nearly all of our body fat, half our protein and survive, but if we lose more than 10% of the water in our body we die. In the USA the average home uses about 185 gallons of water a day for drinking, washing, household needs, cooking and irrigation of lawns. If one includes the water needed for crop irrigation to provide us with food and the water used in manufacturing industry to provide electricity we each use, this rises to 1650 gallons per individual per day! The total fresh water supplies in the world is perhaps 9.8×10^{15} gallons, of which we could expect to exploit only about 30%. Hence, it would seem that we can ultimately support 5×10^9 people or roughly the current world population. Yet the world population is rising, and rising most quickly in those regions of the world where the supplies of fresh water are most limited (Asia and Africa). Hence, many scientists predict that in the twenty-first century one of the greatest problems facing humanity will be the adequate provision of water.

The efficient use of water is therefore a priority for both the global and urban environments. As Chapter 8 describes the global water cycle, the discussion will be limited here to the water needs in the urban environment. As a town grows so the need for water grows until it exceeds the local supply. Either the town has now reached its maximum size or new supplies of water must be found. The Romans faced with just such a problem built the first water transportation systems, piping water from rivers and lakes, often hundreds of miles away, through aqueducts to provide towns with new supplies. Most modern cities are fed from long-distance reservoirs, and the transportation (pumping) of such water supplies is another major energy consumption process.

4.4.1 Sewage

Once used, how do we dispose of the now polluted water? Once again the Romans faced with this problem developed a fairly extensive and effective sewage system. The Romans had toilets flushed with water, and the larger cities had sewage systems that carried the waste away from the city. However, with the collapse of the Roman Empire humanity reverted to unsanitary practices that led to the deaths of millions of people from the disease that such neglect of sanitation inevitably brings. Throughout the Middle Ages in Europe human waste was thrown out of windows into the street where open gutters eventually took it into the nearby river; often the same river that served as the drinking water supply. Not until the nineteenth century were cities again provided with public sanitation facilities to transport and treat the waste. Even today in many less well developed countries, sanitation problems remain and cholera and typhoid are still killing hundreds of thousands every year.

What is in sewage and how may it be treated? Urban sewage is 99.9% water and about 0.1% sand, grit and organic material. Modern sewage plants (Figure 4.10) are designed with primary and secondary stages of treatment. In the primary stage floating debris (usually organic) is removed by skimmers and the suspended inorganic matter (sand and grit) is filtered out; the 'solid waste' collected by these processes is then disposed of by burial or incineration.

Secondary processes involve chemical treatment of the sewage to remove:

1 Inorganic materials, such as salts that can pass through the first stage of the process unchanged, e.g. nitrates and phosphates.
2 Dissolved organic matter.
3 Bacteria and other disease carrying micro-organisms.

Figure 4.10 Modern sewage treatment plant under construction in south-west England.

In (2) the sewage is subject to biochemical action and undergoes decomposition by bacteria in large tanks. The process usually takes place in water in which there is no free oxygen and hence is an *anaerobic* process. In the case of sugars, it is a fermentation process, and with proteins it is putrefaction. The anaerobic process produces volatile effluent, e.g. carbon dioxide, ammonium ions, methane, the characteristic rotten egg-smelling hydrogen sulphide (H_2S) and 'sludge'. Methane accounts for some 72% of the volatile products and may be used in the future as a fuel, while the sludge is often dried and either burnt for heat or used as a fertilizer.

Bacteria and the micro-organisms are killed by addition of chlorine to the effluent before it is allowed to leave the sewage plant and is returned to the river or lakes system from where it may be drawn upon again and recycled. Indeed, in London it has been estimated that the drinking water from the domestic tap has been recycled four or five times.

Today industrialized countries are demanding ever better quality water supplies with legislation being passed to determine what levels of pollutant are acceptable in the drinking water. This has led to the development of refined sewage plants in which a tertiary stage of treatment ensures that such quality targets are met but these are extremely expensive to construct and it will be many years before they are available world-wide. Nor do most of these tertiary systems produce water that many believe to be fit to drink and, therefore, water is treated still further before it is allowed to go through to the domestic tap. Indeed, it has been felt for many years that the deliberate addition of some chemicals (such as fluoride) might be beneficial to human health. Fluoride helps to protect teeth against the decay arising from the increasing amount of sugar eaten in modern diets. However, the addition of fluoride to water has met with considerable public opposition in some countries as a tiny minority are believed to have suffered some side-effects.

4.5   Lighting

Light allows us to carry out tasks and activities safely and easily. The development of artificial light sources has allowed humanity to expand the working day from the sunlit hours, (which limited the hours our ancestors worked) to a 24-h day. Today artificial lighting accounts for a significant proportion of the energy used in the urban environment.

The need for lighting in towns and cities has been recognized by town planners since the earliest times. Legislation providing adequate lighting for urban populations have been in force for over a century. Should anyone wish to build an extension to their house, they (or the builder) must obtain 'planning permission'. If the extension restricts the provision of light to a neighbouring property the owner can object to the local authorities and if they find that this home will suffer from a significant loss of light, planning permission will be refused. Similarly, in the office, school, lecture theatre and classroom there are recommendations for the amount of light that should be provided for safe working practice.

The light received on any surface can be measured in terms of the energy density (W m^{-2}) in the same way as any other type of electromagnetic radiation, but the effect of light on a human eye is dependent on the sensitivity of the eye and, therefore, special units have been developed for the measurement of light.

Luminous intensity (I) is the power of a light source, or illuminated surface, to emit light in a particular direction, the unit of which is the *candela* (cd), a basic unit of the SI system. *Luminous flux* (F or Φ) is the flow of light energy and is measured in lumen (lm), while the *illuminance* (E) is the density of luminous flux reaching a surface measured in lux (lx). Typical values of illuminance experienced are 0.1 lx in moonlight, 5000 lx when under an overcast sky and 50 000 lx when standing in bright sunshine. For design purposes, *natural light* is assumed to come from an overcast sky and *not* from direct sunlight, the natural light through windows into a room is usually only a small fraction of the total light available from a complete sky. Table 4.3 shows the recommended values of illuminance while working in a classroom or office.

Table 4.3 Typical lighting measurements (lx).

Office desk	500
Drawing board	750
Assembly work	500
Electronic work	1000

Another quantity used when designing the lighting of a house or street is the *daylight factor*. This is defined as the ratio between the actual illuminance at a point inside the room and the illuminance possible from the sky. Thus, the daylight factor outside a building is 100% while inside a building near a window daylight factors of 20% are typical. While working at a desk area, daylight factors should be 2% minimum and 5% is recommended. Should this factor not be met, there is obviously a need for additional light sources in the room.

4.5.1 Sources of light

Today *artificial light* is generated by one of two types of electric lamp. Incandescent lamps, which produce light by heating substances to a high temperature at which they are luminous, and discharge lamps, which produce light by passing an electric current through a low-pressure gas or vapour until they are ionized and can conduct, forming a luminous arc or discharge between the electrodes.

The most familiar lamp is the tungsten filament lamp, or 'light bulb'. The light bulb is an incandescent lamp in which a coil of tungsten wire is heated to around 3000°C by its resistance to the electric current. The filament is situated within a glass bulb filled with low-pressure inert gases (helium or argon), which prevent the filament from burning away. It is sometimes mixed with a small amount of halogen (bromine or iodine). They are easy to install and cheap to produce. The modern light bulb differs only slightly from the earliest design of Thomas Edison's in 1879.

In contrast, the light given off by a gas discharge seems unpromising in its colour quality but the rate of light output, in lumens, is high for a low rate of electrical power. In 1910 the French chemist Georges Claude produced the first commercial discharge lamp by passing current through neon gas, producing the orange-red light now so apparent in advertising. The most popular discharge lamp in use today is the familiar sodium street lamp which permits 15 times more light per watt input than a tungsten filament lamp, but the colour of the emitted light is the harsh yellow wavelengths characteristic of the sodium atoms; a glow that many increasingly find unsuitable, particularly as they are a major source of global light 'pollution'.

Fluorescent gas discharge lamps were introduced in 1939. A gas discharge, usually mercury, emits a high proportion of invisible ultraviolet radiation. A coating of fluorescent powder on the inside of the glass absorbs the UV radiation and re-radiates this energy in the visible part of the spectrum. The fluorescent coating therefore increases the efficiency of the system and allows the colour quality of the light to be controlled.

Given the immense market for artificial lighting, it is not surprising that there is considerable research into the development of more efficient and longer-life lamps. The ability of a lamp to convert electrical energy to light energy is measured by its efficacy. The *luminous efficacy* is given by luminous flux output/electrical power input and is measured in lumens/watt. The luminous efficacy of a lamp decreases with time and for a discharge lamp may fall by as much as 50% before the lamp fails. The nominal life of a lamp is usually determined by the manufacturer considering the failure rate of a particular model combined with its fall in light output. In a large installation, it is economically desirable that all the lamps are replaced at the same time on a specific maintenance schedule.

Table 4.4 shows the characteristics of typical modern electric lamps. The higher wattage lamps generally have the higher efficiency and longer life. Today, energy-saving bulbs are small fluorescent tubes that consume less electricity and last much longer than ordinary incandescent bulbs (Figure 4.11).

Table 4.4 Characteristics of electric lamps.

Lamp type	Typical efficacy (lm W^{-1})	Nominal lifetime (h)	Applications
Tungsten filaments	12	1000	homes, restaurants, hotels
Tungsten halogen	21	2000–4000	display lighting
Compact fluorescent	60	8000	homes, offices and public buildings
Tubular fluorescent	60	8000	offices and shops
Mercury fluorescent	60	8000	factories and roads
Mercury halide	70	8000	factories and shops
Low-pressure sodium	180 (at 180 W)	8000	roadways and area lighting
High-pressure sodium	125 (at 400 W)	8000	factories and roadways

Figure 4.11 A modern energy efficient lamp.

4.6   Urban pollution

Living in an urban environment has always been regarded as having one essential disadvantage; the urban environment is not 'natural' but is an artificial environment designed imperfectly by people. The many advantages of living in a town or city are offset by the consequence of placing a large population together in a confined space. The resulting pollution has implications for human health. It is often said that urban dwellers go to the countryside to breath in the 'clean air', suggesting that urban air is somehow 'dirty'. Air is never perfectly 'clean' even in the countryside, in the sense that it contains only nitrogen, oxygen, carbon dioxide, water vapour and the rare gases. Even before humanity made its impact on the planet there were many sources of trace gases, which are now described as 'pollutants'. 'Natural air pollution' could be caused by dramatic events such as volcanic eruptions releasing huge amounts of ash and gases (e.g. sulphur dioxide [SO_2]) into the atmosphere or by local forest fires with resultant soot particulates. However, with the development of industry new and ever-increasing amounts of pollutants are being introduced into the Earth's atmosphere with serious consequences for the global climate, e.g. increased carbon dioxide [CO_2] leading to global warming (see Section 7.8) and release of CFCs leading to ozone depletion (see Section 7.6).

The average adult male uses some 13 kg of air per day compared with the consumption of 1.2 kg of food and 2 kg of water. The quality of the air we breath is, therefore, at least as important as the quality of the food or water we consume. Hence, the presence of pollutants within the air we breath is seen to be a growing threat to our lives. Air pollutants can affect us directly by their toxicity or indirectly through their ability to trigger other illnesses (e.g. asthma and other respiratory problems). Determining what is an acceptable level of pollution and therefore enacting legislation to restrict the emissions is extremely difficult to do since it is obviously unacceptable to expose individuals in clinical trials to pollution levels which cause illness or fatality! Thus, data on the effect of pollution must be drawn from statistical correlations between mortality or illness and exposure to low doses, a study which is known as *epidemiology*. However, the natural variety inherent within the human population ensures that it is hard to ascertain any specific level deemed 'safe'. The safe exposure limit may be dependent on age, fitness, ancestry and social background. One of the most hazardous types of pollution is often self-inflicted by smoking tobacco. Cigarettes contain carcinogens and nicotine as well as producing sufficient carbon monoxide to lead to greater susceptibility to heart diseases and respiratory problems. Any survey of the effects of industrial pollution on the average person may crucially be dependent on whether they have previously smoked! Nevertheless, in the past two decades the importance of air pollution to human health has been appreciated and attempts made to ascertain the impact of industrial emissions on society. This has resulted in recommendations of limits for acceptable levels of some pollutants. Commensurate with these recommendations have been the development of methods to monitor pollution. Most cities now have a pollution monitoring network so that warnings may be given when the levels of nitric oxides, ozone and particulates exceed nominated levels. Such warnings allow the most vulnerable to take preventative measures (e.g. diagnosed asthmatics can carry an inhaler).

4.6.1   Urban pollutants

Pollutants can be divided into two categories, primary and secondary. Primary pollutants are the chemical species emitted directly, while secondary are those formed from primary pollutants by local chemistry. The greatest damage often comes from the secondary pollutants rather than the primary, for example while sulphur dioxide might be emitted from a power station its conversion into sulphuric acid is more damaging to the local environment.

Table 4.5 lists the estimated global emission of primary pollutants produced by global industrialization compared with emissions from natural processes. In can be seen that for most compounds natural processes are the larger contributor. Thus, in judging the effect of humanity on the planet, we are looking at small changes on a large baseline of natural processes with unknown fluctuations.

Nevertheless, in any urban environment it is possible to monitor increases in local atmospheric pollution and seek to determine at what concentrations they will provide a significant hazard to human health. The major hazardous pollutants in urban conurbations are carbon monoxide, nitric oxides, sulphur dioxide, ozone and particulate matter.

Table 4.5 Global emission of primary pollutants in 1976 (10^9 kg year^{-1}).

Species	Man-made	Natural
CO_2	2×10^4	10^6
CH_4	188	1800
CO	600	2500
SO_2	207	~10
H_2S	2	50
NO	90	1200
NH_3	7	1200

Carbon monoxide

Most air pollution arises either directly or indirectly from the combustion of fuels. The formation of carbon dioxide is the major consequence of burning fossil fuels and has led to claims of major changes in global climate due to global warming (see Section 7.8), but since CO_2 is relatively chemically inert its production does not lead to important consequences for human health. Incomplete combustion of carbonaceous materials can, however, lead to the formation of carbon monoxide (CO). A prime source of CO is the internal combustion engine. In London, the four million cars arriving daily can emit over 10 000 tonnes of CO or 2.5 kg per car per day!

CO is an odourless, colourless gas that is toxic, since it combines with the haemoglobin in the red blood cells more readily than oxygen. Hence, CO tends to block the distribution of oxygen around the body and can lead eventually to suffocation. Concentrations of 100 parts per million (ppm) in air for 10 hours will lead to headaches and reduced ability to think clearly, while 300 ppm will result in nausea and possible loss of consciousness, and 1000 ppm for just one hour can result in death within 4 hours. Fortunately, in most cities the average carbon monoxide levels are a few parts per million, but in periods of unusual weather conditions or by major road intersections levels of over 100 ppm have been reported and new regulations are to be introduced to limit CO emissions from cars and trucks such that maximum concentrations of 10 ppm over 8 h and 35 ppm in any single hour (the 'rush hours') are met in most urban environments.

Nitric oxides

In any nitrogen–oxygen mixture heated to above 1100°C nitric oxide (NO) is formed. Should the gas cool slowly then the NO may decompose to nitrogen and oxygen, but if the cooling takes place rapidly, as in most internal combustion engines, then the NO does not have time to decompose but is emitted into the car exhaust at concentrations of about 4000 ppm. Nitrogen oxide is a colourless gas that is toxic in high concentrations, but its toxicity is minor compared with the secondary chemical, nitrogen dioxide (NO_2), which it may form in an urban environment. NO_2 is also formed in the engine, but in small quantities. The majority of the urban NO_2 is formed by the reaction of NO with O_2, such that within 10 hours of emission 50% of the NO has been converted to NO_2. A reddish-brown gas, at about 0.1 ppm, it can be smelt and at 5 ppm long-term exposure can affect the respiratory system. At concentrations of 20–50 ppm damage to heart, lungs and liver can occur, and at 150 ppm serious (and potentially irreversible) damage to lungs can occur within 3 h of exposure. Alerts in cities are made once concentrations reach 4 ppm and the recommended standard is set at only 0.5 ppm!

Sulphur dioxide

Since sulphur is present in all fossil fuels (0.5–5% by weight in most coals, usually as iron sulphide, FeS_2) it is not surprising that upon being burnt such fuels release the sulphur in the form of sulphur dioxide (SO_2). The sulphur dioxide in the atmosphere is largely due to industry since there are only a few natural mechanisms that release sulphur, being mainly decay of organic matter and in volcanic eruptions. Most of the sulphur dioxide emissions arise from the burning of coal in power stations although in some instances sulphur emissions from vehicles may be important to local pollution levels.

On being released into the atmosphere the sulphur oxide is rapidly transformed into sulphuric acid, H_2SO_4, in the form of an *aerosol*. An aerosol is a suspension of particles in a liquid or gas the most obvious example of which is the liquid suspension in a spray can. Similarly, in the atmosphere the sulphuric acid particles are suspended in the air and quickly dissolve into water droplets such that they may act as condensation nuclei (see Chapter 8). Such particles are therefore one form of *acid rain* (see Section 4.8), which is responsible for the damage of trees and the decline of fish stocks in many freshwater lakes (Figure 4.12).

Buildings are very vulnerable to sulphur dioxide pollution since the carbonates of which marble, limestone and mortar are composed are readily attacked by acid rain (dilute sulphuric acid) to form soluble sulphates that are washed away by the rain. Hence, many old buildings and fine statues have been damaged by the increased production of sulphur dioxide until it has become necessary to remove many great works of art from the façades of buildings and city squares and replace them with copies to avoid their destruction. Similarly, sulphur dioxide can attack oxide coatings on metals and paints, thus requiring the frequent restoration of surfaces; perhaps easy on a domestic house but consider the effect on the Forth Road Bridge or the Golden Gate of San Francisco!

Since it is hard to differentiate between the effects of sulphur dioxide and sulphuric

Figure 4.12 Industrial emissions from a factory have been shown to cause damage to trees and people across wide geographical areas.

acid, setting limits for the acceptable emissions of sulphur dioxide is difficult. However, in combination with particulate emissions (see Section 4.6.2) measurable increases in respiratory illness and mortality rates have been found in cities where concentrations of SO_2 exceed 0.1–0.2 ppm. A survey in The Netherlands has suggested that there is a definite increase in the mortality rate from lung cancer and bronchitis for concentrations >0.4 ppm., while plants have been shown to suffer appreciable damage if exposed to 0.01 ppm for a year and 1 ppm for one hour. Hence, although in contrast to nitric oxides and carbon monoxide, sulphur dioxide emissions are small their effect on local health may be much greater.

To reduce global sulphur emissions, it has been necessary to reconsider the role of coal in the electricity production industry. Low sulphur-containing coals may be used in preference to local supplies (although with subsequent loss of local jobs); gas or oil can replace coal since they contain less sulphur, or the coal-burning plants can be modified to prevent the release of sulphur into the atmosphere. In general, low sulphur content coals are less efficient energy producers than higher sulphur-bearing coals and the extra transport costs may make their adoption unfeasible. Thus, some countries have fitted 'flue gas scrubbers' in their power stations to remove the sulphur dioxide from effluent; limestone, dolomite or some other alkaline substance being used to transform the sulphur dioxide into calcium sulphate ($CaSO_4$), a solid waste which is easily collected.

4.6.2 Particulates

Not all pollutants come in the gaseous form; indeed the most dangerous to human health may be in the form of *particulates*. These suspended particles in the atmosphere may come from a variety of sources both man-made and natural. Natural sources include sea salt, dust caused by erosion, soot from forest fires and volcanic eruptions. The presence of these particles are vital to the formation of clouds (see Chapter 8) since they are the nuclei around which the cloud droplets form.

Globally these sources produce annually 15 times more than the estimated man-made sources (mainly caused by combustion) and, occasionally, such as in a dramatic volcanic eruption, can lead to both local and global climate change. The eruption of Tambora in Indonesia in 1815 led to the year 'without summer' with subsequent crop failures and starvation across Europe and the Middle East, while after Krakatoa's eruption in 1883 all over the world people marvelled at the spectacular sunsets. It is now known that both of these events were a consequence of the expulsion of enormous amounts of particulate matter into the upper atmosphere that scattered the solar radiation and reduced the amount of sunlight reaching the ground. More recently examination of the effects of Mount St Helens in the USA, El Chichon in Mexico and Mount Pinatubo in the Philippines has shown that global surface temperatures may be reduced for several years after a large volcanic event (Figure 4.13).

The consequences of industrially-produced particulates are, however, usually more local and are limited to those areas where the particulates are produced. The greatest source of industrial particulate (>90%) is 'fly ash', made during coal combustion while petrol combustion only produces some 5% of the total particulate flux world-wide. The chemical content of the particulate is, however, fairly constant, being predominately carbon (soot) such that the health risk arises from the size of the particulates rather than their chemical composition. The particulates enter the human body through the nose where the larger particles might be removed by nasal hair or trapped in the mucus membrane of the airways. However, the smaller particles (<10 μm) can reach the lungs and there may line

Figure 4.13 Volcanic eruptions produce considerable amounts of particulate matter which may be detrimental to health of flora and fauna.

the air sacs (alveoli); directly reducing the efficiency of the lungs or through the release of chemicals stored within the particulates cause bronchitis and emphysema.

National monitoring of particulates of $\leqslant 10\,\mu$m (i.e. PM_{10}) is being conducted in many industrialized countries and the effect of long-term exposure to varying concentrations examined. In general, long-term exposures to particulate concentrations $>80\,\mu g\,m^{-3}$ have been found to lead to a general deterioration in health. Hence, national standards recommending that annual mean exposures should not exceed $75\,\mu g\,m^{-3}$ have been introduced in the USA, while the 24-h maximum that should not be exceeded more than once a year has been set at $240\,\mu g\,m^{-3}$. The cities in the USA currently average concentrations of $100\,\mu g\,m^{-3}$ with New York, Chicago and Pittsburg recording between 155 and $180\,\mu g\,m^{-3}$ and local maximum measurements $>2000\,\mu g\,m^{-3}$ have been recorded.

Particulates are removed from the atmosphere by sedimentation (i.e. gravity) but this is usually only effective for the largest particles ($>20\,\mu$m); smaller particles are removed by *impaction* with surfaces (i.e. collision and sticking), while wash-out by rain is also an effective method for the removal of pollutant particulates. Nevertheless, one cannot rely on nature to remove these hazardous substances and it is better to remove them at source. It is now possible to fit filters in the chimneys of power stations and industrial plants to remove most of the particulates. Generally, 99% of the fly-ash must be removed before the gas plume leaves the chimney. Two methods have been developed to meet these targets: one uses electrification to charge the particulates so that they can be collected in electrostatic filters; the other passes the gases through a porous air bag in which all but the smallest particulates are then trapped (like in a vacuum cleaner bag). In the next decade, as legislation of PM_{10} emissions is enacted, new and improved methods for the removal of particulate matter in chimneys and from the car exhaust will be needed.

4.7   Smog

Air pollution from burning coal has been a problem for centuries. In England as early as the fourteenth century Edward I forbade the use of coal because of the smell and smoke it produced, and one violator of that law was actually executed! However, with the expansion of cities and development of industry coal came to play an essential role in the development of civilization. The use of coal and the development of cities led to the production of *smog*. Smog is used to describe the combination of smoke and fog, for which London was so infamous in Victorian times, and which has been featured in many Hollywood films portraying such characters as Sherlock Holmes. British coal, having such a high sulphur content, readily formed aerosol particles which acted as nuclei for fog formation. In 1952, a particularly severe smog settled over London for several days and as a result some 4000 people were estimated to have died as a result of respiratory and coronary failure (see Case study 4.1). Following this dramatic event parliament passed the Clean Air Act (1956), which stipulated the types of fuels that could be burnt and the kinds of smoke that might be emitted in both industrial and domestic use. The ban on all but 'smokeless fuels' has all but removed smog from London and other European cities, but elsewhere in the world 'pea-souper' fogs may still have to be endured by the population.

In 1943 Los Angeles began to experience a new form of smog, caused by photochemistry. It was found that sunlight broke down the nitrogen dioxide NO_2 to form NO and oxygen atoms:

$$NO_2 + sunlight \rightarrow NO + O$$

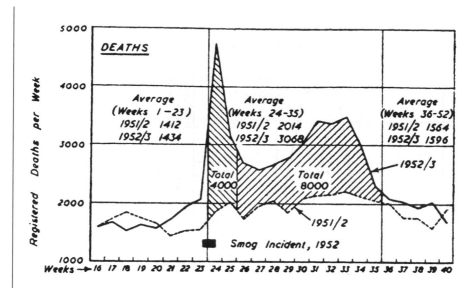

Figure 4.15 Mortality and air pollution over an extended period, 1952–53.

The official figure of 4000 could, therefore, have masked a possible 12 000 deaths. This raises interesting issues about the interpretation of medical statistical data, which are processed within a political context.

4.8   Acid rain

The increased emission of sulphur dioxide and nitric oxide has led to the formation of acids, primarily sulphuric and nitric, which once dissolved in the cloud droplets in the atmosphere may be returned to the ground in the form of *acid rain* (Figure 4.16). This can often fall many hundreds of miles from the pollutant source. Many regard the effects of acid rain on the environment as important as the more well-known problems of global warming and ozone depletion (see Chapter 7), and demand governmental control of such pollutants in the same way that global treaties have been drawn up for CO_2 and CFC regulation.

Before considering the consequences of acid rain, it is important to remember that rain is naturally acidic, having a pH of about 5.6. The pH scale ranges from 0 to 14 with the midpoint, 7, taken as neutral; pH <7 represents an excess of hydrogen ions and, hence, acidity, while pH >7 represents alkalinity. The pH scale is logarithmic, i.e. $pH = -\log_{10}(H^+)$. Hence, a change of 1 on the pH scale corresponds to a change in a factor of 10 in the hydrogen ion concentration (H^+).

Natural rain water is slightly acidic due to the formation of carbonic acid (H_2CO_3), as a result of the solvation of atmospheric CO_2 within the rainwater. Acid rain arises from the further decrease in the pH of the rain arising from the absorption of the sulphur and nitrogen oxides. Acid rain is in fact something of a misnomer since much of the sulphur and nitric oxides can fall to the ground in gaseous or particulate form and do not dissolve in water until they reach the ground. Such *dry deposition* is equally damaging to the environment as acidified raindrops (Figure 4.16).

Figure 4.16 Acid rain damage to statues on the frontage of the Houses of Parliament in London, UK.

In the 1960s and 1970s monitoring of the pH of lakes in the north-eastern USA and across Scandinavia revealed a decrease in the pH of the lakes with consequent loss of fish stocks, while in central Europe trees began to show evidence of severe distress. In some lakes pH levels fell to 4.0 with occasional storms indicating rainfall of pH as low as 2.1. A correlation between the wind patterns blowing from industrial areas to those regions of greatest acidic rainfall was found, suggesting that the acid rain was a direct result of increased industrial emissions. The consequences of such acid rainfall was soon seen in the local ecology. Fish are particularly sensitive to acidity both as a result of direct toxicity and indirectly through damage in the food chain. Fresh water fish stocks were seen to decline in affected lakes and rivers, salmon and trout being especially susceptible. The reduction in the pH of the lake to <5 was found to kill the fish eggs and when <4 few fish could live in the acid waters.

Corrective measures were therefore taken to correct for the acidity of the lakes by the addition of lime to the affected areas, but in some cases the fish stock still continued to decline indicating that other, more subtle, processes were at work. Research discovered that the ecology was dominated by the type of geology surrounding the lake. If granite rocks were dominant the acid rain ran into the lakes unchanged, while if the rain fell on limestone the acidity was partially offset. If the acid rain fell on aluminium-bearing rock, then the consequences for the fish were grave indeed. The liberation of Al ions into the lake and river killed the fish quickly, attacking the fish gills and suffocating them to death. Hence, there were no easy corrective methods that could be used to offset the consequences of acid rain and once again there is a need for legislation to reduce the emissions of both sulphur dioxide and nitric oxides and hence to re-examine the role both of the motor car and the methodologies of electricity production.

4.9   The car as an urban pollutant

The motor car is a major contributor to urban pollution. With the steady increase in the number of cars in cities all over the world, specific attention is being paid to how cars may be designed to limit their pollutant emissions. Table 4.6 lists the major constituents of motor car exhaust fumes.

The potential health problems arising from the unchecked rise in the number of motor cars have led to governments passing legislation to reduce the amount of pollution emitted from car exhausts. Such legislation has led to car manufacturers refining both the design of their car engines and modifying the chemical constituents within the fuel (petrol) used. The adoption of *catalytic converters* to reduce further harmful emissions has

Table 4.6 Pollutants emitted from car exhausts.

Emitted pollutant	Environmental impact
Carbon dioxide	contribution to global warming as a greenhouse gas
Carbon monoxide	toxic and can cause respiratory disease
Nitrogen oxides	formation of acid rain
Sulphur oxides	formation of local smog and acid rain
Benzene, hydrocarbons, aldehydes	toxic and are carcinogens
Lead	toxic and possibly lead to mental abnormalities
Particulates	respiratory disease (e.g. asthma), possible carcinogens

led to a steady fall in the total vehicle emissions globally since the late 1980s. However, with increased prosperity in many previously underdeveloped regions of the world, the total number of cars worldwide is forecast to rise dramatically in the next decade leading to an increase in global pollution from car emissions. Thus, car designers are now developing Ultra-Low Emissions Vehicles (ULEV), producing one-eighth of the emissions of a standard car, and eventually they hope to produce a Zero-Emission Vehicle (ZEV) for commercial exploitation early in the twenty-first century.

4.9.1   Internal combustion engine

The pioneers of the motor car were the German engineers Gottlieb Daimler (1834–1900) and Karl F. Benz (1844–1929). Daimler developed the first lightweight, petrol-powered *internal combustion engine* in 1883, and fitted it first to a bicycle and then, in 1886, to an open carriage. In 1885 Benz developed the first practical car powered by such an engine; the three-wheeled vehicle having a top speed of 8 mph ($13\,km\,h^{-1}$) and, unlike Daimler's 'horseless carriage', was an entirely new vehicle. In 1908 mass car production began when Henry Ford produced the Model T (Figure 4.17). At first, electric cars vied with internal combustion engine models but since their performance and range was much less the battery-driven cars they were soon phased out and the ubiquitous internal combustion engine with all its pollutants came to dominate and change the world.

In the internal combustion engine four cylinders are connected to a single crankshaft such that the movement of one piston initiates motion of the other three pistons. The sequencing of the strokes is the same but not simultaneous (Figure 4.18). In the *intake stroke* the first cylinder sucks in the vaporized petrol–air mixture through the inlet mani-

Figure 4.17 Model T Ford and a modern car.

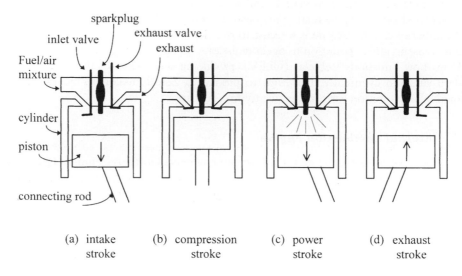

sparkplug
inlet valve   exhaust valve
                   exhaust
Fuel/air
mixture

cylinder

piston

connecting rod

(a) intake   (b) compression   (c) power   (d) exhaust
    stroke   stroke   stroke   stroke

Figure 4.18 Schematic diagram of the operation of an internal combustion engine.

fold, while in the *compression stroke* the second cylinder compresses the mixture. The downward *power stroke* occurs when the mixture in the third cylinder is exploded electrically by the discharge across the sparkplug. Finally, in the *exhaust stroke*, the hot gases (together with toxic chemicals) are ejected through an exhaust manifold in the fourth cylinder. The sequencing continues with the first cylinder now providing the compression stroke, the second cylinder the power stroke, the third cylinder the exhaust stroke and the fourth cylinder the intake stroke. The cycle is then repeated.

The internal combustion engine provides a good example of the First Law of Thermodynamics (see Chapter 2) in action:

$$dU = dQ - dW$$

where the change in internal energy, dU, of the gas in the cylinders is increased by heat input, dQ, and the work done dW.

It is the power stroke that provides the energy required for the other three cylinders to operate. The petrol–air mixture contained in the cylinders expands as the pistons are free to move. Under ideal conditions the pressure in the cylinder will not increase as the piston moves, and the energy transferred to the gas will raise its temperature and do useful work. Thus, dU will be positive because dQ will increase and dW will be positive since work has been done by the gas as it expanded and moved the piston. This is an isobaric change, i.e. change at constant pressure.

In reality, the intake stroke is extremely rapid, and this results in an adiabatic change which is accompanied by a change in the gas pressure. Rather than an isobaric change, an adiabatic change results in a slight decrease in the gas's internal energy (and therefore its temperature) and a loss of the engine's efficiency. As a consequence, the temperature of the fuel mixture must now rapidly increase to the engine's equilibrium temperature so that the dQ is positive. The compression stroke is also an adiabatic process and work is done on the gas raising its temperature.

The thermodynamics of the power stroke is complex and is still the subject of active research. In the ignition stage the fuel (hydrocarbons and air) is converted into CO_2, water and CO. Thermal energy is supplied to the fuel mixture very quickly and ignites the gases, which then expand and cool with work being done by the gas. As the gas expands and cools, thermal energy is dissipated from both the engine and the gas.

In the final exhaust stroke the work is done on the gas and this energy is wasted as it is ejected and increases the inefficiency of the system.

In the 1930s lead was added to the petrol to improve the performance of the engine. The addition of lead reduces the amount of incomplete combustion in the ignition stage, but lead is a highly toxic compound and, therefore, may have harmful health effects. In addition, leaded petrol cannot be used in any car running with a catalytic converter since the platinum catalyst is poisoned by lead (see Section 4.9.3). This led to the development of lead-free petrol. Unfortunately, such petrol has another additive, benzene, which is a carcinogen.

4.9.2   Efficiency of the car engine

In Section 2.2.2 the efficiency (η) for an ideal heat engine can be expressed by:

$$\eta = 1 - \frac{Q_2}{Q_1} \tag{4.1}$$

where Q_1 is the energy supplied and Q_2 is the energy which is rejected by the engine. The Carnot cycle is for an ideal case and the ordinary petrol engine can be described more realistically by the *Otto cycle*. This too has a four-stroke system, and can be expressed in the P-V indicator diagram (Figure 4.19), with the following four features:

From A to B there is a compression stroke in which the working fluid is compressed adiabatically. From B to C thermal energy (Q_1) is taken in under isovolumetric conditions (i.e. conditions of constant volume). This implies that from the First Law of Thermodynamics:

$$Q_1 = dU = mc_v.dT = mc_v.(T_C - T_B) \tag{4.2}$$

where c_v is the specific heat capacity of the fluid at constant volume.

From C to D the power stroke is produced with an adiabatic expansion, and from D to A, energy (Q_2) is released as the temperature falls, again under isovolumetric conditions, where

$$Q_2 = mc_v.dT = mc_v.(T_D - T_A) \tag{4.3}$$

If equations 4.1–4.3 are incorporated, the efficiency of this system will be:

$$\eta = 1 - \frac{Q_2}{Q_1} = 1 - \left\{ \frac{(T_D - T_A)}{(T_C - T_B)} \right\} \tag{4.4}$$

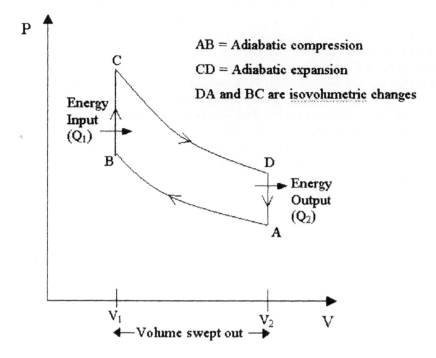

Figure 4.19 The Otto cycle.

This expression suggests that to maximize the efficiency, $(T_D - T_A)/(T_C - T_B)$ must be as small as possible. This can be achieved by making the ratio of the maximum volume contained by the cylinder to the minimum volume contained by the cylinder (i.e. V_2/V_1) as large as possible. This is called the *compression ratio, r*. The relationship between the efficiency of the engine and the compression ratio is then given by:

$$\eta = 1 - \frac{1}{r^{\gamma-1}} \tag{4.5}$$

where γ is the adiabatic index or the ratio of the principal specific heat capacities (c_p/c_v), and γ depends on the atomicity of the fluid used; r is effectively reduced if the ignition is inappropriately timed.

For most modern cars the efficiency is low (between 20 and 30%). Hence, there is considerable research on how to design more fuel-efficient engines. The advantages of a more efficient engine are obvious; less fuel will be required, resulting in the car having a smaller fuel tank and a lighter engine producing fewer emissions.

4.9.3   *Reducing vehicle emissions*

However, such is the magnitude of urban pollution caused by vehicle emissions that governments cannot wait for the design and construction of such new engines but must take measures to reduce the emissions from existing systems. The Clean Air Acts of 1968 and 1970 set restrictive standards on exhaust emissions that must be met by all new motor vehicles. The regulations specify the number of grams of hydrocarbons, CO and NO_x that can be emitted per mile. The amount of emission of any pollutant is a complex relationship between the air–fuel ratio, ignition timing, compression ratio and engine speed. Manufacturers have made impressive advances in engine performance through use of controlled fuel injection and ignition timing both of which may be controlled by on-board computers. The use of computers and chemical sensors, in tandem, to monitor emissions and the operating conditions of the engine have led to the development of experimental vehicles complete with on-board diagnostic systems. Thus it should soon be possible to manufacture cars that routinely monitor emissions and inform the driver when it is necessary to service the engine.

As an alternative to reducing the emissions at source, it is possible chemically to convert the pollutants into less harmful exhaust gases. Such is the role of the *catalytic converter* (Figure 4.20). Catalytic converters are now fitted to most modern cars and seek to convert NO into N_2 and O_2. The decomposition of NO is slow at room temperature but at higher temperatures in the presence of a platinum catalyst the rate is greatly enhanced. However, the platinum catalyst is poisoned by lead contained in the antiknock additive of petrol, so that cars with such catalytic converters must run on lead-free fuel. Similar converters using chromium, iron and copper oxides may be used to oxidize the hydrocarbons and CO.

Alternatively the modern car engine could be redesigned to operate with a different fuel that produces less of the harmful pollutants. Brazil has an excess of biomass (see Section 5.10), which it is attempting to use to supplement fossil fuels for energy production. In more than half a million cars produced in Brazil ethanol, manufactured from sugar cane, is used as a fuel. The emissions from such an ethanol-fuelled car are simply CO_2 and water vapour, thus making these cars 'pollutant free'. However, such a fuel is relatively

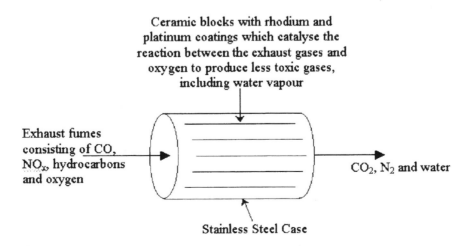

Figure 4.20 Schematic diagram of a modern catalytic converter.

Figure 4.21 A solar-powered car.

inefficient and requires frequent refilling and, while petrol prices remain low, it is unlikely to be adopted on a large scale.

There is also renewed interest in electric cars and even solar-powered cars (Figure 4.21) but the time needed to recharge the batteries of such cars and their short range ensure that they will remain underdeveloped until economic pricing and perhaps even more draconian pollution control allow them to compete with traditional petrol-driven combustion engine vehicles.

4.10   Noise pollution

Noise is not often thought of as a 'pollutant', but unwanted sound (noise) can seriously degrade the quality of life. Even if the sound is comprised of the finest music it can be considered as noise if it occurs in the middle of the night. The acceptance of noise by people obviously depends on the individual, but in the past decade governments have legislated on levels of sound acceptable to the community. The level of noise deemed to be acceptable are dependent upon:

- *The type of environment:* acceptable levels of surrounding noise are affected by the type of activity. A library, for example, has different requirements to those on a factory floor
- *Frequency structure:* different noises contain different frequencies and some frequencies are found to be more annoying or more harmful than others. For example, the high whining frequencies of certain machinery or jet engines are more annoying than lower frequency rumbles
- *Duration:* a short period of high level noise is less likely to annoy than a long period. Such short exposure also causes less damage to hearing.

The measurement and legislation of noise must take all these factors into account and the scale used to assess them should be appropriate to the type of situation. For environmental purposes, sound can be described as a sensation in the human ear and brain produced by pressure variations in the surrounding air. Sound also involves a transfer of energy to the ear, although the hearing system is so sensitive that it reacts to just a few milliwatts; otherwise our ears would overheat! However, relatively small amounts of energy are capable of causing permanent hearing damage, so the protection of hearing is important for our continuing quality of life.

4.10.1   Human ear

The sensitivity of the human ear (Figure 4.22) to sound waves depends on the frequency of the sound and on the intensity of the sound. Different people have different hearing sensitivities, but average values can be measured and provide a 'map' of the sound that the human ear can detect. The *threshold of hearing* is the weakest sound that the average human hearing can detect. The threshold varies slightly with the individual, but it is remarkably low and occurs when the membrane in the ear is deflected by $<10^{-7}$ m. There is also a high threshold, the *threshold of pain*, which is the strongest sound that the human

Table 4.7 Ranges of human hearing.

Range of human hearing	Measured as intensity (I)	Measured as pressure (p)
Threshold of hearing	10^{-12} W m^{-2}	2×10^{-5} Pa
Threshold of pain	100 W m^{-2}	200 Pa

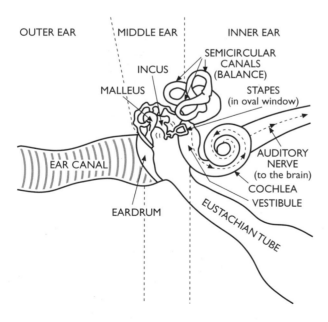

Figure 4.22 Schematic diagram of the human ear.

ear can tolerate (Table 4.7). This is less distinct than the threshold of hearing because the effect of high pressures on the ear changes from sound to pain and then to physical damage to the eardrum.

The human idea of 'loudness' is complicated by the fact that the human ear is more sensitive to some frequencies than others. Sounds with a frequency $<100\,Hz$ (bass notes) need to be of a higher energy before damage occurs than sounds in the most sensitive $1–5\,kHz$ range.

4.10.2 Sound levels

Absolute measurements of sound intensity can be expressed in either $W\,m^{-2}$ or in units of sound pressure, Pa, but such units do not correspond directly to the way in which the human ear responds to sound levels. Since the human ear has a non-linear response to the energy content of sound, a logarithmic scale is used to describe the response of the ear. The energy or pressure measurements associated with a particular sound level are converted to a sound level measured in *decibels*, named after A.G. Bell, inventor of the telephone. The number of decibels for any noise event is then given by:

$$N = 10\log_{10}(I/I_o) \text{ or } N = 20\log_{10}(p/p_o)$$

where I_o and p_o are the values for the threshold of hearing, I and p the intensity and pressure of the sound being measured respectively. This ratio leads to a convenient scale of sound levels that starts at zero (when $I = I_o$ or $p = p_o$). The faintest audible sound (at $1000\,Hz$) is rated as $0\,dB$ but in reality a $3\,dB$ change is the smallest difference that is normally considered significant. Normal speech is $50\,dB$ and an aircraft engine at close range is some $120\,dB$. Table 4.8 lists the noise levels of other typical sounds that may be encountered during the day.

It is important to recognize that a $3\,dB$ change corresponds to a doubling (or halving) of the sound energy but it does not sound twice as loud. It requires a $10\,dB$ increase to

Table 4.8 Sound levels experienced in everyday life.

Sound	Intensity	
	$W m^{-2}$	dB
Threshold of hearing	10^{-12}	0
Whisper	10^{-10}	20
Quiet room in a house	10^{-8}	40
Normal conversation	$10^{-6}–10^{-5}$	60–70
Road traffic	10^{-5}	70
Factory floor	10^{-2}	100
Ghetto-blaster	10^{-1}	110
Jet taking off 60 m away	1	120
Threshold of pain	1	120
Space rocket lift off	10^{8}	200

Worked example 4.2   Combination of two sources of noise pollutants.

If one jet causes a sound level of 120 dB on take-off, what is the sound level of three such jets taking off simultaneously?

Solution
For one jet:

$$120\,dB = 10\log I/(1 \times 10^{-12}), \text{ where } I/1 \times 10^{-12} = 10^{12}$$
$$\text{Thus, } I = 1\,W\,m^{-2}.$$

For three jets:

$$I_{total} = 3\,W\,m^{-2}$$
$$dB = 10\log 3/1 \times 10^{-12} = 124.8\,dB.$$

make the sound seem approximately twice as loud, while an increase of 20 dB seems four times as loud. This may seem like good news for peace and quiet, but working in reverse, one has to reduce the energy of an annoying sound to one-tenth of its energy level before it sounds half as loud. Noise control, therefore, involves large reductions in sound energy.

4.10.3   Hearing loss

Unfortunately, the damage to hearing is directly proportional to the energy, and so a tenfold increase in energy can increase the risk of lasting damage to one's hearing by a factor of 10, but will only sound twice as loud. Therefore, when there is an increase in the volume setting of a personal stereo headset it may create a personal environment which is damaging to hearing and which would be considered unacceptable in an industrial environment.

Most types of deafness involve a loss of sensitivity over certain frequency ranges. *Conductive deafness* is the presence of 'mechanical' faults in the system which transmits sound vibrations to the inner ear. It can be caused by such defects as a broken eardrum, blockage by wax or the stiffening of the bones of the inner ear. Various medical remedies exist for this type of deafness. In contrast, *nerve deafness* is seldom reversible. Nerve deafness is the result of damage to the nerves carrying sound information to the brain. This type of deafness can be caused by infections, head injuries and by exposure to excessive noise.

Hearing loss caused by excessive exposure may be divided into two main types: *temporary threshold shift* (TTS) is the temporary loss of hearing which recovers in 1 or 2 days after exposure to the excessive noise; and *permanent threshold shift* (PTS) is the permanent loss of hearing caused by longer exposure to excessive noise.

Experience of several TTS may lead to a gradual loss of sensitivity to hearing over certain frequency ranges such that eventually PTS occurs. Indeed, as one gets older there is a gradual loss of hearing sensitivity called *presbyacusis*, in the higher tones around 4000 Hz, which remains largely unnoticed except for the misunderstanding of sibilant sounds such as 's'. However, loss of hearing caused by age adds to the loss of hearing caused by noise exposure earlier in life and the combination of the two effects may in the future lead to the increased occurrence of deafness among middle-aged people.

4.11   Summary

In this chapter, the consequences of living in a town or city have been reviewed. As a result of the rising environmental consciousness of the general public, governments world-wide are addressing the issues that surround the slow but subtle move towards sustainable cities. A major priority of these cities is the improvement of human health, and with it the quality of life that the growing global urban population is to enjoy. As energy sources and production are prerequisites for this to occur, we will now consider how in the future energy may be supplied without detriment to the environment.

References

Urban environment

Hall, P., *Cities in Civilization*. London: Orion, 1998.
Hibbert, C., *Cities and Civilizations*. London: Weidenfeld and Nicolson, 1996.
Hough, M., *Cities and Natural Process*. London: Routledge, 1995.
Mumford, L., *The City in History*. Harmondsworth: Penguin, 1966.

Transportation

Ogelsby, C. H., *Highway Engineering*. New York: Wiley, 1975.
Wright, P. H. and Paquette, R. J., *Highway Engineering*. New York: Wiley, 1979.

Water and its impurities

Camp, T. R., *Water and its Impurities*. New York: Reinhold Publishing, 1963.
Dinar, A. and Loehman, E. T., eds, *Water Quantity/Quality Management and Conflict Resolution*. Westport, Connecticut: Pregaer Press, 1995.
Dunne, T. and Leopold, L. B., *Water and Environmental Planning*. W. H. Freeman + Co. San Francisco: 1978.
James, W. and Neimczynowicz, J., eds, *Water, Development and the Environment*. Boca Raton: CRC Press, 1992.

Atmospheric pollution

Boubel, R. W., Fox, D. L., Turner, D. B. and Stern, A. C., *Fundamentals of Air Pollution*, 3rd edn. New York: Academic Press, 1994.

Hodges, L., *Environmental Pollution*. New York: Holt, Rinehart and Winston, 1973.

Mason, B. J., *Acid Rain*. Oxford: Oxford University Press, 1992.

Seinfeld, J. H., *Atmospheric Chemistry and Physics of Air Pollution*. New York: McGraw-Hill, 1986.

Noise pollution

Bragdon, C. R., *Noise Pollution*. Philadelphia: University of Pennsylvania Press, 1971.

Taylor, R., *Noise*. Harmondsworth: Penguin Books, 1970.

Physics of environmental health

McMichael, A. J., Haines, A., Sloof, R. and Kovats, S., eds, *Climate Change and Human Health*. Geneva: World Health Organization, 1996.

Schnoor, J. L., *Environmental Modeling: Fate and Transport of Pollutants in Water, Air and Soil*. New York: Wiley, 1996.

Turiel, I., *Physics, The Environment and Man*. Englewood Cliffs: Prentice-Hall, 1975.

Discussion questions

1   What are the basic differences between the London Smog and that found in Los Angeles?
2   (a)   What are some of the useful by-products of a municipal sewage treatment plant in addition to the purified water output?
     (b)   Why is the combining of sewage disposal and storm sewers such poor practice?
3   Investigate sources and levels of pollution in your (a) home, (b) place of work and (c) local environment. What might be done to reduce these levels?
4   Urban planners are increasingly taking environmental issues into account. Discuss the themes that they might be addressing.
5   You have been put in charge of a large city's transport system, with the remit to revolutionize its ailing condition. What would be your masterplan and, in particular, what policies could be employed to alleviate congestion and to reduce pollution?
6   It has been suggested that cities can be unhealthy places. What evidence is there to support this?
7   Discuss how varieties of pollution can lead to the generation of environments in which people can be at risk.

Quantitative questions

1   If carbon monoxide (CO) is in the air at 1 ppm what are the number of CO molecules in $1 \, m^3$ and what is the mass of CO in $1 \, m^3$?
2   Consider a 1000 MW coal-burning power station operating 24 h a day at an efficiency of 33%. Given that 1 tonne of coal produces 7800 KWh of energy, calculate (a) how much coal is burnt daily and (b) if the coal has 1% of sulphur content how many tonnes of sulphur dioxide are released daily by the plant.

3   A car travels 12 km for every litre of petrol consumed and has an engine efficiency of 20%. The energy output of the car during an 8 hour journey of 400 km is 180 MJ.
  (a) Determine the mean energy output from the engine per km.
  (b) Calculate the average input energy to the engine per km.
  (c) Determine the energy provided per litre of the fuel.
  (d) Consider the power stroke in the car's internal combustion engine. When the engine operates at 1800 revolutions per minute, calculate the energy generated by each of these strokes.

4   The temperature of the catalyst (heat capacity $C = 2.5\,JK^{-1}$) in a catalytic converter is raised by 400 K in 20 s when starting the car engine using the 12 V supply derived from the car battery. Assuming the heating process was only 1% efficient, what was the total energy drawn from the battery? Hence, calculate the electric current drawn from the car battery.

5   (a) A car engine raises the temperature of the fuel to about 600°C and expels combustion products at about 80°C. Calculate the efficiency of the engine if it achieves 70% of the maximum theoretical efficiency of a 'perfect' heat engine working between these temperatures.

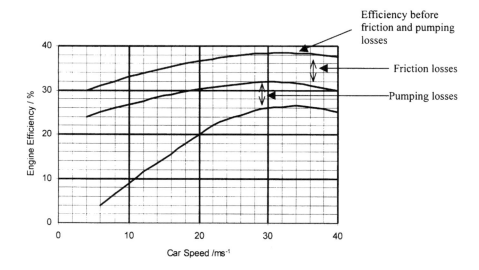

  (b) The diagram above shows the engine efficiency as a function of car speed, for a car in top gear on a level road
  Describe the variation of frictional losses and the pumping losses with speed. Describe the causes of frictional losses and pumping losses in the engine.
                    (London University: 'A' Level Physics: Energy Option: January 1990).

6   Show that (a) increasing by 10 dB increases the sound intensity by 10 times and (b) adding four identical sources of noise together adds 6.0 to the decibel scale.

7   A washing machine generating 90 dB of noise is turned on at the same time as a ghetto-blaster generating 100 dB is on in the room. What is the total noise level in the room?

8   How thick should a sound barrier be made if it is to attenuate effectively sound transmitted at (a) 100 Hz and (b) 10 000 Hz?

9   What is the power per unit area in watts per square metre ($W\,m^{-2}$) if the sound intensity on the decibel scale is 65 dB? A trumpet player generates a sound of 95 dB in an adjacent flat. What attenuation is needed, in decibels, of the intercommunicating wall to bring the sound level to an acceptable level of 65 dB?

proportion of annual income spent upon energy is roughly the same in the developed and developing countries at around 5%.

Approximately 40% of the global energy is used by industry, 40% by offices and domestic consumers and 20% in transport. About one-third of primary energy usage is used to provide electricity of which about half is used by industry and half by domestic and commercial activities. However, these figures as stated may not be reflective of the actual energy use.

The efficiency of generating electricity from thermal combustion is often low, e.g. from coal about 33% of the primary energy arrives as delivered electricity. Thus, *multipliers* are needed to deduce the primary energy corresponding to the delivered energy, e.g. 33% efficiency corresponds to a multiplier of 100/33 = 3. The delivery of energy always requires the use of energy, so multipliers are needed for all processes. For instance, if 130 energy units of primary oil are needed to produce and deliver 100 energy units of processed fuel, the efficiency will be 100/130 = 77%, and the multiplier is 130/100 = 1.3.

Similarly, transport energy per person depends greatly on how many people travel together. By introducing a modifying factor into individual energy use for transportation, it can be seen that three hundred passengers on a train will use less energy per capita than a single person in a car going to work.

Energy is also used to produce food and goods. For food, about 3 MJ of energy per kg of delivered food are used to grow, produce and transport food. For goods: about $100\,MJ\,kg^{-1}$ are needed to produce manufactured goods. Thus, the per capita use of energy cannot be simply calculated and any individual energy audit (see Question 1 and Appendix 3) may, therefore, greatly underestimate the actual energy used.

5.1.2 World energy supplies

At present the world generates most of its energy from the fossil fuels – coal, oil and gas. These are finite resources and ultimately will be exhausted. Table 5.2 lists the current estimates of global energy resources and the period for which they may be exploited at the present rate of use. Known reserves will allow fossil fuels to provide energy until the mid-twenty-first century after which (and before if demand continues to grow), supplies will begin to come under pressure and lead to increases in their cost. Further exploration and new technology will undoubtedly allow new reserves to be found and exploited but ultimately, within a century, commercially viable reserves of oil and gas will have been exhausted and a return to coal is likely. There are several hundred years of coal and lignite stocks exploitable. Beyond this, supplies of uranium (for nuclear power) may provide energy supplies for many centuries but with the commensurate risk of potentially

Table 5.2 Estimate of global energy reserves.

Fuel	Proven reserves (Gtoe)	Ultimate reserves (Gtoe)	Years for exploitation
Coal and lignite	696	3400	400
Oil	137	200	40
Gas	108	220	60

disastrous nuclear accidents. Hence, there is a growing interest in the development of renewable energy sources. These are energy sources that are not depleted by their use, and include

- Solar energy
- Hydroelectricity
- Wind
- Geothermal (heat from radioactive decay in the Earth and from near-surface heat transfer).

At present renewable energies provide only 7% of the total global energy demand, mostly through hydroelectric power. However, it has been proposed that by the mid-twenty-first century this may be more than doubled until 20% of the total global energy budget is met by adopting such resources. Nevertheless, fossil fuels and nuclear power will remain the predominant energy sources in the twenty-first century.

The main forms of energy supply, together with the basic physics, essential data and environmental impact criteria will now be described.

5.2   Fossil fuels

At present, the major energy sources are derived from *fossil fuels*. By definition, fossil fuels are carbon materials that have been removed from active ecology (fossil), yet capable, when extracted by combustion with oxygen, to produce thermal energy (fuels). Fossil fuels are formed as solids (coals), liquids (oils) and gases (natural gas). No source of fossil fuel is obtained as a single molecular form, and there are many impurities and complexities. In particular, sulphur is present in many coals and oils, so combustion produces sulphur dioxide (SO_2) as a pollutant, which in turn leads to the problem of acid rain (see Section 4.8). In particular, oil, gas and coal are all derived from *biomass* (see Section 5.10), i.e. from plants and animals via photosynthesis and metabolism. The energy released in the combustion of fossil fuels is given in Table 5.3, ranging from coal at about $25\,MJ\,kg^{-1}$ to natural gas (predominantly methane) at $55\,MJ\,kg^{-1}$.

These values can be compared with the energy available from human labour. A fit person can sustain work at about 20–30 W over an 8-h day; thus producing 160–240 Wh day^{-1}, equal to 0.6–0.9 MJ day^{-1}. A litre of petroleum, costing 15 pence (without government tax), therefore, can provide energy equivalent to about 50 labourers for a day (costing about £2000). Thus, the use of fuels has relieved humanity from exhausting labour, freeing it for other more intellectual and innovative work.

Table 5.3 Energy released from fossil fuels.

Fuel	Energy content ($MJ\,kg^{-1}$)
Coal	
Anthracite	33
Lignite	15–25
Crude oil	45
Methane (natural gas)	55

Nevertheless, the use of fossil fuels has disadvantages in their environmental impact. These include:

- Destroying productive land and ecological systems in open caste mines and spoil tips
- Producing polluting emissions of gases (e.g. SO_2) and particulates (PM_{10}'s) causing ill-health and possibly premature death (see Chapter 4)
- Producing greenhouse gases that affect climate change (see Chapter 7).

These impacts and the finite lifetime for using these resources have therefore led to considerable discussion as to the development and viability of alternative 'clean' energy technologies.

5.3   Nuclear power

Since the 1960s a growing proportion of the world's energy supply has been derived from the use of nuclear power. In 1990, 5% or 0.45 Gtoe of the global energy demand was met by nuclear power, with some countries (notably France and Russia) having a far larger proportion of its energy supply met by nuclear installations. Remaining known reserves of uranium (the nuclear fuel) are such that nuclear power can provide 3000 Gtoe and possibly up to 9000 Gtoe, or many centuries of global energy demand, far more than the reserves of fossil fuels. However, use of nuclear power is clouded by the threat to human health and the environment from nuclear accidents and nuclear waste.

5.3.1   Nuclear fission

As we move through the periodic table the ratio of the neutron number (N) to proton number (Z) for each element increases. This results in increasing instability. Radioactivity is the process whereby unstable nuclei attempt to become more stable by the emission of energy in the form of either α-particles, β-particles or γ-radiation.

Fission is the break-up of heavy nuclei that are already likely to release energy as shown by their natural radioactivity; these are the 'fuels' of conventional nuclear power stations and basic nuclear weapons. The nuclei of elements with highest nuclear mass (e.g. plutonium and uranium) have extremely large energy potentials and are in states of thermodynamic non-equilibrium. This means that if the nuclei are disrupted, they decay into more stable forms and release energy as radiation and thermal energy. The energy released is so great that the combined masses of the residue nuclei are measurably less than the masses of the initial nuclei, i.e. mass is significantly transformed into energy. This is explained by Einstein's mass defect formula, probably the most famous equation in science:

$$\Delta E = \Delta m.c^2 \tag{5.1}$$

where ΔE is the energy released, Δm is the mass deficit and c is the speed of light.

Fission only occurs if the elements with the radioactive nuclei, e.g. the isotope uranium 235, are concentrated. In the natural environment such radioactive nuclei are dispersed in geological substrata and are present, if at all, in very low concentrations of about 0.05% or less. Therefore, when fission sources are mined, the open-caste pit is very large and >99.9% of the material has to be discarded in spoil tips. Moreover, the concentration process uses much water and other materials, and there is much effluent, including polluted water and radioactive radon gas.

After concentration in several stages, the radioactive fuels, typically uranium of nuclear mass about 230–240, are transported to thermal nuclear power stations. Here the radioactive elements are concentrated so that fission reactions occur. The energy emission per fission of ^{235}U is about 3.2×10^{-11} J/atom (or 200 MeV per atom), which equates to about 80 million MJ kg^{-1} of ^{235}U, and 80 000 MJ kg^{-1} of mined material. In practice, nuclear energy is only used to generate electricity at about 30% efficiency, so that the useful delivered energy is about 30 000 MJ kg^{-1} of mined material. This may be compared with the useful energy from burning methane of 55 MJ kg^{-1}; some 600 times less per unit mass.

5.3.2 Nuclear reactors

In modern nuclear reactors natural uranium is the fissile material. It consists of the isotopes ^{235}U (0.7%) and ^{238}U (99.3%), and it is the ^{235}U that is bombarded by slow thermal neutrons. The condition for fission to occur is for the mass of the parent nucleus and nuclear missile (i.e. the neutron) to be greater than the masses of all the fission-products. The difference in mass then provides the energy dissipated under the Einstein relationship, equation (5.1).

The nuclear reaction that results from the interaction between the slow neutrons and the ^{235}U produces a *chain reaction* in which successions of neutrons initiate additional fissioning; between two and three neutrons per fission are generated. The general reaction can be of the form:

$$U + n \rightarrow X + Y + zn + Q$$

where the left-hand side signifies the reactants, X and Y are the fission-fragments, z is the number of fast neutrons generated per fission and Q is the quantity of energy released in MeV.

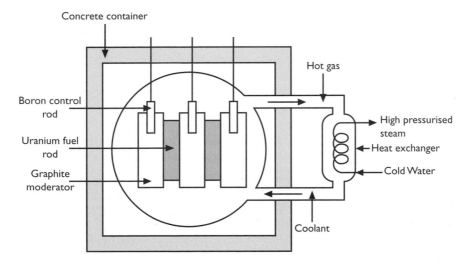

Figure 5.2 An advanced gas-cooled thermal reactor using nuclear fission.

If this process occurs spontaneously a nuclear bomb results. For a nuclear power reactor the process has be to controlled. For the chain reaction to be maintained one or more of these neutrons has to be available to continue the process. The neutron multiplication factor (k) is the net number of effective neutrons per fissioning. With $k > 1$, the chain reaction will continue. Since the ratio (loss of neutrons that are capable of fissioning)/(production of fission neutrons) depends on the surface area/volume ratio $(= 4\pi R^2/4/3.\pi R^3)$, it implies that the ratio depends on the reciprocal of the radius of a spherical specimen of fissile material. Thus, the larger the material, the less chance neutrons will be lost, and the resulting critical mass requires that $k = 1$.

To control the huge amounts of energy released graphite or heavy water *moderators* are used to reduce the speed of the fast neutrons that emerge from each fissioning. Boron or cadmium rods are used to control the nuclear reaction; by raising and lowering these *control rods* the neutron flux can be manipulated.

To extract and use the huge quantities of energy produced a *coolant* (such as water or carbon dioxide) is passed through the system and then through a *heat exchanger*, where the hot coolant transfers energy to water flowing through pipes. This is converted into steam at very high pressure, which then drives turbines which produce electrical energy.

Worked example 5.1   Nuclear fission.

A typical fission reaction includes:

$$^{235}_{92}U + {}^1_0n \rightarrow {}^{95}_{42}Mo + {}^{139}_{57}La + 2{}^1_0n + 7{}_{-1}^0e$$

Calculate the total energy released by 1 g of the uranium undergoing fission by this reaction, neglecting the masses of the electrons. You are provided with the following information:

Mass of neutron $= 1.009u$
Mass of ${}^{95}_{42}Mo$ $= 94.906u$
Mass of ${}^{139}_{57}La$ $= 138.906u$
Mass of ${}^{235}_{92}U$ $= 235.044u$
$1u = 1.66 \times 10^{-27} kg$
Number of atoms in 1 mole of atoms $= 6.02 \times 10^{23}$
Speed of light $= 3 \times 10^8 m s^{-1}$.
(London University: 'A' Level Physics: Paper 2: January 1975)

Solution
Before answering this question, you should be aware of the variety of units used in atomic and nuclear physics. These are the joule (J), the electron-volt (eV), and the unified atomic mass unit (a.m.u or u). $1 eV = 1.6 \times 10^{-19} J$, and $1 amu = 1.66 \times 10^{-27} kg = 931 MeV$.

$1 eV$ is the work done in taking an electron (of charge $1.6 \times 10^{-19} C$) through a potential difference of $1 V$.

The sum of the masses of the reactants is 236.053u, while the mass of the products (ignoring the electrons) is 235.83u.

Thus, the mass defect $= 236.053 - 235.83 = 0.223u = 0.223 \times 1.66 \times 10^{-27} kg = 0.37 \times 10^{-27} kg$.

Using the Einstein relationship, $\Delta E = \Delta m.c^2$, the energy released from one atom $= 0.37 \times 10^{-27} \times (3 \times 10^8)^2 = 3.33 \times 10^{-11} J$.

Because there are 6.02×10^{23} atoms in 235g of ^{235}U, i.e. there are Avogadro's number, N_A, of atoms or molecules in one mole of the substance, there will be $6.02 \times 10^{23}/235$ atoms in 1g (i.e. 2.562×10^{21}).

Thus, the energy released, $\Delta E = 3.33 \times 10^{-11} \times 2.562 \times 10^{21} = 8.53 \times 10^{10} J g^{-1}$.

This is an enormous amount of energy, and would be equivalent to heating 200 000 kg of water to 100°C. The atomic bombs of 1945 contained two separated subcritical pieces of fissile uranium. On being brought together the resultant mass became supercritical, with a mass of about 1 kg. This could generate about $8 \times 10^{13} J$ in 1 μs!

However, major difficulties for nuclear power remain. Nuclear power station processes may be used to produce the further fissionable materials for nuclear weapons. Thus, there is a danger of nuclear weapons proliferation. Every decade since the 1940s has seen new countries detonating nuclear weapons in tests and thereby showing the capability to deploy them in warfare; India and Pakistan in 1998 being the most recent to demonstrate this ability. Since the collapse of Communism in Eastern Europe in the 1990s, there have been reports of criminal and possibly terrorist activity in fissionable material obtained illicitly from nuclear reactors and waste treatment plants, heightening the fear that smaller nuclear weapons may, in the twenty-first century, become part of the terrorists' arsenal and with 'rogue' governments threatening their use in local conflicts.

There is also, at present, no known way to dispose safely of the radioactive waste products, some of which have half-lifes of thousands of years, and there is considerable controversy over the long-term storage of nuclear waste. Some dumping has taken place at sea, for example, in the Arctic and the deep Marianas trench off the Philippines, but an international treaty now forbids this. High-level waste from the chemical reprocessing of spent fuel rods from power stations and decommissioned nuclear weapons must now be stored in specifically designated sites. However, already leaks of radioactive material have been reported, and these may enter the local ground water. It is still not clear what price future generations will pay for our present consumption of cheap electricity. Fears of the potential for serious global pollution from nuclear power systems were highlighted by the disaster at Chernobyl.

In the first phase of a three-step process two protons fuse to produce a deuteron (^2_1D):

$$^1_1H + ^1_1H \rightarrow ^2_1D + _{+1}^0e + \nu$$

The superscript denotes the nucleon or mass number, while the subscript denotes the charge or proton number. In the second step the deuteron then fuses with another proton to produce the helium isotope ^3_2He:

$$^2_1D + ^1_1H \rightarrow ^3_2He + \gamma$$

In the third stage two helium nuclei then fuse to form the helium isotope ^4_2He, which is an α particle:

$$^3_2He + ^3_2He \rightarrow ^4_2He + 2^1_1H$$

The combination of these three reactions is:

$$4^1_1H \rightarrow ^4_2He + 2_{+1}^0e + 2\nu + \gamma$$

where ν represents another fundamental particle known as a neutrino.

This equation shows that in solar thermonuclear fusion four hydrogen atoms are transformed into a single helium atom, but with an energy release explained by Einstein's mass–energy equivalence relationship, $\Delta E = \Delta m.c^2$, of the order of about 27 MeV.

Worked example 5.2   Energy released in nuclear fusion.

Determine the quantity of energy, in joules, released in a single fusion process.

Solution
Using atomic units in which the mass of the oxygen atom is 16, four hydrogen atoms will have a mass of 4.03252 units [u] and one helium atom 4.00386u, there is a mass difference of 0.02866 u. Since one atomic mass unit is 1.66×10^{-27} kg, the energy released per fusion process is $\Delta E = \Delta m.c^2 = 0.02866 \times 1.66 \times 10^{-27} \times (2.9996 \times 10^8)^2 = 4.3 \times 10^{-12}$ J.

Since only a small amount of energy (10^{-13} J) is released by the neutrino, approximately 4×10^{-12} J is available per fusion process. In this process hydrogen is being converted to helium at the rate of 5000 million kg s^{-1}. There is an enormous energy release within the solar core, some of which is transported through the solar interior to the Sun's surface by radiation and hydrogen convection. Visible solar radiation (sunlight) comes from the 'cool' (about 5800 K) outer surface layer of the Sun known as the *photosphere*. This seething cauldron of gas ejects hot gas into the surrounding *chromosphere* where the temperature rises again to about 10000 K, and is the region where relatively dark patches are seen periodically, the so-called *Sun spots*. The outermost layer of the Sun, the corona, is then composed of that matter that has been 'boiled off' from the surface of the chromosphere and is ejected into space as a hot plasma known as the *solar wind*, which travels across the Solar System at typically $1.5 \times 10^6 \, \mathrm{km\,h^{-1}}$. The solar wind interacts with the Earth's magnetic field and upper atmosphere causing the phenomenon known as the *aurorae*, the intensity of which is increased when the Sun is active and solar spots more apparent (see Chapter 7).

Case study 5.2 The nuclear fusion reactor

Nuclear fusion is often considered to be the panacea for future global energy needs; a source of safe and limitless energy. When light nuclei (e.g. hydrogen) are bonded together the binding energy per nucleon of the resulting element increases as does the energy released. This released energy is the source of nuclear fusion-induced energy supplies. Scientists and engineers have been developing prototypes of fusion reactors capable of producing electrical energy for fifty years. However, for this to be achieved a number of factors have to be taken into consideration, summarised by the *Lawson criterion* ($n\tau > 10^{20}$ s.m^{-3}) where n is the density expressed as the number of particles m^{-3} and τ represents the confinement-time of the fusion plasma in seconds. The implication of the Lawson criterion is that very high particle densities, such as 2.5×10^{20} m^{-3} as well as an extended confinement period (typically 1–2 seconds) will be required, which can only be achieved at very high temperatures.

The reactor fuel consists of a 50:50 mixture of the hydrogen isotopes deuterium ($^2_1$D) and tritium ($^3_1$T) which has to be heated to temperatures between 100 and 200 million °C (i.e. at least 5 times more than the temperature at the centre of the Sun). Under these conditions the fuel becomes a plasma and the hydrogen isotopes fuse to form helium together with high energy neutrons ($^1_0$n) and energy of the order of 17.6MeV i.e.

$$^2_1\mathrm{D} + {}^3_1\mathrm{T} \rightarrow {}^4_2\mathrm{He} + {}^1_0\mathrm{n} + 17.6\mathrm{MeV}$$

However, there are two major problems that need to be overcome if such a reactor is to operate. The first relates to the fuel. Deuterium occurs naturally in water and can be generated from hydrogen exchange mechanisms, but tritium, which is radioactive, does not occur naturally. One solution in generating tritium is to line the reaction vessel with lithium (Figure 5.5a) such that a neutron will react with the lithium isotope ^{6}Li to produce tritium and additionally release energy of 4.8MeV:

$$^6\mathrm{Li} + \mathrm{n} \rightarrow {}^3_1\mathrm{T} + {}^4_2\mathrm{He} + 4.8\mathrm{MeV}$$

The second problem relates to the energy losses that occur when the very hot plasma comes into contact with the vessel walls. Powerful magnetic fields are therefore used to achieve *magnetic confinement*, constraining the plasma and keeping it from touching the walls of the reactor. To date the best confinement system has been found to be a hollow vessel called a torus (which is doughnut-shaped) in which the plasma particles are aligned along the magnetic field lines. The *tokamak* (which is Russian for a toroidal-shaped magnetic enclosure) is an advanced design of such a magnetic confinement system (Figure 5.5b). It consists of a toroidal magnetic field generated by coils located at right angles to the torus. This field integrates with that generated by the large current, formed by the transformer, that flows through the plasma. In addition there are podoidal field coils assisting in shaping and positioning the plasma.

Despite the great opportunities that such reactors will bring to the global energy industry nuclear fusion is unlikely to be commercially viable for another 40 years. The very high temperatures required to initiate fusion reactions have only been reached in (a) 'hydrogen bombs', where the energy released is uncontrolled, and in (b) controlled laboratory developmental conditions. Major engineering problems that have yet to be solved include:

(i)   how to sustain the reactions within a vacuum, to absorb the radiation in solid material
(ii) how to remove the thermal energy to produce steam for generating electricity from turbines
(iii) passing the energy from the vacuum to water will produce radioactive materials that have to be contained and eventually disposed of as waste and
(iv) in contrast to the experimental laboratory reactors currently under development, the generating stations would have to be tens of thousands MW capacity in centralised locations distributing the electricity on international grids.

Therefore at present no operating system has been developed and the extremely high costs of research requires an international effort if such reactors are to be developed within a feasible time frame.

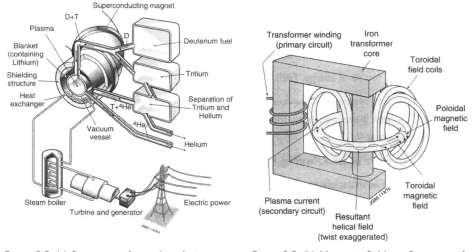

Figure 5.5 (a) Structure of a nuclear fusion power station.

Figure 5.5 (b) Magnetic field confinement of a plasma.

5.4   Renewable energy

Renewable energy is energy obtained from natural and persistent sustainable sources of energy occurring in the environment. Renewable energy sources are energy sources that are not depleted by their use, e.g. solar energy, hydroelectric, wind and geothermal.

Figure 5.6 shows the different types of renewable energy and their energy fluxes. The fluxes range in quantity by a factor of 1:100000, with the dominant flow being the solar radiation incident upon the Earth's surface.

The renewable energy flux densities of solar radiation, wind, wave and usual water flow

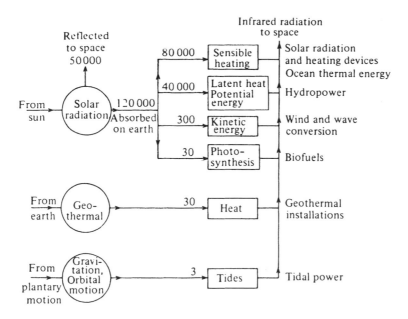

Figure 5.6 Renewable energy sources and their energy fluxes in terawatts.

are in the order of $1\,\text{kW}\,\text{m}^{-2}$. For example, this is the irradiance of bright sunshine at midday (see Chapter 7) or the power per unit area of wind blowing at $6\,\text{m}\,\text{s}^{-1}$. Such power flux densities are about 10–1000 times less than within fossil fuel engines and boilers. Therefore, to capture such energy, quite large structures are needed relative to the power produced. Although the energy in the environment is 'free', efficient capture is needed to make these structures cost-effective and to reduce the visual impact. Hence, at present, renewable energies provide only 7% of the global energy demand with 6% being obtained from hydroelectric plants.

In the following sections the main renewable sources are described in turn and their technological limitations considered. Their possibility for growth as a proportion of global energy supply will also be considered.

5.5   Solar energy

The Sun plays a key role in the development and operation of our global environment (see Chapter 7). The solar flux reaching the Earth is 180 000 million million Watts. Hence, the amount of solar energy reaching the Earth in just 40 min is equal to the total energy used by the world's population in 1 year. How can this enormous reservoir of energy be captured and used to provide the energy that is needed?

Since solar energy has on average a small energy flux density ($<300\,\text{W}\,\text{m}^{-2}$) large collecting devices are required to harness it. However, if captured and retained, the energy falling on an average-sized house can make a significant contribution to heating (or cooling), and provide electricity to heat and light it. Careful design of a house's architecture may allow much of the incident energy to be retained. Large windows, for

The energy gained per unit time by a fixed mass of water (m) is:

$$P_{output} = mc.dT/dt \tag{5.2}$$

where c is the specific heat capacity of the fluid and dT/dt is the rate at which the temperature changes.

If the fluid is flowing through the collector then equation (5.2) must be adapted for the mass flow-rate (dm/dt), and then:

$$P_{output} = \frac{dm}{dt}.c.(T_p - T_a) \tag{5.3}$$

where T_a is the temperature of the fluid entering and T_p is the emergent temperature. To allow for the energy losses the collector can be treated as one whole system, and this argument can be extended to the *capture efficiency* of the system:

Useful energy removed = {energy gain from Sun − energy loss from the structure}.

Thus,

$$P = \{taAG - (T_p - T_a)/R\} \tag{5.4}$$

where R is the thermal resistance for energy loss from the collector.

The capture efficiency (n) of the system is easily measurable, since

$$P = nAG$$

so

$$n = P/(AG) = ta - (T_p - T_a)/(RAG)$$
$$= ta - U[(T_p - T_a)/G] \tag{5.5}$$

This is the *Hottel–Whillier–Bliss equation*, where the energy transfer coefficient, $U = 1/(RA)$. By obtaining the efficiency, (n), the irradiance (G) and the temperatures (T_p and T_a) from experimental data, a plot of this equation gives intercept (ta), and slope (U). These parameters characterize the particular solar water heater.

Solar concentrator

The flat-plate collector is useful for heating water up to 60°C, in all climates including cloudy ones. However, if higher temperatures are to be achieved for the working fluid (such as several hundred degrees), in climates which have considerable direct radiation, other devices are necessary. The *parabolic trough concentrator* can achieve this. It consists of a concentrator (such as a mirror) which focuses the radiant beam on to a receiver, R (Figure 5.9), which lies along the axis of the concentrator. The receiver, which could consist of a conducting material (through which the working fluid flows) absorbs the focused energy.

Using water as the working fluid flowing through the focus of the concentrator it can

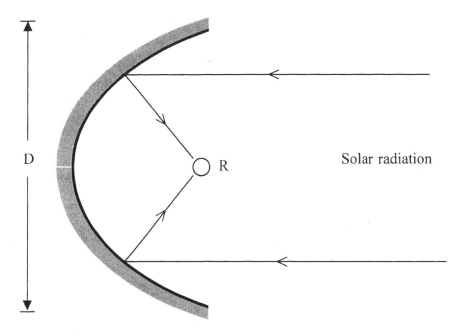

Figure 5.9 Side view of a parabolic concentrator.

be determined whether the silvered parabolic reflector warms water to higher temperatures more effectively than the flat-plate collector, given that the energy absorbed per second (P) by the receiving tube is approximately:

$$P = LD.G \qquad\qquad (5.6)$$

where D is the concentrator's width, L the length of the concentrator, G the solar irradiance R is the focus, and LD the area of the concentrator.

An obvious problem here is to maximize the energy transfer. Owing to the Earth's rotation the efficiency of the device will rapidly decrease once the solar beam is no longer directly facing the concentrator. To ensure that the solar beam is maintained on the concentrator, *tracking* can be used. This is the process whereby the motion of the concentrator is synchronized to follow the Sun's apparent trajectory. This can be achieved by using light-dependent resistors as sensors in a potential divider circuit and two small motors that can rotate the solar collector through two perpendicular directions to monitor the apparent motion of the Sun.

5.5.2   Solar photovoltaic electricity

Sunlight can also be converted directly into electricity by means of photovoltaic (PV) solar cells. These are already in common use as power sources for small calculators and watches.

To understand how sunshine can generate electricity, it is necessary to understand the photon properties of light. Radiation of frequency υ has photon energy $h\upsilon$. A photon of visible light can give its energy to an electron and excite it across a voltage of about 1–2 V. If the excited electron can be trapped at this voltage, then it is possible for the electron to move as an electric current and so give its energy to an electrical load, such as a motor. This is the principle of photovoltaic electricity production from solar cells. The photovoltaic effect was first observed by the French physicist Edmund Becquerel in 1839. To have a better understanding of the photovoltaic effect, it is useful briefly to see how the photoelectric effect works. Though they are not the same process they do have parallels in that they both generate an electric current from the action of light.

Photoelectric effect

The photoelectric effect was discovered by H. Hertz in 1887. He shone ultraviolet light on a spark gap and noted that the electrical discharges across it occurred more easily. Further experimental research by W. Hallwachs (1888) and P. Lenard (1895) produced the following results:

*   Certain metals when illuminated with light of certain wavelengths generated photo-electric currents
*   The kinetic energy, and therefore the velocity, of the photoelectrons did not depend on the intensity of the incident light, but on its wavelength and hence the frequency
*   Photoelectrons were not liberated if the frequency of the light was less than a threshold frequency (v_o) (Figure 5.10).

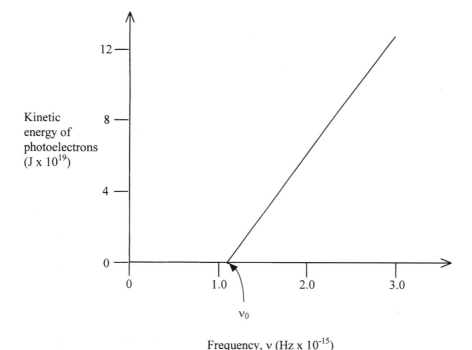

Figure 5.10 The photoelectric effect describes kinetic energy of electrons liberated from a surface by photons with an energy hv where v is photon/light frequency.

The existing classical wave theory of electromagnetic radiation could not explain all these results. In 1905 Einstein applied Planck's Quantum Theory (see Case study 2.1) and argued that light, being radiation with a given wavelength and frequency, could be *quantized*, i.e. it could occur in discrete packets of energy ($= h\nu$) called quanta, where h is Planck's constant ($= 6.6 \times 10^{-34}$ Js) and ν is frequency. These quantized packets were called *photons*, and Einstein suggested that if light consists of photons, then the number of photons incident on a metal per second per metre squared depended on the intensity, while their energy depended on the frequency (and not the intensity).

He considered that each photon of radiation transferred its energy to an electron in a metal. The minimum amount of energy required to release the electron at the surface of the metal was called the *work function* (Φ), and the energy that remained was the kinetic energy of the liberated photoelectron. There would be a range of kinetic energies depending where in the crystal lattice the electrons were initially situated. Those further away from the surface would lose energy in collisions in the process of escaping.

Einstein summarized his ideas in the famous *photoelectric equation*:

$$h\nu = \Phi + \frac{1}{2}.m.v_{max}^{2} \tag{5.7}$$

where Φ is the work function of the metal and $\Phi = h\nu_{o}$, where ν_{o} is the threshold frequency, and $\frac{1}{2}.mv_{max}^{2}$ is then the maximum kinetic energy of the liberated photoelectron. The units of each term can be measured in joules or electron-volts (eV), where $1\,eV = 1.6 \times 10^{-19}$ J. Sodium, for example, has a work function of 2 eV.

Worked example 5.3   Photoelectric effect.

The maximum kinetic energy of the electrons released from a metal surface is 1.6×10^{-19} J, when the frequency of the radiation is 7.6×10^{14} Hz. Determine the minimum frequency of the radiation for which photoelectrons can be released. Assume Planck's constant.

Solution

Apply the Einstein photoelectric equation (5.7), noting that the work function $\Phi = h\nu_{o}$, where ν_{o} is the minimum frequency.

Therefore, $6.63 \times 10^{-34} \times 7.6 \times 10^{14} = 6.63 \times 10^{-34}\nu_{o} + 1.6 \times 10^{-19}$. You should find that the threshold frequency, ν_{o}, is 5.2×10^{14} Hz.

Solar cells

Solar cells consist of materials that are not pure metallic conductors. They consist of semiconductors, in which the majority charge-carriers are either electrons or 'holes', depending on the type of semiconductor. A 'hole' is an electronic vacancy.

Band theory is a useful way of explaining electrical conduction in semiconductors. In semiconductors band theory visualizes three bands that differ in potential energy (Figure 5.11): the valence band, the conduction band and, sandwiched between them, a forbidden band, which is sometimes called the energy gap, denoted by E_g. The width of this gap in energy has an important bearing on the electrical conductivity of the semiconductor. The width of the forbidden band depends on the material that forms the metal or semiconductor. In a metallic conductor there is no forbidden band. This allows electrical conduction by electrons which can be thermally excited from the valence to the conduction band at all temperatures (Figure 5.11a).

Many solar cells use silicon as it is a pure semiconductor. It is *tetravalent*, which implies that it has four electrons in its outermost orbit. These are also referred to as valence electrons. Silicon in its pure state is an effective insulator, which means that the forbidden band can be large enough not to allow electronic transitions between the valence and conduction bands (Figure 5.11b). If, however, the silicon is 'doped', i.e. impurities are added, its electrical properties are modified (Figure 5.11c).

The cell's operation depends on the silicon semiconductor having layers with different dopant 'impurities', so forming a junction or boundary within the solid. In solar cells, the material is constructed in thin sheets with the junction just beneath the top surface (Figure 5.12). Electrical connection is made at the top and bottom, so that many cells can be connected in series to make a module.

In a solar cell the forbidden band is 1.12 eV. Thus, when light illuminates a solar cell electron-hole pairs will be generated, at the junction (which creates a voltage across it), if the photon energy $h\nu \geq E_g$, where E_g is the band gap energy. Any photon arriving with an energy <1.12 eV will not excite an electron into the conduction band but it may heat up the crystal, while if the photon energy is >1.12 eV not only will an electron be excited into the conduction band, but also the excess energy will be used up in heating the crystal.

The crystal itself has its own resistance, and the efficiency of the cell depends on temperature. As the temperature increases, the internal resistance increases and, as a result, the electrical conductivity decreases. At 0°C the best silicon solar cell has an efficiency of about 24%, and this decreases as the temperature rises.

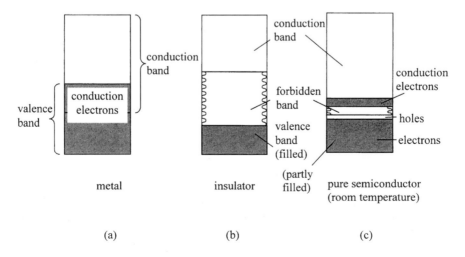

Figure 5.11 Band theory.

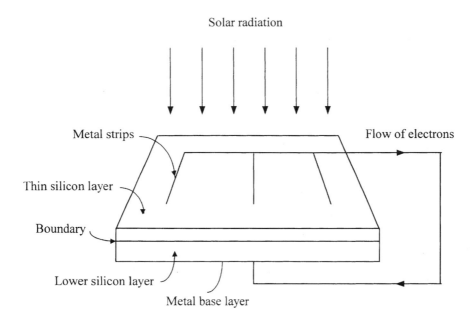

Solar radiation

Metal strips

Flow of electrons

Thin silicon layer

Boundary

Lower silicon layer

Metal base layer

Figure 5.12 Schematic diagram of a solar cell.

Usually about 30 cells are in series, to give a total open circuit voltage of about 15 V direct current (DC). When the Sun shines on it, the module is able to charge a 12 V DC battery; in bright sunshine the current would be about 4 A. The modules are connected in series and parallel to produce DC currents at the required voltage. This DC power can be converted into alternating current (AC) with an electronic invertor. For a silicon solar cell, the potential difference available is about 0.5 V and is generated with a peak power of about $0.75 \, W \, m^{-2}$. Connecting many solar cells in series allows higher voltages and in parallel greater currents to be generated and, hence, greater power outputs.

Solar cells are becoming increasingly popular for a variety of applications. There is no doubt that they can make an important contribution to our domestic and industrial needs. For example, an engineering firm has developed photovoltaics to generate electrical power for hospitals in remote regions of east Africa (Figure 5.13).

In the future, one can anticipate that solar cells will be applied more widely especially in developing countries to provide local sources of energy in remote rural areas. Here the predominant need is for small amounts of power for lighting, for communications, for refrigeration of vaccines and for pumping water. The beneficial aspects of photovoltaic power is that electricity is being produced with no moving parts, no noise, no pollutant emissions and within transportable components of extremely long life-time. The disadvantages are that electricity is only produced in daytime, and therefore has to be stored or integrated with other sources, and that the initial capital cost is presently large in comparison with other sources (about $7 per watt capacity for photovoltaics, compared with about $1 per watt capacity for electricity from combustion of coal or from wind power). Nevertheless, capital costs are decreasing as volume production increases at about 20% per year from the 100 MW per year production of 1998.

Figure 5.13 Solar cells providing electrical power for hospitals in north-east Africa.

5.6   Wind power

Wind energy has been used for thousands of years in providing energy for useful work in both linear and rotational motion, as in wind for sailing ships and windmills respectively. Historically, they were used for pumping water and milling grain. In Holland windmills were used for the drainage of fields, and the multiblade wind-turbine is used world-wide for pumping in isolated areas. Today, electricity is generated from wind in many parts of the world, especially in Europe, India and the USA. However, there is no country (Figure 5.14) where it is not applicable in some form.

Even within Europe there is considerable variability in the wind speed. Britain would seem to be a suitable area for capturing wind energy, because prevailing south-westerly winds enhance the wind-capability of the western-facing coastal and upland areas. Thus, the prospect of using wind power partly to meet energy demands in these areas of high wind has led to the development of new wind turbines. These may be categorized into two types: the horizontal axis wind turbine as illustrated in Figure 5.15, and vertical axis wind turbines as shown in Figure 5.16. Often clustered together such turbines may form a *wind farm*. The disadvantages of such a 'farm' are obvious. Since wind is intermittent electricity generation will be intermittent. Nevertheless, they may provide a local energy supply and are often used to provide power to draw water from deep artesian wells.

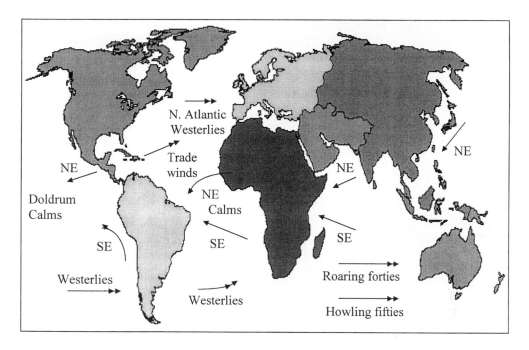

Figure 5.14 Global prevailing winds.

Figure 5.15 A wind farm in Friesland, North Germany consisting of three-bladed wind generators.

(a)

(b)

Figure 5.16 Examples of wind turbines (a) Darrieus (or 'egg-beater') and (b) Savonius rotor.

5.6.1   *Average power of a moving mass of fluid*

The unequal heating of the Earth and its atmosphere gives rise to pressure and density differences which power convection currents and these constitute winds (see Chapter 9). Winds are therefore moving air masses, with kinetic energy. This implies that they follow the rules of fluid dynamics (see Section 5.6.4).

   The following derivation is equally applicable to water flowing in a cylindrical pipe and air flowing towards and through a wind-rotor. The fluid, of density ρ, is flowing through a cylindrical, transparent pipe, 1 m long (L) and with a speed v. Then, the power of the moving fluid, P, will be kinetic energy/time:

$$P = \frac{1}{2}.mv^2/t$$

If this time is replaced by an expression for the speed of the fluid and the length of the cylinder of fluid, $t = L/v$, then the power becomes:

$$P = \frac{1}{2}.mv^3/L$$

A cylinder with a cross-sectional area (A) will have a volume of $A \times L$. Thus, the mass of the moving fluid, $m = V \times \rho = AL\rho$.

This will give an equation for the available power of the moving fluid:

$$P = \frac{1}{2}.\rho.A.v^3 \tag{5.8}$$

This equation suggests that if the speed of the fluid doubles then the theoretical power will increase eightfold.

Worked example 5.4   Available wind-power.

The average speed of Atlantic winds at about 10 m height is $10\,\mathrm{ms}^{-1}$. If the radius of a wind-turbine is 30 m, calculate the available wind-power, given that the density of air, ρ, is $1.2\,\mathrm{kg\,m}^{-3}$.

Solution
Apply equation (5.8), where the cross-sectional area is $\pi r^2 = 2826\,\mathrm{m}^2$. $P = 1700\,\mathrm{kW}$ (1.7 MW).

However, this does not take into account the efficiency of the turbine. A typical coefficient of performance is 40%; hence, the turbine would generate about 0.7 MW. In a $20\,\mathrm{ms}^{-1}$ wind, since the power increases as the cube of the wind speed, the same turbine could produce 5.6 MW. In practice the generated power is limited to the rated power of the electricity generator to avoid over-heating, so generation in less powerful winds may have to be limited by changing the pitch of the blades.

5.6.2   *Bernoulli's theorem and the aerofoil*

Wind-turbine propeller blades have a design similar to that of an aircraft wing (the aero-foil), so that the motion of the blades can obey the rules of *aerodynamics*. The theory of the aerofoil was developed by the British engineer Frederick Lanchester (1868–1946). When an aircraft is moving through the air the wind provides lift in the following way. The wing is designed such that the wind flowing above the wing is travelling faster than that below it. The air travelling faster over the wing produces a reduction in pressure compared with the air travelling below the wing, which is moving at a lower speed and whose pressure is not reduced as much. A pressure difference therefore results and it is this that provides the lift. This lift is the product of the pressure difference and the area over which it acts. This is an expression of *Bernoulli's theorem*.

Consider a fluid flowing in a tube of flow which has a varying cross-sectional area as it goes from X to Y (Figure 5.17). If there is a difference in pressures at the ends of the pipe, the fluid will flow along the pipe and the difference in pressure divided by the length of the pipe will give the pressure gradient (dP/L). Assuming that the fluid is incompressible, the density at X is the same as the density at Y, this implies that over a given time interval, dt, the mass of fluid entering the tube at X should equal the mass flowing out at Y. Hence, the mass flow rate at X = mass flow rate at Y.

Daniel Bernoulli (1700–82) was a member of a prodigious Swiss family. His father and uncle helped to develop the methodology and applications of differential equations. Daniel, too, was a gifted mathematician. He held professorships in both Botany and then Physics, but his main interest was in the field of partial differential equations. He conducted pioneering work in fluid dynamics and developed mathematical models of the transmission of infectious diseases.

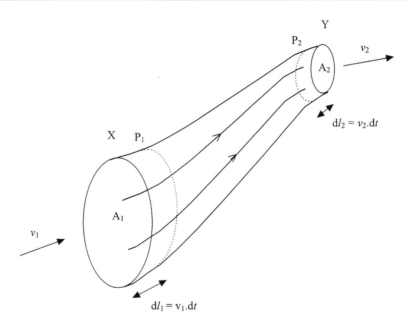

Figure 5.17 The equation of continuity.

Now the mass of liquid = volume $(V) \times$ density (ρ), thus the mass rates can be expressed as:

$$V_1.\rho_1/dt = V_2.\rho_2/dt$$

Since $V_1 = A_1.dl_1$ and $V_2 = A_2.dl_2$ and as the liquid flowing is incompressible the density remains constant, such that

$$A_1.v_1 = A_2.v_2 \tag{5.9}$$

This is the *Equation of Continuity*, and states that as the cross-sectional area at a point increases the velocity of the fluid decreases at that point. This implies that as the velocity

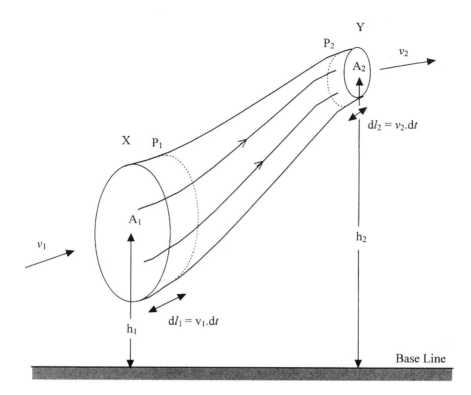

Figure 5.18 Bernoulli's theorem.

changes a resultant force acts on the fluid, and this force has its origin in the pressure difference.

If the pressure at X is P_1 and at Y it is P_2 (Figure 5.18), for the fluid to flow in the specified direction P_1 must be greater than P_2. Consider an element of fluid cross-sectional area (A_1) and length (dl_1) at X, and area (A_2) and length (dl_2) at Y. Since it is incompressible and the mass-rate at X is the same as at Y, then the net work done in taking a volume element of the fluid from X to Y is

$$(F_1 \times dl_1) - (F_2 \times dl_2) = (P_1.A_1.dl_1) - (P_2.A_2.dl_2) = P_1.V - P_2.V = (P_1 - P_2).V$$

where V is the volume.

This work done is equal to the corresponding change in the kinetic and potential energies of the element. Hence:

$$(P_1 - P_2).V = (\tfrac{1}{2}.mv_2^2 - \tfrac{1}{2}.mv_1^2) + (mgh_2 - mgh_1)$$

Dividing both sides by V and separating the terms for X and Y respectively:

$$P_1 + \tfrac{1}{2}.\rho v_1^2 + \rho gh_1 = P_2 + \tfrac{1}{2}.\rho v_2^2 + \rho gh_2 \tag{5.10}$$

This is the mathematical formulation of *Bernoulli's theorem*, and it is sometimes expressed as:

$$P + \frac{1}{2}.\rho v^2 + \rho g h = \text{a constant} \tag{5.11}$$

Bernoulli's theorem is therefore an expression of the Principle of the Conservation of Energy. It shows that for a fluid flowing in a tube of flow, whose cross-sectional area may vary and whose height may also vary above an arbitrary base-line, that the sum of the pressure, the kinetic energy per unit volume and the potential energy per unit volume, at a point on a flow-line, is constant.

Bernoulli's theorem not only has applications in physics and engineering, as can be seen in wind energy and hydro-power, it is valid for blood flow in arteries and veins and in the transport of phloem in plants. The theorem explains how the pressure difference between the top and the underside of an aircraft wing provides the lift force, and how through the process of 'tacking' a sailing boat can be made to travel faster than the wind-speed.

Worked example 5.5   Lift force on an aircraft wing.

Air is flowing across an aircraft wing with a velocity of $100\,\mathrm{m\,s^{-1}}$ above the wing, and $80\,\mathrm{m\,s^{-1}}$ below it. Determine the force required to provide lift, given that the total wing area is $30\,\mathrm{m^2}$, and the density of air is $1.29\,\mathrm{kg\,m^{-3}}$.

Solution
Apply Bernoulli's theorem, assuming that the potential energy component is constant, and that the fluid is incompressible.

Thus, $P_1 + \frac{1}{2}.\rho v_1^2 = P_2 + \frac{1}{2}.\rho v_2^2$

Rearranging to obtain the pressure difference, $dP = P_1 - P_2 = \frac{1}{2}.\rho(v_2^2 - v_1^2)$. The lift force, $F = dP \times A = \frac{1}{2} \times 1.29\,(100^2 - 80^2) \times 30 = 7 \times 10^4\,\mathrm{N}$.

5.6.3   Forces acting on wind-turbine propeller blades

Consider a propeller linked to a stationary wind-turbine, with the wind rushing through it (Figure 5.19). (The wind in the diagram is moving into the paper.) The blades are designed such that the pressure difference resulting from the difference in wind speed on either side of the blade provides a deflecting lift. Thus, for blade A the lift is upwards, while for blade B it is downwards. Assuming the lift forces are equal and opposite the couple which is produced generates a turning effect (a torque).

Tip–speed ratio

A blade of fixed shape requires wind to approach it at an optimum angle ϕ. At the end tip of the blade, the approaching wind is relative to the moving blade tip, and the cotangent of the angle of approach gives an expression for the tip–speed ratio:

$$\cotan \phi = v/u = \lambda = \text{the 'tip–speed ratio' where } \cotan \phi = \frac{1}{\tan \phi} \tag{5.12}$$

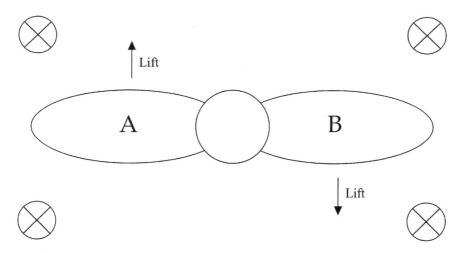

Figure 5.19 Wind flowing past a propeller blade.

where v is the speed of the tip and u the speed of the wind. Keeping ϕ constant implies that v should be proportional to u at all times. This would require the rotational speed and frequency to vary continuously as the fluctuating wind changes speed. Some turbines attempt to do this, but more commonly the meteorological conditions of each site are investigated to determine the best average value for a fixed rotational rate.

The tip–speed ratio (λ) is designed to be as large as possible for electricity generating wind-turbines (in practice about 10) to give maximum speed of rotation for the generator with as small a gear box as possible. This also requires there should be as few blades as possible (usually two or three), which has the benefit of reducing the amount of expensive material required.

Worked example 5.6   Tip–speed ratio.

An aeronautical engineer is designing a wind-turbine to achieve optimum performance in a wind of speed $15\,\mathrm{m\,s^{-1}}$, a blade radius of $30\,\mathrm{m}$ and a blade shape with a tip–speed ratio of 10. Determine (a) the angular frequency and (b) the period of rotation.

Solution
The speed of the blade tip needs to be $10 \times 15 = 150\,\mathrm{m\,s^{-1}}$. Then:

(a) Angular frequency, $\omega = v/r = (150\,\mathrm{m\,s^{-1}})/(30\,\mathrm{m}) = 5$ radian $\mathrm{s^{-1}}$. The frequency (f) will then be $\omega/2\pi = 5/(2\pi)\,\mathrm{s^{-1}} = 0.8\,\mathrm{Hz}$.
(b) Rotational period $= (1/0.8)\,\mathrm{s} = 1.25\,\mathrm{s}$.

Thus, the larger the wind turbine, the slower it rotates to maintain the optimum tip–speed ratio.

Wind turbines for mechanical work, such as milling or water-pumping, are designed to produce large torque at low speeds. This is in contrast to large rotational speeds at low torque for electricity production. Therefore, 'wind mills' and 'wind pumps' have many blades to maximize the turning force on each blade.

Case study 5.3   Betz limit.

Equation (5.8) showed that the available power of the wind can be expressed as $P = \frac{1}{2}.\rho A v^3$. In reality, the actual power that can be extracted is a fraction of this. The extracted power can, therefore, be expressed as:

$$P = \tfrac{1}{2}.C_p.\rho A v^3, \tag{5.13}$$

where C_p is called the *coefficient of performance*. How is this factor derived?

Consider a cylinder of wind with an unperturbed upstream speed of v_1. In applying the equation of continuity (equation 5.9) it is assumed that the density remains constant and that the fluid is incompressible. The propeller-rotor functions as a 'actuator disc', in which as wind energy is extracted there is a pressure drop across the disc. Hence, the linear momentum of the wind also decreases.

In passing through the rotor the moving air expands and this affects the efficiency. The speed of the wind (v_3) is also slower downstream than at the rotor (v_2) or upstream (v_1). These are shown in Figure 5.20.

If A_1 and A_3 are the cross-sectional areas of the ends representing the tube of flow that passes through A_2 (which represents the area swept out by the actuator) and it is assumed that a constant mass of air sweeps through A_2 and that $v_1 > v_2 > v_3$,

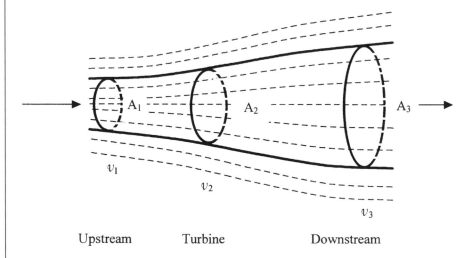

Figure 5.20 Wind-pattern through a rotor.

from Newton's Second Law the force of the wind acting on the turbine equals the rate of change of momentum:

$$F = (mv_1 - mv_3)/t$$

Therefore, the power extracted by the rotor:

$$P_r = F.v_2 = (mv_1 - mv_3).v_2/t \tag{5.14}$$

The difference in kinetic energy per second of the wind upstream and downwind is the extracted power $P_w = KE_{upwind} - KE_{downwind} = \frac{1}{2}.mv_1^2 - \frac{1}{2}.mv_3^2$.

$$\text{Therefore, } P_w = \frac{1}{2}.\frac{m}{t}.(v_1^2 - v_3^2) \tag{5.15}$$

From equations (5.14) and (5.15), an expression for the speed of the wind passing through the rotor is obtained by equating P_r and P_w:

$$v_2 = (v_1 + v_3)/2 \tag{5.16}$$

If one now applies the equation of continuity for an incompressible fluid, then the mass flow-rate can be written as:

$$m/t = \rho.A_2.v_2 \tag{5.17}$$

By incorporating equations (5.14) and (5.16) an expression can be obtained for the power extracted by the rotor:

$$P_r = 2\rho.A_2.v_2^2.(v_1 - v_2) \tag{5.18}$$

At this point the *interference or perturbation factor* (a) is introduced. This is an expression for the fractional wind velocity drop at the rotor:

$$a = (v_1 - v_2)/v_1 \text{ and } a = (v_1 - v_3)/2v_1 \tag{5.19}$$

Using these terms and equation (5.18), one can now formulate a model for the extracted power in terms of the initial upwind velocity (v_1) and the interference factor:

$$P_r = \frac{1}{2}.\rho.A_2.v_1^3.\{4a(1 - a)^2\} \tag{5.20}$$

or $P = \frac{1}{2}.C_p\rho Av^3$, where $C_p = 4a(1 - a)^2$ is the coefficient of performance and P is the power of the upwind. C_p, therefore, represents the fraction of the power that can be extracted, and is also called the power coefficient. It is a measure of the efficiency with which a rotor can extract energy from the wind, and it is called the *Betz limit* (after the German engineer Albert Betz who developed this idea during the 1920s) because it places a limit on the maximum power that can be extracted. It suggests that the kinetic energy of the upwind cannot all be captured by the rotor and that some remains to travel downwind initially at a lower speed. Thus, from the equation of continuity, the tube of flow would expand. The expression suggests that a maximum of just over a half of the wind's power can be extracted. C_p has a maximum of 16/27 or about 59%. In practice, a modern wind-turbine might extract about 40% of the power of the wind.

5.6.4   Laminar and turbulent flow

Wind, like water, is a fluid, and in Section 2.5.3 the effect of wind on the body was examined. Different types of air-flow can affect the efficiency of power generated from wind-turbines and *fluid dynamics*, which is the branch of physics concerned with fluids in motion, has made a valuable contribution to our understanding of such processes.

Fluids can either flow in streamlines (*laminar* flow), or with eddies and vortices (*turbulent* flow) (Figure 5.21).

In laminar flow the frictional force is proportional to the velocity (v), because the planar streamlines have uniform velocities. In turbulent flow, where there are accelerating bodies of fluid subjected to centripetal forces, the frictional force is proportional to the velocity squared (v^2). Turbulence is an example of non-linear dynamics and can be modelled using the mathematics of chaos. Turbulent air flow can increase the drag on propeller blades and thereby reduce efficiency.

In a series of classic experiments in 1883, Osborne Reynolds, Professor of Engineering at Manchester University, investigated laminar and turbulent flow in pipes. Under laminar conditions, as the velocity of flow was gradually increased streamlines were created. This ensured that the velocity of the water particles in each particular streamline remained constant. As the flow-rate was increased, there was a point when there was a sudden transition, at the *critical velocity* (v_c), to the eddies and vortices characteristic of turbulent flow. Using the method of dimensions it can be shown that the critical velocity depends on the dynamic viscosity (η), the density of the fluid (ρ) and the radius of the pipe (r):

$$v_c = k\eta/r\rho \tag{5.21}$$

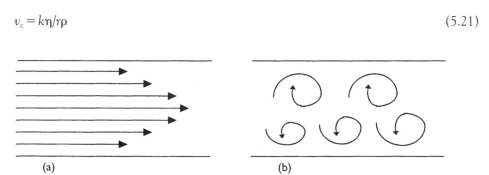

(a)   (b)

Figure 5.21 (a) Laminar flow and (b) turbulent flow.

where, $k = R$, where R is a dimensionless quantity called the *Reynolds' number*. When a fluid is flowing R can be expressed as the ratio of two forces which are operational within the system. The first is the inertial force that derives from Newton's first and second laws of motion, and the drag or viscous force, which is a product of the shear stress and the surface area over which a layer operates. R can then be expressed as:

R = inertial force/viscous force

This shows that when the inertial forces are small R is low, and when the viscous forces are small R is high. This number is a useful indicator of the type of flow for a given velocity of the fluid. For a circular tube the type of flow associated with the following approximate ranges of the Reynolds number are:

Laminar: $1 < R < 1000$
Transition: $1000 < R < 10\,000$
Turbulent: $R > 10\,000$

It has been suggested that for a pipe the critical region in which laminar flow changes into turbulent is defined by $2000 < R < 3500$, whereas for external flow over a body the transition region is much higher, $10^5 < R < 2 \times 10^6$.

5.7   Hydroelectric power

The energy of moving water has been harnessed for centuries, in such appliances as water mills, indeed the utilization of water power was a crucial factor in Britain's economic development during and following the Industrial Revolution.

Today, hydroelectric power is still competitive economically with electricity generated by other means. Applications are from very small scale (e.g. a 2 kW electricity generator for a single house beside a stream) to extremely large structures. Two of the world's largest hydroelectric plants, each with over 10 000 MW capacity, are in South America at Guri in Venezuela and at Itaipu on the borders of Brazil and Paraguay. China has commenced a major hydroelectric programme on the Yellow River to produce 10 GW scale to supply a utility national grid (1 GW = 10^6 kW). The world total hydroelectricity capacity is about 550 GW, with an average load factor of 42% (i.e. the annual electricity produced is 42% of what would have been produced if the generators were always at full capacity). On a world scale there is potential for four times as much hydroelectricity, but this would require extensive loss of land and habitat through the erection of dams. As a consequence, there is considerable concern when ecologically important valley land is threatened in this way.

A variety of hydropower is pumped storage. Hydroelectric power is often used to provide additional electrical power at times of peak consumption. Water is pumped up from low levels into a holding reservoir at periods of low priced electricity, and then allowed back through rapidly started turbines to generate electricity at times of peak demand. Thus, hydropower cannot only give very fast response of a few tens of seconds, but also it has the capacity for long-term interseasonal storage.

5.7.1   Water moving through a cylindrical tube

Water of volume flow per unit time (Q), density ρ (fresh water $1000\,\mathrm{kg\,m^{-3}}$) and relative height ($H$) above a datum-plane has a potential energy (P) per unit time of:

$$P = \rho Q g H \tag{5.21}$$

where g is the acceleration of gravity, $9.8\,\mathrm{m\,s^{-2}}$. If this water falls through a turbine, a large proportion of the potential energy can be converted to kinetic energy of the rotating machinery. This 'shaft power' has been used historically for milling grain and sawing timber, and is used now for generating electricity.

Thus, when water can be either dammed or stored in mountainous areas its gravitational potential energy can be transformed into kinetic energy and then via turbines into electrical energy (Figure 5.22). This is the mechanism of *hydroelectric power*.

The total efficiency of converting the potential energy to electricity at end use, is obtained by considering the losses at all stages, namely: (1) pipe friction loss of 5% in large systems and perhaps 30% in small systems; (2) turbine losses between 5 and 10%; (3) electricity generator 5%; and (4) transmission and distribution loss about 10%.

5.8   Tidal power

Large oceans and seas rise and fall in surface height at daily (diurnal) and half daily periods. Far from shallow coasts, the range is only about 1 m and of no value for power generation. However, near to shallow coasts and especially in certain estuaries the range is

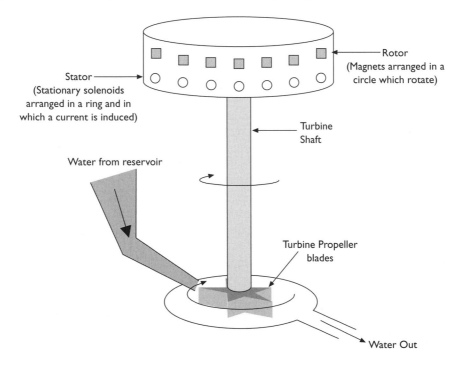

Figure 5.22 A hydroelectric power station.

considerably enhanced and gives the potential for the extraction of power. The gravitational attraction between the Earth, the Moon and the Sun generates the tides. As a result an Earth entirely covered by sea would have two 'bulges' of equatorial water (Figure 5.23a).

Bulge 1, facing the Moon, occurs because the Moon's gravity pulls the water towards itself. Bulge 2, on the opposite side from the Moon, occurs because the Earth–Moon system (Figure 5.23b) is like a weight-lifter's dumb-bell with unequal weights. The centre of mass of the combined system lies 1600 km below the Earth's surface. Not only does the Moon orbit this centre of mass, but so does the Earth, with a period of 27.3 days. This motion is in addition to the Earth's daily rotation and its annual orbit of the Sun. These orbital patterns help us to decipher the forces acting between the Earth and Moon. When the 'dumb-bell' system rotates in the plane of the Earth and the Moon, the centripetal forces on the Earth's seas are different. Water on the side away from the Moon has a slightly smaller centripetal force than the water facing the Moon, resulting in bulge 2.

As the Earth spins on its axis, an observer from a satellite in space would see the ocean surfaces trying to maintain these two bulges. Yet on the Earth, land masses pass through the positions of the bulges, so an observer on a coast notices that the sea rises and falls as the day and night progress. In practice there are many complications, including those of (1) the Sun- and Moon-induced effects that move in and out of phase through each lunar month, (2) the Earth's axis of spin is not perpendicular to the solar and lunar plane, (3) the depth of the sea and shape of the coasts affecting the tidal wave motion, and (4) the effect of wind forces.

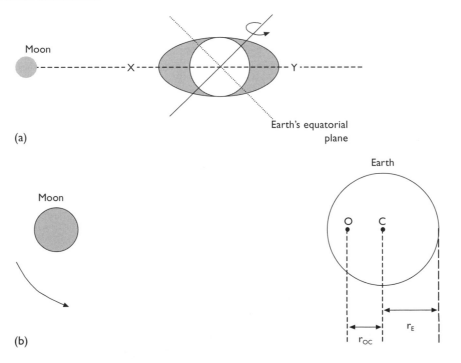

Figure 5.23 (a) Tidal bulges
(b) Rotation of the Earth and the Moon about the Earth–Moon centre of mass (O), where $r_E = 6380$km and $r_{oc} = 4670$ km

Tidal flow power

The strongest tidal currents are near to the shore and, especially, in the channels around islands. Turbines placed in the extended flow of moving water may operate in a similar manner to wind turbines in air, and power may be extracted in a similar way. Historically there were many such small tidal mills and at La Rance in France a tidal power station generates 240 MW.

The enhancement by river estuaries occurs because the estuary becomes a resonating channel, similar to a wind instrument such as a horn giving its fundamental musical note when in resonance with the lips of the player. If an estuary of length (L) and depth (h) is in resonance with a half-daily tide then:

$$L^2 = (36\,000\,m)^2.h \tag{5.22}$$

This condition is met by the Severn Estuary in England (length about 200 km, depth about 30 m) and so the 10–14 m tidal range is one of the largest in the world.

A tidal power plant consists of a barrier built across an estuary to form an enclosed basin. The barrier has gates and turbine ducts to allow sea water to enter or leave the basin as required. Normally as the tide rises, water is allowed into the basin until high tide is reached. Then the gates are closed. As the tide recedes, the water level on the sea-side becomes several metres lower than in the basin. Towards low tide, the trapped basin water is allowed out through turbines, and thereby generates electricity.

With water in a basin of area A, trapped at height R above low tide, the maximum volume of water passing through the turbines all at low tide is AR. With density of 1025 kg m^{-3} seawater will have a mass $AR\rho$. The centre of gravity of this mass is $R/2$ above the low tide level, so the potential energy loss (E) as the water runs out at low tide is:

$$E = (\rho AR)g.R/2 = (\rho AgR^2)/2 \tag{5.23}$$

This potential energy is available for every tidal period, i.e. every 12 h. Of course, there are additional factors. Trapped water cannot be let out instantaneously at low tide. The tidal range is not constant, varying monthly between a maximum spring tide and a minimum neap tide. The tidal oscillations follow the lunar month, not the solar day. Upstream tidal mud flats are disturbed and reduced in extent, so perturbing the ecology. This may be especially serious for migratory birds searching the mud flats for food.

Nevertheless, there is considerable attraction for tidal power inspite of the large initial capital costs. For instance, a tidal power plant across the Severn estuary in south-west Britain could provide about 15% of total electricity for England and Wales.

5.9   Wave energy

Wave energy is an indirect product of solar energy. Solar energy generates winds (see Chapter 9) which in turn sweep over the sea surface generating waves. The quantity of energy imparted by the wind to the sea depends on the strength of the wind and the length of its track over the ocean. The British Isles are fortunate in having coastlines with energetic waves. This is particularly true of the Atlantic-facing coasts, where every metre of wave-front generates tens of kilowatts. Figure 5.24 shows the average wave energy contours around Britain and Ireland. The data shows the annual energy (MWh) and the bracketed figures are the power intensities (kW m^{-1}).

Several attempts have been made to harness wave energy during the past 200 years. There are probably now over 100 different methods proposed to extract power from sea waves. However, relatively few commercial scale demonstrations have been constructed. Devices vary by type of site and type of technology. Sites are (1) in deep water, (2) near shore and (3) on shore. The greatest wave power potential is in deep water, but the storm forces and electricity transmission challenges are formidable. A power plant on shore is the easiest to erect and operate, but the waves have to enter the device by a channel.

The simplest method, and the one with the longest operational experience, is the shore-based 'tapered channel' near Toftestallen in Norway. Here a site was selected where there was a natural focusing of waves between a rocky gully. From this, a constructed channel leads to a reservoir behind a barrier of about 2 m height above mean sea level.

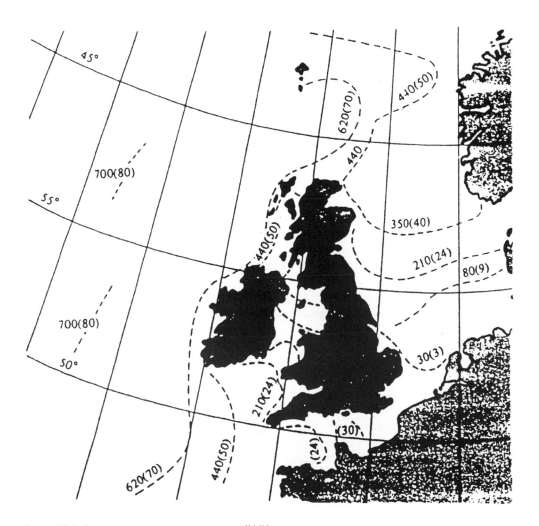

Figure 5.24 Average wave energy contours off UK coasts.

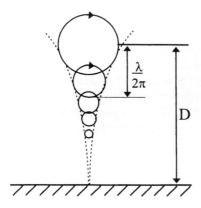

Figure 5.25 Deep water wave profile.

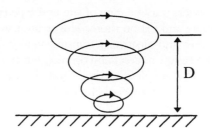

Figure 5.26 Shallow water wave profile.

Waves run up the channel and break into the reservoir, filling it. The trapped water is released back to the sea through a conventional hydropower turbine generator.

Water waves have both kinetic energy, by virtue of their motion, and gravitational potential energy by virtue of the height of the wave. Since the nature of water waves is a means by which wave energy can be harnessed, it should be appreciated that the behaviour of the water particles in waves varies with depth.

A *deep water wave* is defined as one in which the depth is greater than half the wave's wavelength. If one looks at the profile of such a wave (Figure 5.25), it is observed that in the upper zone the particles follow a circular orbit. In contrast, for *shallow coastal waves* (where the depth is less than half the wavelength) the power of the waves attenuate rapidly because the lower orbits are affected by friction with the sea-bed (Figure 5.26). In shallow waves the motion of the water particles are different. Further down the orbit becomes more elliptical, and deeper still, the particle motion is linear. Thus, the position and type of *transducer* (i.e. a means of energy conversion) will have to match the behaviour of the water itself.

Many different types of energy converting devices have been designed for wave energy, the most well known of which are *Salter's 'duck'* and the *oscillating air column*.

Salter's duck

Professor Stephen Salter of Edinburgh University investigated the harnessing of the energy of water waves using a mechanical–hydraulic device. It is called a 'duck' because of its nodding oscillatory behaviour. It moves with the motion of the waves and was designed to 'capture' the wave's energy as it passes. The ducks consisted of a line of cone-shaped vanes (Figure 5.27) fixed to a central shaft, and the oscillatory motion of the device drove a rotary-pump which powered a generator

Oscillating air column

For several years Professor Trevor Whitaker's team, based at Queen's University, Belfast, has been researching the potential of harnessing wave energy using a structure called an

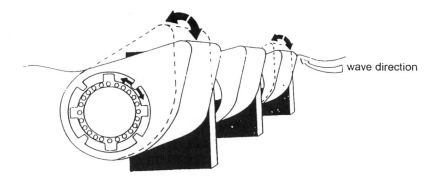

Figure 5.27 Salter's duck.

oscillating air column, which has its base in the sea and its top in the air. The water inside the column rises as the crest of the wave rises (Figure 5.28(a)). The air above this water is compressed and is forced past the turbine from left to right. When the water inside the column falls air is sucked in from the top (5.28(b)). Again the air flows unidirectionally from left to right to power a coupled generator. The large frequency of high waves off the north-west coast of Scotland make it a particularly attractive area in which to site wave energy converters. Whitaker's team have constructed such a device (Figure 5.28) on the Hebridean island of Islay, which can generate up to 75 kW.

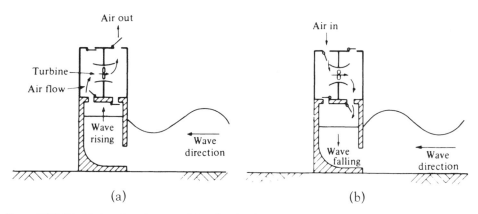

Figure 5.28 Oscillating air column on Islay, Scotland.

5.9.1   *Mathematics of wave power*

At the simplest level a model of a sinusoidal wave can be taken that represents each half of the wave by a cuboid (Figure 5.29).

As one particle of water rotates, it supplies energy in the direction of the wave for continued particle motion, so that the wave as a whole moves forward with energy and as a travelling wave. Taking the wavefront-width as 1 m, the expression for the potential energy component is $mg.\Delta h$, where Δh = change in vertical height between X and Y.

Now, $mg.\Delta h = V\rho.g.\Delta h$, where V is the volume of the cuboid, and $\Delta h =$ amplitude (A) of the waves. Therefore, the potential energy is:

$$V\rho g.\Delta h = A \times 1 \times \frac{\lambda}{2}.\rho g.A = \frac{1}{2}.\lambda.A^2.\rho g \qquad (5.24)$$

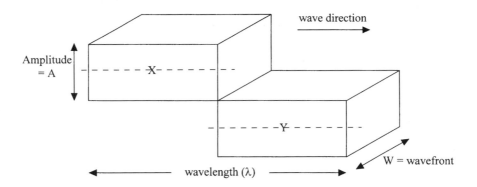

Figure 5.29 Deriving the power of waves by approximation.

If all this potential energy can be converted into kinetic energy then the power generated would be kinetic energy/time, and if this time corresponds to the period (T), then the power available would be:

$$P = \frac{1}{2}.\frac{\lambda}{T}.A^2.\rho g = \frac{1}{2}.\rho g.A^2.v \qquad (5.25)$$

It can also be shown that the power carried forward by a wave of amplitude (A), period (T), wavelength (λ) and per unit length across the top of the wave is:

$$P = \rho g^2 A^2 T/8\pi \qquad (5.26)$$

where

$$T^2 = 2\pi\lambda/g. \qquad (5.27)$$

A typical ocean wave has wavelength about $100\,\text{m}$, period about $8\,\text{s}$, and amplitude about $1.5\,\text{m}$; so that the power carried forward is about $70\,\text{kW}\,\text{m}^{-1}$. In practice, the state of ocean waves is complex, with waves of different wavelength and direction coming together. It has been suggested that, to take into account the variability of the heights of the waves, the wave-power (P) can be expressed by the more sophisticated model:

$$P = W.\rho g^2.TH^2/64\pi \qquad (5.28)$$

where W is the width of the wave-front, ρ the density of sea-water $= 1030\,\text{kg}\,\text{m}^{-3}$, T the wave period and H the 'significant wave height', which is the mean height of the highest third of the waves. Height in this case is the vertical distance between trough and crest. For example, if the heights of 900 successive waves were measured, the top one-third would be isolated, i.e. 300. H would be the arithmetic mean of these 300 heights.

Worked example 5.8   Power generated from waves.

A series of waves impinge on a wave-converting device which is 1000 m long. The period is 10 s, and the heights (i.e. trough to crest) are: 1.6, 2.4, 1.8, 2.8, 1.4 and 2 m. Determine the power generated.

Solution
The top third are 2.4 and 2.8; whose mean is 2.6 m. Then, the power generated, P, using equation 5.28 is:

$$P = 1000 \times 1030 \times 9.81^2 \times 10 \times 2.6^2/64\pi = 3.3 \times 10^7 \, W = 3.3 \times 10^4 \, kW = 33 \, kW \, m^{-1}.$$

5.10   Biomass and biofuels

Biomass material is predominantly organic carbon-based material that reacts with oxygen in combustion and natural metabolic processes to release heat. The initial material may be transformed by chemical and biological processes to produce *biofuels*, e.g. methane (natural gas) or liquid (oil) or solid (coal). The initial energy given to the biomass is derived from the Sun as photosynthesis (see Chapter 11). When released in combustion, the energy stored in the biofuel is dissipated, but the elements of the material remain. Hence, the carbon is often released as carbon dioxide and passes into the atmosphere where it acts as a greenhouse gas and then leads to global warming (see Chapter 7). However, not all biomass material forms biofuels, indeed over the past 3 billion years the bulk of the carboniferous material has been trapped in the earth as carbonates, such as chalk ($CaCO_3$), bicarbonate ions (HCO_3^-) in water and the sea. Furthermore for every carbon atom removed from the atmosphere in this way (carbon burial) a molecule of oxygen remains in the atmosphere. Therefore, the formation of biomass was part of the process of forming our present oxygen-rich atmosphere that maintains life on Earth.

Biomass as a fuel lies second to hydroelectric power as a 'renewable energy source'. However, time and extreme conditions have made the chemical composition of fossil fuels, especially coal and oil, varied and complex. In contrast, biofuels are of simpler and more uniform chemical composition. A benefit of the great majority of biofuels is their very low sulphur content. Examples of biofuels are:

Gaseous biofuels: used for heating and cooking, and in engines for electricity and heat generation, and occasionally for transport, e.g.
- Biogas (methane, CH_4 and CO_2) from anaerobic digestion of plant and animal wastes

- Producer gas (CO and H_2) from gasification of plants, wood and wastes.

Liquid biofuels: used mainly for transport fuels, e.g.
- Oils from crop seeds (e.g. rape, sunflower)
- Esters produced from such oil
- Ethanol from fermentation and distillation
- Methanol from acidification and distillation of woody crops.

Solid biofuels, e.g.
- Wood from plantations, forest cuttings, timber yards and other wastes
- Charcoal from pyrolysis
- Refuse-derived fuels, e.g. compressed pellets.

The total world growth of new biomass is about $250 \times 10^9\,t\,year^{-1}$. The energy released by the combustion of biomass and biofuels is the heat of combustion, and this depends strongly on the moisture content of the biomass material. The heat of combustion of green wood is about $8\,MJ\,kg^{-1}$ and the combustion is poor; dry wood usually burns well at $15\,MJ\,kg^{-1}$; natural fats and oils burn at about $40\,MJ\,kg^{-1}$, and methane gas at $55\,MJ\,kg^{-1}$.

There are many forms of biological material that may be used as fuel; however, methods to use them as energy sources can be arranged into three categories of seven types:

(1) Thermochemical: processes with the application of heat:
- *Combustion:* burning in an ample supply of oxygen. The less oxygen and the more hydrogen atoms in the molecules, the greater the heat of combustion per unit dry matter, e.g. in units of $MJ\,kg^{-1}$: methane 55, petrol 47, ethanol 30, charcoal 32, average coal 27, peat 12–15.
- *Pyrolysis and gasification:* involves heating at various temperatures, in the absence of air or in reduced air flow. Usually the biomass is not dry, and water may be intentionally present. The products are extremely varied, including gases, vapours, liquids, oils and solid char. These products may have value as chemicals and/or as fuels. If the main purpose is to produce gaseous fuel, the process is called gasification; the fuel is producer gas, which incorporates CO, H_2, N_2, water vapour and some other gases.
- *Other thermochemical processes:* there may be a wide range of chemical and heat processes in industry to pretreat the raw biomass material. For instance, acid may be used to break down cellulose and starches into sugars, which in turn can be fermented.

(2) Biochemical processes: these are processes near ambient temperatures, involving biological organisms or biochemistry. There are many such processes in nature from which plant breeding and genetic engineering can improve yields.
- *Alcoholic fermentation:* micro-organisms, in the presence of oxygen in air, produce ethanol from sugars by fermentation. The alcohol can be distilled off and used as a fuel, in place of, or mixed with, petroleum. A common crop is sugar cane, from which the sugars are extracted by compression; the residue (bagasse) is used as a combustion fuel for raising steam for process heat and electricity generation.
- *Anaerobic digestion:* other micro-organisms, in the absence of oxygen, produce methane (CH_4) and CO_2, which are trapped as biogas. The heat of combustion of

the mixture (about 60% CH_4, 40% CO_2) is about $28 \, MJ \, kg^{-1}$. Biogas can be used to burn in a flame for heat, or to be an engine fuel with spark ignition.

- *Biophotolysis*: photolysis is the splitting of water by light to produce hydrogen and oxygen. Biophotolysis is the emission of hydrogen by the metabolism of certain biological organisms. Such effects are uncommon and have not been developed commercially.

(3) Agrochemical fuel extraction: some plants exude liquid or waxy materials that can be collected by cutting (tapping) the trunks or by crushing harvested stems, or more likely, as a chemical base. Examples are natural rubber, turpentine and eucalyptus oil. The extrudates are used as a chemical base or fuel.

Biomass and biofuels are being used in many countries world-wide but the largest use is in Brazil. Since the 1970s, large plantations of sugar cane have been grown to produce biomass fuel suitable for gasification, known as *biogasse* and *barbojo*. Biogasse is the residue from crushing the cane; while barbojo consists of the tops and leaves of the cane plant. The alcohol distilled from the cane is used mainly in transport and generates considerably less pollution than traditional fossil fuels. The most efficient use of biomass is to turn it into biogasse, which can then be burnt in a traditional gas turbine to produce electricity. However, the low efficiency of conversion form biomass to fuel and the large areas that would be needed to grow sufficient material has slowed the introduction of such biofuels. At present, the basic cost of electricity produced in this way is about twice that of fossil fuels and biofuel for transport is also twice as expensive per km than diesel and petrol.

Case study 5.3   Fuel cells.

Fuel cells are a method of converting chemical into electrical energy directly. Unlike the internal combustion engine producing exhaust fumes from cars and other vehicles, fuel cells are non-polluting, releasing only water. The fuel is normally hydrogen, though methanol and natural gas can also be used, and this interacts with oxygen from the air electrochemically to generate electrical energy in the form of direct current (DC). Though the idea of fuel cells was developed in the nineteenth century, it was in the space programmes of the 1970s that they came to be utilized. They consist (Figure 5.30) of two carbon electrodes and an electrolyte.

Consisting of either phosphoric acid or potassium hydroxide, the chemical reactions involve:

1   Hydrogen fuel input:

$$H_2 \rightarrow 2H^+ + 2e$$

2   Production of water from the oxidation of the hydrogen ions:

$$\tfrac{1}{2}.O_2 + 2H^+ + 2e \rightarrow H_2O$$

The electrons migrate from the negative (cathode) to the positive electrode (anode) via an external load resistor (R). At the anode hydrogen atoms are ionized, while at the cathode the hydrogen ions interact with oxygen atoms and electrons to produce water.

The conversion of DC into AC can be achieved using an inverter. These cells can have as high as 60% efficiency, and they have a large power/mass ratio.

Even though fuel cells have an efficiency twice that of an internal combustion engine and are non-polluting they have not as yet been widely deployed. Why not? Until the present time it has not been in the interest of the oil companies to see them developed, so they remain expensive (about $3000 per kilowatt!) and their lifetimes are uncertain, but future research and the gradual decline in global oil supplies should ensure their wider adoption.

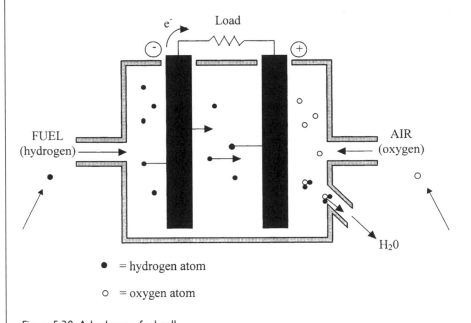

Figure 5.30 A hydrogen fuel cell.

5.11 Geothermal power

Geothermal energy is the energy derived from deep in the Earth's crust. The temperature profile within the Earth is shown in Figure 5.31. Within the first 2 m, the temperature changes with time due to seasonal variation at the surface (see Chapter 10). After that, the temperature increases with depth. The usual rate of increase is about 20–30°C km^{-1}, so making working conditions in deep mines unpleasantly hot. The centre of the Earth is at about 4000°C. However, the average outward flux of energy at the surface is only 0.06 W m^{-2}; a quantity far too small to be useful. The source of this energy is not from the cooling of an initially hot Earth (as Lord Kelvin, the great physicist and engineer, mistakenly thought), but from the radioactive decay of dispersed elements in the Earth.

The temperature gradient between the core and the surface (dT/dz) is the driving force for the thermal conduction of energy outwards. This is defined by a form of Fourier's law (see Section 2.4.1): dQ/dt = $k.A.$dT/dz, where k is the average thermal conductivity of the Earth's crust.

Despite the normal outward geothermal thermal energy flux being very small, there are locations where conditions are very different. Near to the Earth's plate tectonic boundaries the temperature gradient below the surface boundary can be more than 80°C km^{-1}, so that water extracted from deep bore holes may provide steam at high pressure. This steam can be used directly or indirectly to generate electricity and heat, as in Larderello in Tuscany, where geothermal power was first generated in 1904.

In other locations there may be anomalies of high heat flux between 40 to 80°C km^{-1}. Hot subterranean aquifers can be tapped for hot water, e.g. as in Paris and Southampton where the energy is used for piped district heating of buildings. Large accumulations of subterranean granite, a poorly conducting rock of high thermal capacity, may be at elevated temperatures (as in Cornwall); drilling such granite and forcing water through extracts the heat. Attempts have been made to use such 'hot rock' sources for electricity generation, but failure to control the heat exchange water within the cracked rock has produced difficulties.

It should be noted that where the rate of energy replenishment by the Earth is less than the rate at which it is being extracted, the energy source, at least in the short-term, becomes non-renewable.

Figure 5.31 Temperature inside the Earth as a function of depth.

5.12   Summary

Humanity's ability to discover, invent and innovate is boundless. The quest for imaginative sources of energy that are effective, efficient, clean and sustainable are on-going. Deep oceanic thermal energy conversion and hydrogen fuel cells are just two possible sources. At the global level there is a slow but subtle move towards sustainable societies. There is an increasing use of solar collectors and photovoltaic cells in buildings, and in some cities combined heat and power is providing energy for hotels and sports centres. In Denmark, for example, 50% of houses are warmed through district heating. Both public and private transport is a major factor in global energy consumption, and therefore requires that integrated energy policies will also entail increasing use of systems that can reduce pollution in all its forms.

Projections of the demand for energy in the twenty-first century, particularly accompanying the increasing industrialization of countries in the developing world, suggest that it will outstrip energy production. There will then be an energy gap and as a consequence a potential energy crisis. With the eventual decline in energy production from fossil fuels, and the environmental problems arising from their continuing use, such as acid-rain and air pollution, the search for new sources of energy, which are non-polluting and energy efficient, will become ever more urgent. Future predictions suggest that renewable energies will play an increasingly important role but nuclear energy probably remains the most viable option for closing this energy gap, in spite of the inherent safety issues involved.

References

Renewable energy and energy studies

Boyle, G., ed., *Renewable Energy: Power for a Sustainable Future*. Oxford: Oxford University Press, 1996.

Elliott, D., *Energy, Society, and Environment: Technology for a Sustainable Future*. London: Routledge, 1997.

Foley, G., *The Energy Question*, 3rd edn. Harmondsworth: Penguin, 1987.

Howes, R. and Fainberg, A., eds, *The Energy Sourcebook: A Guide to Technology, Resources and Policy*. New York: American Institute of Physics, 1991.

Masters, G. M., *Introduction to Environmental Engineering and Science*. Prentice-Hall: New Jersey 1991.

McMullan, J. T., Morgan, R. and Murray, R. B., *Energy Resources*, 2nd edn. London: Arnold, 1983.

Patel, M. R., *Wind and Solar Power Systems*. London: CRC Press, 1999.

Ramage, J., *Energy: A Guidebook*, 2nd edn. Oxford: Oxford University Press, 1997.

Twidell, J. W. and Weir, A. D., *Renewable Energy Resources*. E. and F. N. Spon/Routledge, London 1996. The full justification for all the analysis of Chapter 5 is given in this comprehensive text.

Web sites

http://www.eren.doe.gov
Energy efficiency and renewable energy network, US Department of Energy
http://www.ise.fhg.de
Solar energy and renewable energy-related servers
http://www.cat.org.uk

The National Centre for Alternative Technology, Machynlleth, Powys, North Wales
http://www.brookes.ac.uk/other/uk-ises/home.html
The UK Solar Energy Society
http://www.bwea.com/contents.html
The British Wind Energy Association
http://www.ewea.org
Web-site of the European Wind Energy Association

Discussion questions

1   Conduct your own energy audit and compare it with the table in Appendix 3. You might like to consider your domestic heating, lighting, cooking, travel, holidays, energy consumed in your working life and leisure activities. How does your energy usage compare with that of the average European shown in Table 5.1?

2   For each of the renewable energy sources cited in the text, discuss their: (a) seasonal variation, (b) efficiency, (c) advantages and disadvantages, and (d) economic feasibility.

3   Discuss the criteria for the economic viability of renewable energy sources. In your analysis you may wish to include the following issues: (a) research and development, (b) initial capital investment, (c) maintenance, running and labour costs, and (d) the de-commissioning process. You might like to compare your sources with the life-history of a nuclear power station.

4   Investigate what contribution could be made by renewable energy sources to your school's or college's energy requirements. You might like to consider heating, lighting, transport, electrical power and cooking.

5   (a) 'With few exceptions, mankind derives all its energy ultimately from the Sun.'
      (i)   Discuss briefly three methods by which energy is derived from the Sun, explaining the role of the Sun in each process. Your choices should be as diverse as possible. Classify each energy resource you mention as renewable or non-renewable.
      (ii) State two energy resources which are independent of solar energy.
      (iii) Indicate the approximate percentage of Britain's energy need which is supplied by each resource to which you have referred in parts (i) and (ii).
   (b) Discuss quantitatively the feasibility of siting a 600MW solar power station on land, at latitude 51°N, in close proximity to areas of high population density. (The maximum conversion efficiency from solar energy to electrical energy is about 10%).
                                    (London Board: 'A' Level Physics: Energy Option, June 1988)

6   It is suggested that the available power (P) to a wind-turbine is proportional to the density of the air (ρ), the area swept out by the propeller blade (A) and the wind speed (v). Apply dimensional analysis to derive an expression for this power.

7   Distinguish between the purpose and mode of action of a *flat plate collector* and a *photovoltaic cell*, each of which has been designed for exposure to solar radiation. Select *one* of these devices and draw a carefully labelled diagram to illustrate its structure. Add a scale to your diagram.
                                    (London Board: 'A' Level Physics: Energy Option: January 1991)

8   The term *combined heat and power (CHP)* is often used when referring to district heating systems. Explain what is meant by CHP and discuss the problems associated with moving from the existing pattern of electrical power production to one in which district heating using CHP is employed.
                                    (London Board: 'A' Level Physics: Energy Option: June 1993)

9   (a) Explain with the aid of diagrams what is meant by (i) nuclear fission and (ii) a nuclear chain reaction.
   (b) Draw a labelled block diagram showing the main parts of a nuclear-fission electrical generating plant. Do not include fuel reprocessing.
   (c) Outline the roles of the *moderator*, the *coolant*, and the *control rods* in a nuclear reactor and

state a suitable material for each in (i) an advanced gas cooled reactor, (ii) a pressurized-water reactor.

At the present time about 20% of the UK's electrical power is generated in nuclear power stations. It has been suggested that the use of such stations be phased out. Discuss briefly two alternative sources of energy for such power stations, paying particular attention to (i) the physics of the energy conversion, (ii) the problems of exploiting the source on a large scale.

(London Board: AS Level Physics: June 1990)

10   The Department of Energy in 1988 described as 'Promising but uncertain' the following renewable energy technologies: tidal energy, shoreline wave energy and land-based wind energy. Choose any one of these. Describe briefly, with the aid of a sketch, the way in which the energy is harnessed and give an approximate calculation to illustrate the average power which might be available using the system you have described.

(London Board: 'A' Level Physics: Energy Option: January 1992)

11   Where will Kyoto lead?

The Kyoto conference in December 1997 was a ground-making event in that it set the first legally binding targets for CO_2 emissions, to be reached globally by 2010. What target percentage reduction figure did the EU agree at the negotiations and what did the US agree? Describe the difficulties encountered in reaching these conclusions and the reasons for them.

What is the additional morally binding percentage target which the UK government has set for this country?

As far as the UK is concerned, energy prices have dropped dramatically in recent years. Current advertising suggests, for example, that gas prices will drop a further 15% in the next few months. The Government is soon to publish its strategy for meeting its reduction targets, against this falling price background. Discuss the problems it faces, in particular with respect to the domestic, commercial, industrial and transport sectors. The wider social and fiscal issues should be recognized.

Suggest and discuss the actions it might take to address these problems and move towards the target, again recognizing the sectoral differences.

(City University: BEng Mechanical Engineering and Energy Management: May 1997)

12   Discuss the features and the modes of operation of a national energy policy.

13   In the emergent 'global village' of the twenty-first century, discuss the role of international cooperation in influencing the development of sustainable communities.

Quantitative questions

1   In the 19th century, coal was the principal fuel used to provide domestic hot water. Suggest how, in a sunny climate, solar power might be used to provide hot water for a small isolated community of about 20 people. Your answer should include simple diagrams and, wherever possible, calculations for which you may have to make estimates as well as make use of the following data:

Solar flux at Earth's surface at midday $\sim 1\,kW/m^2$

Specific heat capacity of water $\sim 4kJ/kg.K$

Average volume of hot water needed per day per person ~ 10 litres

State any assumptions made in arriving at your estimates.

(London Board: 'A' Level Physics: Energy Option: June 1990)

2   (a)   An insulated metal surface is given a negative charge. It is then illuminated with ultra-violet light. Explain why it loses its negative charge and becomes positively charged.

(b)   Is it true that the photoelectric equation obeys the conservation of energy principle?

(c)   The work function of caesium is 1.89 eV. Determine:

(i)   The maximum wavelength capable of generating the liberation of photoelectrons from the metal's surface.

(ii) Using Einstein's photoelectric equation calculate the maximum speed possessed by the photoelectrons, if the light incident on the surface is 540 nm.

(iii) Calculate the electromotive force (EMF) generated by a caesium photocell if solar energy, expressed by a hypothetical average of 545 nm, is incident on it.

(iv) If the power per unit area generated by the photocell, at 545 nm, is 300 W m^{-2}, determine the current per unit area from the metallic surface.

(v) Suggest what might be the limitations with photocells, as they are at present, in providing electrical power.

(Assume h, c, e, mass of an electron and that $1\,\text{eV} = 1.6 \times 10^{-19}\,\text{C}$.)

3   The solar constant has a value which varies from 1420 W/m^2 in December to 1330 W/m^2 in June.

(i) Explain what is meant by the term *solar constant*.

(ii) Suggest why the solar constant is different in December from June.

(iii) Explain why the mean temperature in England is lower than the mean temperature in the West Indies, although the solar constant is the same in both places. (The West Indies are closer to the equator than England).

(iv) State two factors, apart from those already mentioned, which affect the amount of power which a place receives from the Sun at any particular time.

(Cambridge University: 'A' Level Physics: Environmental Physics Option: June 1997)

4   Differentiate the expression for the coefficient of performance, $C_p = 4a(1-a)^2$ with respect to a, and show that C_p is a maximum at 16/27 for $a = 1/3$.

5   You might like to use the software package Excel to plot a graph of C_p against a (for $0 < a < 1$). Take a from 0 to 0.6 in 0.05 intervals, and calculate C_p using $C_p = 4a(1-a)^2$. You should obtain a reversed parabola.

6   (a) Actuator disk theory was originally developed for propellers. Describe the conceptual principles of the theory as it applies to wind turbines. What is the fundamental difference between propellers and a wind turbine application?

(b) The force exerted by the wind on an actuator disk can be written as:

$$F = 2\rho A_D U_\infty^2 . a(1-a)$$

where ρ is the density of air, A_D is the area of the disk, U_∞ is the upstream wind speed, and 'a' is the axial inflow factor described by $a = 1 - (U_D/U_\infty)$ in which U_D is the wind speed through the disk.

Write down expressions for the power developed by the actuator disk, the power coefficient and the thrust coefficient of the disk.

(c) Calculate the maximum value of the thrust coefficient and the value of 'a' at which it occurs. What is the significance of the axial inflow factor exceeding this value?

(d) By a similar calculation it can be shown that the theoretical maximum value of power coefficient is 16/27. Compare this with typical values for real wind turbines and explain any differences.

(e) The axial inflow factor describes the slowing down of the air as it approaches and passes through an actuator disk or wind turbine rotor. What physical features of a real turbine rotor would affect the value of the axial inflow factor?

(City University: BSc/BEng Aeronautical and Mechanical Engineering: Wind Energy: May 1998)

Chapter 6

Revealing the planet

6.1　Introduction

A thousand years ago, most people had little experience of the world outside their village. Few would travel during their lifetime more than a few miles from where they were born and thus people around the world were largely ignorant of the existence of other continents and nations. Then, in the fifteenth century, Europeans began to venture across the oceans establishing contacts (and empires) across the Asian, African and American continents. Maps were constructed, which when printed ensured that, for the first time, humanity became aware of the geography of the Earth. Drawn carefully from topographical surveys conducted by intrepid explorers the map makers of the early twentieth century could still not be sure that their maps were correct in every detail; rivers might be off course and towns placed some miles from their real place. Indeed, early in the twentieth century the Soviet Union deliberately produced incorrect maps relocating whole cities to confuse and mislead any invader! The Soviet generals could not imagine that by the end of that century no part of the Earth's surface would be free of scrutiny and that every movement of an army could be recorded and reported to its opponent in minutes!

Two hundred years ago communications were at the speed of the horse and sailing ship, a slowness that fashioned many historical events. British and American armies would fight the Battle of New Orleans on 8 January 1815, ignorant that the Treaty of Ghent was signed on Christmas eve 1814, ending the US–British 'War of 1812', for only on 13 February would the news reach the Americas by sailing ship. Two hundred years later communications are instantaneous as telephones and satellites link every part of the globe.

Such advances in communications and observations of the Earth's surface arise from one of the greatest technological revolutions – the Space Age. With the launch of the first satellite the planet could be viewed in a new, beautiful and unique way (Figure 6.1). Today, satellites also explore all aspects of the Earth's environment, monitoring climate change, measuring deforestation and desertification, and even locating mineral deposits. Satellites are 'revealing the planet'.

6.2　Remote sensing

Remote sensing is observation at a distance. It is the process by which it is possible to derive information about an object without being directly in contact with it. It is something we do continuously when our eyes or ears sense things around us. In these cases electromagnetic waves (light) and sound waves are used respectively to receive information

Figure 6.1 The Earth as viewed from a satellite.

about a remote object. Thus, by its nature, remote sensing requires a 'field' to carry the information to the observer. Other fields that can be used to transmit information are gravitational and magnetic fields.

Satellites use electromagnetic fields to receive and transmit information. The particular wavelength used depends on the specific application and how strongly that wavelength is absorbed by the atmosphere. To study the composition of the Earth's atmosphere it is necessary to know how each of the components of the Earth's atmosphere absorb particular wavelengths. Since each atom or molecule has a distinctive 'finger print', by selecting one of these characteristic wavelengths it is possible to selectively detect specific molecules in the atmosphere and to study how they change with time and conditions. For

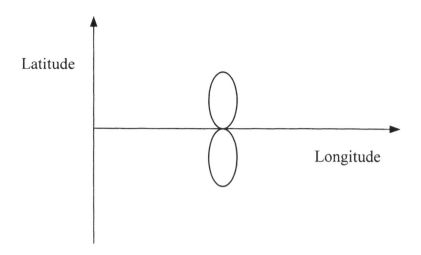

Figure 6.3 Schematic diagram of the path of a satellite in a geosynchronous orbit.

Since $v = 2\pi R/T$, where T is 24 h (i.e. the period of one revolution)

$$T^2 = 4\pi^2 R^3 / GM \tag{6.4}$$

Hence, T^2 is proportional to R^3. This is *Kepler's Third Law* of planetary motion and holds for a satellite around the Earth as it does for a planet around the Sun.

Geostationary orbits are, however, limited to the equator and the same part of the Earth's surface may not always wish to be viewed, e.g. a military satellite will want to examine different parts of the Earth's surface as conflicts and battle fronts develop. Thus, alternative orbiting patterns may be used.

In *geosynchronous orbits* the orbital period is the same as in a geostationary orbit but the orbit traces out a figure of eight pattern in latitude centred about the equator (Figure 6.3).

In a *Sun-synchronous* orbit the satellite procedes around the Earth at the same angular speed as the Earth goes around the Sun. In this case, the satellite always passes over a particular point on the Earth's surface at the same local time. Many satellites use this type of orbit. The altitude of such orbits ranges from 500 to 1500 km. Hence, from Kepler's Third Law the period is about 100 minutes; such that there are some 14 orbits around the Earth in any 24 h. The pattern that this type of orbit follows across the Earth's surface is shown in Figure 6.4. Careful adjustment of the orbital parameters makes it possible to ensure that the orbital cycle is exactly repeated over several days (typically 3–35 days). This is useful for monitoring temporal changes in surface properties while providing global coverage. One such example is the American National and Oceanographic and Atmospheric Administration (NOAA) polar-orbiting satellite used to monitor weather patterns.

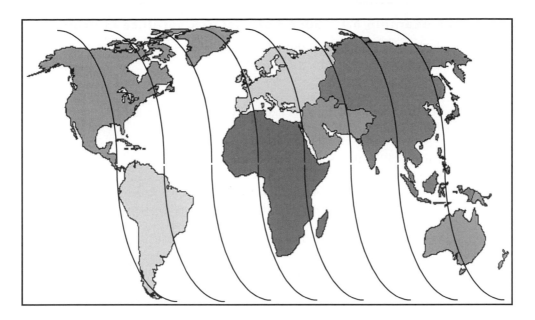

Figure 6.4 Pattern of ground tracks by a satellite in a Sun-synchronous orbit with a 3-day repeat period.

6.4   Resolution of satellite images

Resolution is a measure of the level of detail which a 'picture' taken from the satellite can reveal. The level of detail *required* for monitoring the progress of a weather front is obviously very different from that needed to count the number of tanks deployed on a battlefield. Meteosat, the weather satellite, has a resolution at the equator of 2.5 km for the visible wavelengths and 5 km for longer wavelengths. The level of resolution falls away at higher latitudes and is particularly poor for Europe but remains sufficiently high to give clear indications of coastlines and therefore can locate the position of weather fronts with respect to geographic features (Figure 6.5).

In general, the longer the wavelength the lower the resolution. Therefore, visible sensors have a high resolution and provide fine-scale detail whereas microwave sensors have a lower resolution and are more suited to studies of large uniform areas such as the oceans and ice sheets. The exception to this is the *Synthetic Aperture Radar* (SAR), which uses an ingenious technique to create very large 'synthetic' apertures to provide high resolution imagery in the microwave region (see Section 6.5).

For a detailed exploration of the Earth's surface at a higher resolution a satellite must be as close to the Earth's surface as possible, while not too close so as to experience frictional forces from the atmosphere. Above 650 km the Earth's atmosphere is sufficiently thin so as not to affect the satellite. Therefore, many satellites are placed in orbits between 650 and 900 km and are Sun-synchronous to obtain maximum surface coverage.

The data are then collected as the satellite passes over the Earth's surface, each point of the surface being recorded in several 'pictures', the number depending on the rate at which the pictures are taken and processed. The ground speed of the satellite is, of course, high (see Worked example 6.1).

Worked example 6.1   Speed of a NOAA satellite in orbit.

What is the tangential speed of the NOAA satellite placed 820 km above the Earth's surface and with a period $T = 101$ min?

Solution
The speed of the satellite in an orbit radius R ($=$ radius of the Earth + the height above the Earth) and period T is given by $v = 2\pi R/T$:

$$v = 2\pi(6.37 \times 10^3 + 0.82 \times 10^3)/101 \times 60 = 7.45\,\text{km}\,\text{s}^{-1}$$

The speed of the satellite's shadow V_s across the Earth is then its ground speed multiplied by the ratio of the radius of the Earth to the radius of the satellite orbit:

$$V_s = 7.45 \times (6.37 \times 10^3/7.19 \times 10^3) = 6.60\,\text{km}\,\text{s}^{-1}$$

Thus, if the satellite was only to take one 'picture' every $1/20$ s, objects some 330 m apart might appear only in one or two 'photographs' and hence would be the maximum resolution achieved by such a satellite.

6.4.1   Image processing

A satellite does not collect moving pictures in the same way as a camera but scans and stores data digitally in a similar way to a computer screen (Figure 6.5). A computer screen typically has 1024 dots (pixels) from side to side and 768 from top to bottom. Each pixel can appear in a range of colours. The information displayed in the picture on the computer screen is digitized and fed to each pixel and is repeated about 20 times per second. Similarly, in a satellite the data are collected and stored in an array. In the case of the NOAA satellite, each digital array has a resolution of 1 km. Thus, as NOAA passes over the Earth each strip under each orbit is scanned (Figure 6.6).

The width of the strip over which NOAA passes is 2800 km and is often known as the 'swath width'. If the picture is 1 km square, then the track might be divided into 2800 squares and since NOAA's shadow takes $1/6.60\,\text{km}\,\text{s}^{-1} = 0.15$ s to pass over 1 km, the instrumentation must record and store all the information required from that 1 km^2 in 0.15 s. If the resolution was to be improved to 0.1 km, then there would be a ten-fold increase in data transmission. Hence, the time taken to process the data would be required to be 10 times faster should the data transfer rate be fixed. The increased resolution can only be achieved by allowing the satellite to move more slowly across the Earth's surface (impractical since this is set by the radius or orbit) or by reducing the field of view of the satellite. This allows global coverage once every several days instead of once a day. Thus, there always has to be a balance between resolution and picture frequency!

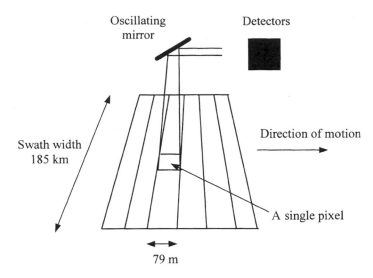

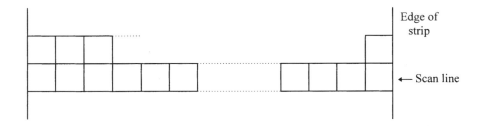

Figure 6.5 Schematic diagram demonstrating resolution achievable from a satellite.

Figure 6.6 NOAA strip.

6.5   Radar

The principle of RADAR (RAdio Detection And Ranging) is simple (Figure 6.7). A short pulse of radiation is transmitted from an aerial and the reflection of that pulse of radiation from targets is recorded by the same aerial. Measurement of the time delay between the emitted and recorded pulse determines the range of the target. Radar was developed in Britain in the 1930s and played an important role in the conduct of the Second World War; radar stations being used to detect aircraft up to 160 km away. Today, radar is widely used in air traffic control systems to monitor the position of aircraft, police to gauge the speed of traffic, by astronomers to probe the surfaces of other planets and meteorologists to detect rain.

A wide range of wavelengths is used in modern radar, extending from 0.8 to 30 cm (Table 6.1) all of which is not greatly affected by atmospheric aerosols (dust), or clouds.

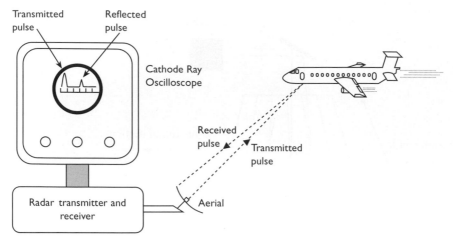

Figure 6.7 The principle of radar.

Table 6.1 Wavelengths used in modern radar systems.

Band	Wavelength range (cm)
K_a	0.8–1.1
K	1.1–1.7
X	2.4–3.8
C	3.8–7.5
S	7.5–10
L	15–30

In the simplest application of radar to remote sensing a pulse (duration τ) of radio waves is emitted vertically downwards from the satellite and the return pulse is timed to provide an accurate measurement of the altitude of the satellite (Figure 6.8). The time taken for the pulse to travel to the surface and back is measured. Then, the range from the satellite can be calculated using $h = ct/2$, where c is the velocity of light.

The US satellite Seasat uses radar to measure the height of the ocean surface to within a few centimetres. Since water is less dense than rock and the gravitational field intensity for $1\,m^3$ of water is less than that from $1\,m^3$ of rock, the gravitational field at the surface of the oceans is lower where the water is deeper, thus causing a slight build up of water over the shallow areas. The contour of the oceans as measured by Seasat is, therefore, an inverse contour of ocean depth. Hence, Seasat can remotely map the topography of the seabed locating the ocean trenches and ridges which by any alternative method (e.g. use of a submersible) would be prohibitively expensive. Seasat has also been used to map the elevation of the two great ice sheets, Antarctica and Greenland, the very remoteness of which had precluded their mapping.

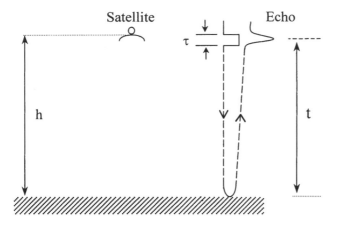

Figure 6.8 The use of radar to measure the altitude of a satellite.

Now consider radar from a moving platform, e.g. an aerial mounted on the underside of an aeroplane. Recall what you hear as an ambulance, police car or fire engine passes by. The pitch of the siren changes, being higher as the vehicle approaches the observer and lower as it recedes. This is due to the *Doppler effect*. See Appendix 4 for a mathematical treatment of this effect. The change in pitch may be used to measure the speed of the source. The same phenomena occurs with stars; the *red shift* of the emitted light being used to determine the speed with which outer galaxies are receding from the Earth. Similarly for flying aircraft; the detected frequency of the radiation at a point in front of the aircraft is raised, compared with that emitted, and the frequency of the radiation detected behind the aircraft is lowered (Figure 6.9). An analysis of the timing and phase of the return signal when combined with the height and speed of the aircraft provides an accurate picture of the terrain and targets around the aircraft. Hence, by flying over the terrain, the aeroplane acts as a large aperture detector. This principle of *synthetic radar* is widely used

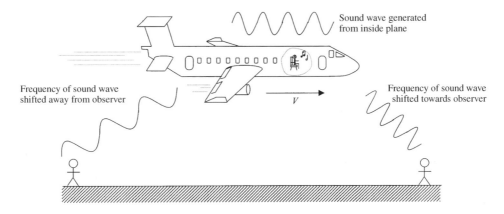

Figure 6.9 Doppler effect used in environmental monitoring.

and is also incorporated into the design of satellites. For example, the European Research Satellite ERS-1 was launched by the European Space Agency in July 1991 to collect information about the oceans, sea surface temperature (see Plate 7), coastal water and land use. B and C microwaves are used for synthetic aperture radar, the return beam being collected for a 800 m long 100 km wide strip with a resolution of 30 m on the ground. When operating in such a mode ESR-1 produces 100 Mbits of data for the ground-based computers to process!

Christian Doppler (1803–53) was a mathematician and physicist who was born in Salzburg, Austria. Between 1829 and 1833 he produced his first research papers in mathematics and electricity. In 1842, while Professor of Mathematics at the Technical Academy in Prague, he developed a theory to explain the apparent change in frequency of the whistle produced by an approaching and receding train. The resulting Doppler effect can be applied to both sound and light waves. The effect has wide applications, e.g. in medical physics such as in studying blood flow patterns, and in cosmological and astrophysical contexts, such as in the development of the Big Bang theory through the red-shift evidence for the recession of galaxies. It formed the basis for E.
Hubble's formulation of the velocity of recession of stars, and played an important role in the development of Einstein's theory of relativity. The effect is used in acoustics, especially in ultrasound, and by the police in radar speed-traps. Doppler returned to Vienna and in 1850 became Director of the Physical Institute and Professor of Experimental Physics at the Imperial University. He died from tuberculosis 3 years later.

Worked example 6.2   Doppler effect.

Many bats use the Doppler effect for detecting obstacles and prey. One species sends out high-frequency sound waves and locates the objects in front of it from an analysis of the reflected waves. If the bat flies at a steady speed of $4\,\mathrm{m\,s^{-1}}$ and emits waves of frequency 90,000 Hz, what is the frequency of the wave detected by the bat after reflection from a stationary obstacle directly ahead of the bat? Assume the velocity of sound $= 340\,\mathrm{m\,s^{-1}}$.

(Joint Matriculation Board: 'A' Level Physics)

Solution

Two situations are presented in this problem. In the first case the bat is the moving source of the waves and the stationary object, the reflector, is 'the observer'. Then

$$f_{\mathrm{o}} = \left(\frac{c}{c - v_{\mathrm{s}}} \right) . f_{\mathrm{s}} = \left(\frac{340}{340 - 4} \right) 90\,000 = 91\,071\,\mathrm{Hz}.$$

In the second case, the stationary object now becomes a stationary source, and the bat now becomes an approaching moving observer. Therefore, if 91 071 Hz is reflected, then the bat will receive a signal of frequency:

$$f_o = \left(1 + \frac{v_o}{c}\right).f_s = 91\,071\left(1 + \frac{4}{340}\right) = 92\,142 = 92.1\,\text{kHz}$$

6.6   Applications of remote sensing data

Several applications have already been discussed and include measurements of sea surface temperature by monitoring infrared radiation using microwave radiometers and the mapping of the topography of oceans and ice sheets using radar altimeters. You will be familiar with the use of photographic images from 'spy planes' used to identify the types and location of military installations. During the 1991 Gulf War the massive plumes of smoke produced by the burning oil wells in Kuwait were clearly visible on the images of several satellites.

Current operational satellites may be broadly divided into five categories:

Communication
Earth-surface mapping
Navigation
Military
Weather.

The operational characteristics of military satellites is, of course, largely secret but information on the others is freely available and many images from these satellites are displayed on the World Wide Web. Some web sites are listed at the end of this chapter. Navigational satellites can inform ships and planes or even individuals of their position on the Earth's surface to an accuracy of a few metres. The *global positioning system* is already being placed in some motor cars so that motorists may soon no longer need a large road atlas. Communication satellites transmit into homes news and sporting events live from all over the Earth as well as providing the opportunity to talk to relatives and friends wherever they may be. This chapter will be concluded with a few examples to demonstrate the versatility of the satellites in operation today.

6.6.1   Meteorological satellites

Meteorological satellites are used in weather forecasting and have become a familiar part of everyday lives as their images are displayed every day on television weather forecasts. In November 1977, the European Space Agency launched the first of its Meteosat satellites, Meteosat 1. Unfortunately this failed in 1979, but Meteosat 2 was launched in 1981 and is still operational, while Meteosat 3 now provides most of the images seen on the European weather forecasts (Figure 6.10). Three wavelength regions are used by the Meteosat

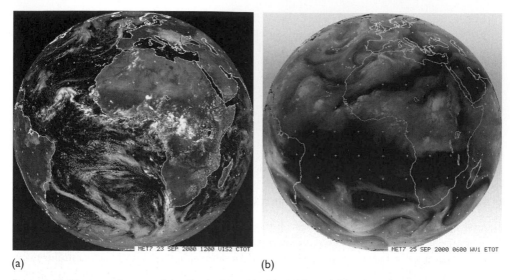

(a)   (b)

Figure 6.10 Meteosat images of the Earth viewed in (a) visible and (b) infrared wavelengths.

system; the visible 0.4–1.1 μm to locate cloud patterns and weather fronts; the 5.7–7.1 μm region to monitor water vapour in the troposphere and the infrared wavelengths 10.5–12.5 μm to monitor temperatures.

Among the instruments included on modern satellites is the *Advanced Very High Resolution Radiometer* (AVHRR). This is a five-band scanning radiometer that operates from the visible to infrared regions. The AVHRR has a resolution of 1.1 km and because it has a large swath width of 3000 km it can provide global coverage every day. Figure 6.11 is an example of a visible image of Europe at night and clearly shows the problem of 'light pollution'. Lighting from street illuminations, advertising signs and houses can be clearly seen by the satellite. Astronomers have been worried for many years about such stray light which, as background lighting, 'pollutes' their observations and deprives millions around the world of the opportunity to view the stars.

AVHRR is widely used in the US NOAA weather satellites. Visible and near infrared wavelengths are used to study cloud patterns, coast lines, snow and ice, and vegetation. Thermal infrared wavelengths monitor land, sea and cloud surface temperatures, while the middle infrared wavelengths record the water vapour in the atmosphere (Figure 6.10).

6.6.2 Landsat

The Landsat satellites are used primarily for investigative work on land use (Plate 6). In fact the EU is using the data to see if farmers really are growing the crops that they claim subsidies for. Data from the Landsat Thematic Mapper is also used for assessing the size of crop harvests, examining the extent of forests and monitoring the spread of deserts. In such cases the Earth's reflectance is monitored at different wavelengths (Table 6.2). The percentage surface reflectance is measured and equals (reflected illuminance/incident illuminance) × 100. The illuminance is measured using a light meter.

Figure 6.11 Light pollution in Europe as observed by satellite.

Resolutions of <30 m are achievable for bands 1–5 and 7, much smaller than most agricultural fields, while resolution in band 6 is about 120 m.

Electronic sensors collect data for each of the seven wavelength bands and are transmitted to the Earth to be combined to provide various types of image. For example, a 'true colour image', i.e. an image that shows the real colours of the observed surface, may be produced by combining signals from bands 1 (blue), 2 (green) and 3 (red). Such images can be directly compared with photographs and are often of a poorer quality. This reflects the poorer pixel quality of computer images and, thus, in the visible region, photographic images are still often superior to those generated by computer. However, in other

Table 6.2 Wavelength bands used in Landsat satellites.

Band	Wavelength range (μm)	Applications
1	0.45–0.52	coastal water mapping, soil/vegetation differentiation
2	0.52–0.60	green reflectance by healthy vegetation
3	0.63–0.69	chlorophyll absorption for plant species differentiation
4	0.76–0.90	biomass surveys
5	1.55–1.75	vegetation moisture, snow/cloud discrimination
6	2.08–2.35	thermal mapping including plant stress
7	1.7–10.4	vegetation moisture and geological mapping

wavelength regions the computer-generated images are superior and provide the only method for global monitoring.

Specific crops may be identified by the infrared wavelengths that they reflect and band 4 is characteristic of the radiation reflected from leaves, and maps of band 4 may be used to identify regions that are cultivating specific crops. In Europe farmers are paid by the European community *not* to grow certain crops on particular areas of land (so-called 'set aside land'). Landsat may, therefore, be used to monitor land use and crop growth. Landsat's resolution of 30 m allows individual fields to be identified and hence those farmers growing crops illegally are discovered by the 'spy in the sky'.

6.7   Summary

The deployment of satellites to explore the planet has led to discoveries that, for their historical impact, can be only compared with the discovery of the Americas by Columbus and other feats of European naval exploration in the fifteenth and sixteenth centuries. Satellites have revolutionized humanity's understanding of the world and in less than 30 years have led to the concept of the *global village*, first proposed by the Canadian sociologist Marshall McLuhan (1911–80). In the global village the progress and speed of modern satellite communications ensures that, today, events in one part of the world can be recorded, studied and acted on with a speed that our ancestors would not even contemplate. The British and American armies fighting at New Orleans in 1815 today would know of the Treaty of Ghent as soon as it was signed and would not have to wait 2 months for the news to cross the Atlantic. Today the world's financial systems react instantly to global news; stock markets in Tokyo are linked to those in London and New York. Television shows incidents as they happen: from sporting events, such as the Football World Cup and the Olympics, to the battlefront. One can literally see history unfold in the home.

The same technology that has enabled us to appreciate the beauty of the Earth also shows the fragility of that world. The Earth is a small planet orbiting a not very special star, which itself is only one of billions in the Universe, and yet here life has evolved – in perhaps – a unique way. The conditions for the evolution of life are tenuous and small changes in any one parameter may have left the Earth a barren rock such as the other planets in the Solar System. Satellites have allowed the study of the ecology of the Earth and also show how the human race has adapted to the physical conditions prevalent upon it. However, they are also revealing how humanity is changing the Earth and perhaps upsetting the delicate balance which has allowed life to flourish. In the second part of this book the *global environment* will be considered, how it operates and how humanity might now be changing it.

References

Bader, M. J., Forbes, G. S., Grant, J. R., Lilley, R. B. E. and Waters, A. J., eds, *Images in Weather Forecasting: A Practical Guide for Interpreting Satellite and Radar Imagery*. Cambridge: Cambridge University Press, 1995.

Baker, D. J., *Planet Earth: The View from Space*. Cambridge, MA: Harvard University Press, 1990.

Barrett, E. C. and Curtis, L. F., *Introduction to Environmental Remote Sensing*, 4th edn. Cheltenham: Thornes, 1999.

Campbell, J. B., *Introduction to Remote Sensing*, 2nd edn. London: Taylor and Francis, 1996.

Cracknell, A. P. and Hayes, L. W. B., *Introduction to Remote Sensing*. London: Taylor and Francis, 1993.

Danson, F. M. and Plummer, S. E., *Advances in Environmental Remote Sensing*. Chichester: Wiley, 1996.

Drury, S. A., *Images of the Earth: A Guide to Remote Sensing*, 2nd edn. Oxford: Oxford University Press, 1998.

Rees, W. G., *Physical Principles of Remote Sensing*. Cambridge: Cambridge University Press, 1996.

Sabins, F. F., *Remote Sensing: Principles and Interpretation*, 3rd edn. New York: Freeman, 1997.

Schott, J. R., *Remote Sensing: The Image Chain Approach*. Oxford: Oxford University Press, 1997.

Schowengerdt, R. A., *Remote Sensing: Models and Methods for Image Processing*. San Diego: Academic, 1997.

Verbyla, D. L., *Satellite Remote Sensing of Natural Resources*. New York: Lewis, 1995.

Vincent, R. K., *Fundamentals of Geological and Environmental Remote Sensing*. Englewood Cliffs: Prentice-Hall, 1997.

Williams, J., *Geographic Information from Space: Processing and Applications of Geocoded Satellite Images*. Chichester: Praxis, 1995.

Web sites

Images of the Earth from space as viewed by satellites may be found on many web sites including: **Nottingham University Remote Sensing web-site**
http://www.ccc.nottingham.ac.uk/pub/sat-images/x2.JPG

Meteosat

http://www.bishnet.free-online.co.uk/weather/meteo/meteofr1.html
http://www.liv.ac.uk/~mark/main.htm
http://www.nottingham.ac.uk/meteosat/infos.shtml

Landsat

http://copac.ac.uk/maps/landsat/
http://www.mimas.ac.uk/maps/landsat/
http://www.stile.lut.ac.uk/~gydrw/STILE/t0050025.html

NOAA

http://www.websites.noaa.gov

Discussion questions

1   Show that the height of a geostationary satellite orbiting in the equatorial plane above the Earth's surface is approximately 36 000 km.
2   How can Doppler radar imaging be used in charting the course of moving rain clouds?
3   Outline a method by which the rate of tropical deforestation (e.g. in the Amazon) can be remotely sensed.

4   Compare and contrast the effectiveness of geosynchronous and Sun-synchronous remote-sensing satellites.

5   Meteosat is a satellite in a geostationary orbit above the equator. It has sensors which detect electromagnetic radiation in the following ranges:
400 nm–1100 nm
5.7 μm–7.1 μm
10.5 μm–12.5 μm
To which regions of the electromagnetic spectrum does each of these ranges belong?
   State and explain which of these sensors would be most suitable for:
(a) mapping cloud formations near the Earth,
(b) mapping the temperature variation of the Earth's surface.
How are satellites such as NOAA used to complement the information from Meteosat?
                    (London University: 'A' Level Physics: Earth and Atmosphere: June 1999)

Quantitative questions

1   The Army has asked the Air Force to place a spy satellite in a circular orbit round the Earth such that it passes over a particular country at intervals of exactly one hour. The required orbit is of radius 6.48×10^6 m, placing the satellite just above the Earth's atmosphere. The Air Force reply that it is impossible to establish the satellite in such an orbit. Why?
                    (Cambridge University: 'A' Level Physics: Further Physics: November 1997)

2   Geostationary satellites orbit at a height of 36000 km above the Earth's surface. The radius of the Earth is 6400 km. Calculate the period (T) of a polar orbiting satellite such as NOAA at a height of 800 km above the Earth's surface, given that:

    (period of orbit)^2 = constant $\times$ (radius of orbit)^3.

In what way do polar orbiting and geostationary satellites complement each other for weather forecasting purposes?
                    (London University: 'A' Level Physics: Earth and Atmosphere: Specimen paper)

3   What is the difference between an active and a passive satellite.
   LANDSAT, a polar orbiting satellite, scans a path of width 185km and completes one orbit in 99 minutes. Calculate the number of orbits made by LANDSAT in one day.
   The circumference of the Earth at the equator is about 40,000km. Show that while this satellite makes one orbit of the Earth, a point on the Earth's equator will travel approximately 2800km as the Earth spins on its axis.
   Explain with the aid of a diagram why it takes several days for LANDSAT to complete a scan of the Earth.
   One use of such a satellite is to monitor the vigour of crops. State how a satellite image can distinguish healthy crops from dying ones.
   Data from the satellite must be calibrated by comparing it with data collected on the ground. State two problems which occur when making this comparison.
                    (London University: 'A' Level Physics: Earth and Atmosphere: January 1997)

4   The sensors on board a geostationary satellite have a *low spatial resolution* but *good temporal resolution*. What is meant by the terms in italics?
   Explain why the spatial resolution of sensors on board geostationary satellites is low.
   One particular sensor is quoted as having a spatial resolution of 5km. Is this better or poorer than a resolution of 25km? Explain your answer.
   An Along Track Scanning Radiometer is able to measure sea surface temperatures to an accuracy of better than ±0.5K because its readings can be corrected for atmospheric absorption of infra-red radiation. Describe how this correction is achieved.
                    (London University: 'A' Level Physics: Earth and Atmosphere: January 1998)

5   Explain why it is necessary in acoustics to consider the separate velocities of the observer and of the source while only their relative velocity need to be considered when the source is emitting electromagnetic radiation. In what circumstances may the equations applicable to the Doppler effect in acoustics be applied to the effect with electromagnetic radiation?

A satellite, emitting a radio signal of frequency 6×10^7Hz, passes directly over an observer on the ground. When the satellite is first observed as it comes over the horizon the frequency is found to be 2000Hz above the value found when the satellite is overhead. The maximum rate of change of frequency is 200Hz/s. Calculate the height of the satellite assuming that this is small compared with the radius of the Earth. (Assume c is 3×10^8 m/s).

(London University: 'A' Level Physics: Special Paper: Summer 1972)

Chapter 7

The Sun and the atmosphere

7.1 Introduction

The Sun has a direct and important influence on many physical processes within the Earth's atmosphere. Sunlight provides the energy for photosynthesis within plants, which in turn creates the atmospheric oxygen required for us to breathe. The solar daily cycle regulates our lives, and the solar annual cycle determines the seasons and hence the agricultural cycle. Yet the role of the Sun in determining the state and behaviour of the Earth's climate is much greater than just a simple observation of its warmth would lead us to suppose. The Earth's atmosphere is literally solar-powered, the Sun being the primary cause of all the atmospheric processes, including the formation of clouds and the generation of both local and global wind patterns.

In this chapter, the mechanisms for the transfer of solar energy through the Earth's atmosphere on to the Earth's surface will be examined. In addition, the conversion of incident solar radiation into emitted terrestrial radiation from the Earth's surface will be described and the role of the 'greenhouse effect' in sustaining life on Earth will be discussed. The increasing influence of industrialization on the climate will be reviewed and the environmental problems of ozone depletion and global warming examined.

In describing these mechanisms, the structure and composition of the Earth's atmosphere will be explored, the role of ozone explained and the causes and effects of ozone depletion by industrial emissions discussed. The physical principles of pressure, the gas laws, escape velocity and black-body radiation will be invoked to explain the structure and retention of the Earth's atmosphere and radiation transport through it.

7.2 Solar energy

7.2.1 Solar output

The Sun, at the centre of the Solar System, is a typical star 1 392 000 km in diameter, with a mass roughly 1000 times that of the rest of the Solar System combined. Like most stars it is composed mainly of hydrogen (70%), with most of the remainder being helium. The Sun generates its heat and sustains itself from nuclear fusion reactions in the core (see Section 5.3.3) where the temperature reaches some 15 million °C, and in which an estimated 600 million tonnes of hydrogen are converted into helium every second. The Sun's visible surface, the *photosphere*, with a temperature of 5500°C is a seething cauldron of gas that sends off jets of hot gas into the surrounding *chromosphere* and where relatively dark patches, known as *sunspots*, are observed periodically.

The total solar output into space is $2.33 \times 10^{25}\,\text{kJ min}^{-1}$, but only a tiny fraction ($1/2\,000\,000\,000$), i.e. one two thousand millionth, of this is actually intercepted by the Earth since the energy received by any planet is inversely proportional to its distance from the Sun. The rate of flow of energy per unit area, the *irradiance* (I) is related to the *solar emittance* (E_s) by:

$$I = E_s (R_s/R)^2 \qquad (7.1)$$

where R is the distance from the centre of the Sun and R_s is the solar radius (Figure 7.1). Since the Earth orbits the Sun at a distance of about 200 solar radii the amount of energy falling on the top of the Earth's atmosphere is only about $(1/200)^2$ of the solar emittance. This is known as the *solar constant* (S), and is equal to $82.4\,\text{kJ m}^{-2}\text{min}^{-1}$ or $1353\,\text{Wm}^{-2}$ ($\pm 1.6\%$). The total power being received by the Earth from the Sun is then equal to this value multiplied by the cross-section of the Earth pointing towards the Sun, πR_E^2, where R_E is the radius of the Earth. If this power is averaged over the whole surface of the Earth, the power received per unit surface area is given by:

$$S\pi R_E^2/4\pi R_E^2 = 338\,\text{Wm}^{-2}$$

Obviously, the solar radiation is not received uniformly over the Earth. For instance, the equatorial regions receive more annual solar energy than the polar regions. It follows that there is a surplus energy for the lower latitudes and a deficit for the higher latitudes. As the tropics do not become progressively hotter or the higher latitudes colder, there must be a transport of energy from the lower to higher latitudes to compensate for the energy imbalance. This is important in connection with global atmospheric motions and will be discussed in Chapter 9.

The altitude of the Sun (i.e. the angle between the incident solar rays and a tangent to the Earth's surface at the point of observation) also affects the amount of solar radiation received at the Earth's surface. The greater the Sun's altitude, the greater the solar flux received at the Earth's surface. Thus, the solar flux at any site on the Earth's surface is dependent on the latitude of the site, the time of day, the season and the extent of cloud cover.

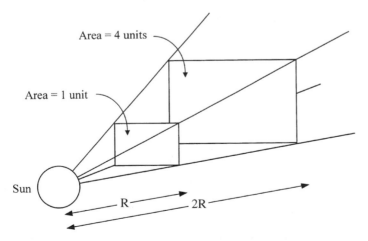

Figure 7.1 Solar irradiance follows an inverse square law.

7.2.2 Rhythm of the seasons

Owing to the *eccentricity* of the Earth's orbit around the Sun, the receipt of solar energy on a surface normal to the Sun is 7% more on 3 January at the *perihelion* (i.e. when the Earth is closest to the Sun) than on 4 July at the *aphelion* (when the Earth is furthest from the Sun) (Figure 7.2). This difference should produce an increase in the effective January Earth surface temperatures of some 4°C over those of July and should make Northern Hemisphere summers warmer than those in the Southern Hemisphere and the Southern Hemisphere winters warmer than those in the north. In practice, once again, atmospheric circulation patterns mask this effect and in fact the actual seasonal contrast between the hemispheres is reversed. Hence, the seasons cannot be ascribed to differences in the Earth's distance from the Sun but are due to the Earth's axis not being at right angles to its orbit but at 23.5° from the perpendicular to a line joining the centres of the Earth and the Sun. This causes first one hemisphere and then the other to point a little towards the Sun, such that in summer the Sun is high in the sky and in the winter it is low.

This makes the proportion of daylight a greater fraction of the 24 hours in the summer, reaching an extreme in the polar summers where the 'lands of the midnight sun' have sunlight for nearly the whole 24 hours (although in compensation in winter it is nearly always night). Also, in summer the Sun's rays are more vertical at the local noon concentrating the heating effect rather than spreading its effect over a slanting path across the Earth's surface. In winter the opposite happens, the Sun rises low above the horizon and its heating effect is dissipated over a slanting path.

Although in the Northern Hemisphere the Sun rises highest in the sky on 22 June, the longest day, few would regard June as high summer. The peaks of summer heat and the troughs of winter cold therefore lag behind the apparent position of the Sun as viewed from Earth. This is due to the finite time it takes for the Earth's surface to respond to the change in the amount of heat arriving from the Sun. Before June much of the solar radiation is used to warm the Northern Hemisphere (the temperate and Arctic zones) after the previous winter. Thus, only as the days get shorter and the Sun rises lower in the sky

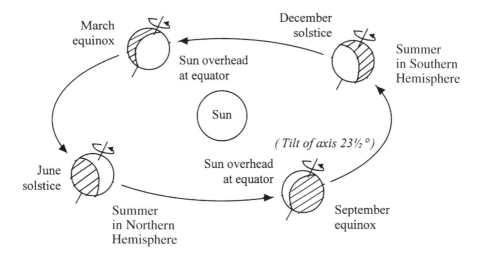

Figure 7.2 The rhythm of the seasons.

throughout July and August is 'high summer' reached. Similarly, through autumn and winter it takes a finite time for the Earth's surface to lose the excess heat stored in the hemisphere and thus the shortest day, 22 December, is seldom the coldest experienced during the winter.

Local climate patterns also greatly influence the seasonal temperature. Thus, the presence of the Gulf Stream (see Chapter 9) ensures that Britain has warmer winters than Canada and Russia, despite parts of all three countries lying at the same latitude and thus receiving the same solar radiation flux.

7.2.3 Solar cycles and climate change

Some scientists have sought to explain variations in the Earth's climate in terms of changes of the solar output. Such suggestions must be regarded as somewhat speculative since no measurements of solar output exist prior to 1978 when satellites outside the Earth's atmosphere could measure solar irradiation free of atmospheric effects. One theory put forward to explain changes in the Earth's climate was that the climate responded to changes in the solar flux during periods of greater and lesser sunspot activity. Sunspot activity has been recorded for many centuries and show a quasi 11-year cycle with longer-term fluctuations superimposed. The colder years between 1650 and 1700 AD, commonly known as the 'little ice age' (when, for example, the Thames froze every winter) seems to have corresponded to a period when there were fewer sunspots, seemingly providing some support for such a theory. However, satellite measurements during the 1980s showed that while the solar output changes during the course of a solar cycle, being reduced when sunspot numbers reached a minimum, the solar constant was only reduced by $1.5\,Wm^{-2}$ or $<0.1\%$, which would change the mean global temperature by only about $0.06°C$.

Case study 7.1 The Milankovitch hypothesis

Many theories have tried to explain the causes of climate change. These include:

- variations in solar energy
- changes in the Earth–Sun geometry
- atmospheric changes due to aerosol particulates resulting from volcanic eruptions
- the mobility of the tectonic plates and
- anthropogenic behaviours.

Here we will consider the role of changes in the solar irradiance.

In 1864 the Scotsman James Croll discovered that changes in the Earth–Sun geometry, caused by three cyclic perturbations, could be responsible for variations in the Earth's orbit. As a result, the Earth would receive different amounts of solar radiation during these cycles. However, it was not until the 1920s that the Yugoslav geophysicist Milutin Milankovitch (1879–1958), at the University of Belgrade, developed a mathematical basis for these ideas. It has therefore become known as the *Croll–Milankovitch hypothesis* and provides a basis for explaining climate change over long time periods and, in particular, provided an explanation for the cycle of ice ages.

The Croll–Milankovitch hypothesis may be summarized as follows. The gravitational attraction between the Earth, the Sun and the planets, especially the Moon, generate three long term cyclical perturbations. When these three cycles are synchronised marked changes in global climate can result.

The first perturbation arises from changes in the distance between the Sun and the Earth and hence a change in the solar flux (irradiance) reaching the Earth's surface. Kepler's First Law (Chapter 6) shows that the Earth revolves around the Sun in an elliptical orbit. One of the consequences of this is that the distance between the two bodies varies and, as a result, the Earth is nearer to and further from the Sun at different times of the year. The Earth is closest to the Sun at the beginning of January, when the solar irradiance reaching the Earth is about 7% greater than when the Earth is furthest away at the beginning of July (Figure 7.2). However, over thousands of years it appears that the shape of the Earth's orbit about the Sun changes from an elliptical one to one which is slightly more circular. This implies that the eccentricity of this orbit varies periodically and it takes about 96 000 years to go through one cycle.

The second cycle concerns changes in the angle of tilt (or *obliquity*) of the Earth's axis. This angle varies between 21.8° and 24.4°, and is at present 23.5°. The period of this cycle is about 40 000 years (Figure 7.3a).

The third cycle arises from the precession of the equinoxes, i.e. the manner in which the Earth 'wobbles' as it spins on its axis. In this cycle there is another

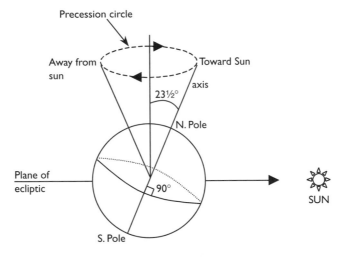

(a)

Figure 7.3. The Milankovich cycles and Earth climate: (a) Orbital parameters of the Earth's tilt and precession. (b) Variation of the orbital parameters in the Milankovitch hypothesis for a period of 300 000 years. (c) Glacial and interglacial oscillations over the past 600 000 years. 0 represents the present time and the positive figures the next 100 000 years.

variation in the distance between the Earth and the Sun once again changing the solar irradiance at the Earth's surface. At the present time the nearest distance (the perihelion) coincides with January, and the furthest distance (the aphelion) in June/July. The period of oscillation of these distances is about 21 000 years (Figure 7.3a).

When these three cycles coincide (i.e. are in phase), Figure 7.3b, it has been found that correlations exist between glacial (interglacial) periods and low (high)

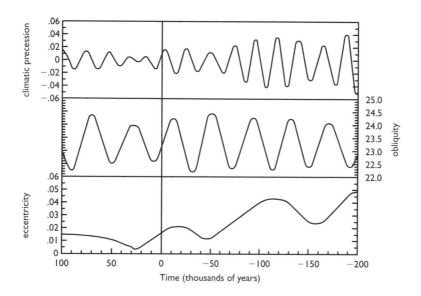

(b)

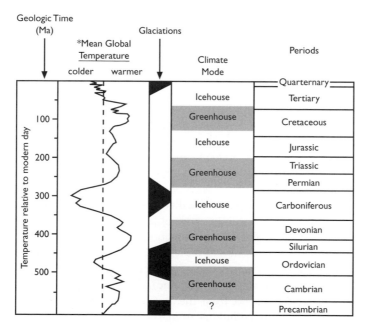

(c)

levels of incoming solar radiation at specific latitudes (Figure 7.3c). Thus the Croll–Milankovitch hypothesis may explain paleoclimatology.

Whether changes in solar irradiance play any significant role in current climatic patterns is less certain. The role of sunspots and solar flares in influencing the earth's weather remains a controversial topic since the change in solar irradiance is small ($\approx$1%). However, some scientists contest that changes in solar flux may lead to changes in the rate of cloud growth at high altitudes which may affect global albedo levels, but to date such a hypothesis remains largely untested.

Milankovitch hypothesis

Over longer time scales the solar output may itself vary (see Case study 7.1). The Earth's orbit although nearly circular, is actually an ellipse (see Section 6.3). The ellipticity of the Earth's orbit itself changes on a 100000-year cycle during which the solar input to the atmosphere may change by 30% of the current global average. In addition, the tilt of the Earth's axis relative to the Sun changes over a 40000-year cycle, varying between 21.6 and 24.4°. Currently it is 23.5°. Such variations, are named after the Yugoslav mathematician Milutin Milankovitch, who examined the evidence for correlation between the ice ages and the ellipticity of the Earth's orbit and the change in the Earth's tilt (Figure 7.3). Of the variance in the climate record of global ice coverage, 60% might be correlated with changes in the solar flux but the actual changes in surface temperature were larger than the changing solar flux would suggest. Other processes must be involved that can enhance the global cooling or warming, one of which may be changes in the atmospheric composition.

7.3　Structure and composition of the Earth's atmosphere

7.3.1　Structure of the atmosphere

The Earth's atmosphere is a gaseous envelope, retained by gravity, surrounding the planet. Most dense at the Earth's surface, 90% of the mass is contained in the first 20 km and 99.9% of the mass within the first 50 km. The atmosphere becomes thinner with increasing height until at some 1000 km it merges indistinguishably with interstellar space. The Earth's atmosphere is, therefore, only a thin shell around the Earth.

The Earth's atmosphere can be divided conveniently into layers characterized by their temperature (Figure 7.4). Each layer is called a *sphere* and the boundary between layers is called a *pause*.

The lowest layer of the atmosphere is called the *troposphere*. Extending some 10 km above the Earth's surface and containing 80% of the atmosphere's mass, it is a turbulent layer in which the weather is generated. Throughout this layer the temperature generally decreases with altitude at a mean rate of 6.5°C km^{-1}, up to a minimum of between −50 and −55°C at the *tropopause*. The tropopause is a temperature *inversion level* (i.e. where a layer of relatively warm air lies above a colder one); the inversion thus acts as a 'lid' limiting both convection and transport from the troposphere into the higher layers of the Earth's atmosphere. Industrial pollutants from the Earth's surface are, therefore, only slowly diffused into the upper atmosphere from the troposphere.

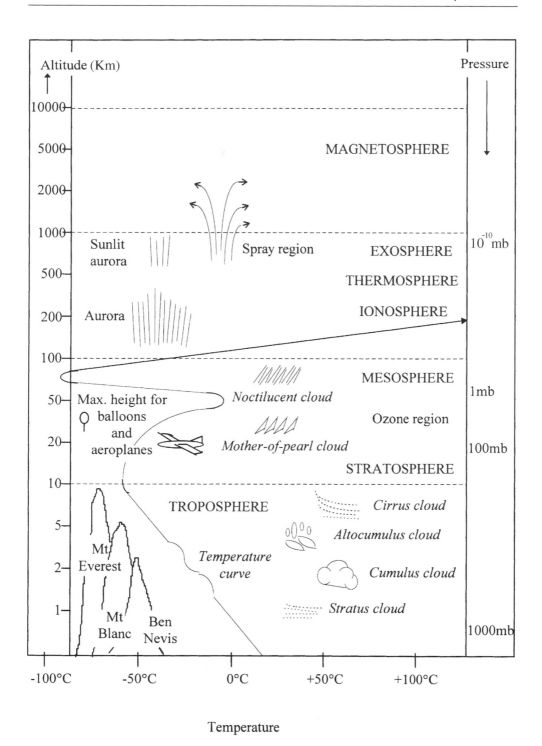

Figure 7.4 The structure of the Earth's atmosphere.

Above the tropopause lies the *stratosphere*, which extends to about 50 km. Throughout the stratosphere the temperature gradually increases with altitude up to a maximum of 0°C at 50 km, at the *stratopause*. This maximum is the result of the absorption of the Sun's ultraviolet rays by ozone. Although only small amounts of ozone are in this region, its presence is essential for the survival of life on Earth, since it filters out much of the biologically harmful solar ultraviolet radiation (see Section 7.6).

Above the stratopause the temperature falls again throughout the region known as the *mesosphere* (or middle atmosphere). This is the coldest region of the atmosphere with a minimum of −100°C (=173 K) at altitudes between 80 and 85 km. Above 80 km absorption by molecular oxygen and ozone causes temperatures to rise again with altitude. This inversion is known as the *mesopause*.

Above the mesosphere the temperature increases very rapidly with altitude and can range from as low as 200°C to >2000°C depending on the time of day, the latitude and changes in the energy emitted by the Sun. Daily variations of 500–800°C may occur with a minimum near sunrise and a maximum around 14:00 hours. This region is known as the *thermosphere*. However, at such altitudes atmospheric densities are extremely low and these temperatures are theoretical, being more a measure of the velocity of the atoms and molecules than the temperature acquired by any satellite passing through such a region since the low pressures do not permit any appreciable heat transfer; the object is actually in a high vacuum.

Above 100 km the shorter wavelengths of the incident solar radiation (cosmic rays, X-rays, high-energy UV radiation) ionize atoms and molecules producing a region of positive ions and free electrons called the *ionosphere* (see Section 7.5.2). Such ionized layers reflect radio signals and are thus of great importance in communications, although higher frequency VHF and UHF signals may pass through without reflection and thus can be received by orbiting communications satellites. The ionosphere is also the region of the *aurorae*: the Aurora Borealis in the Northern Hemisphere and the Aurora Australis in the

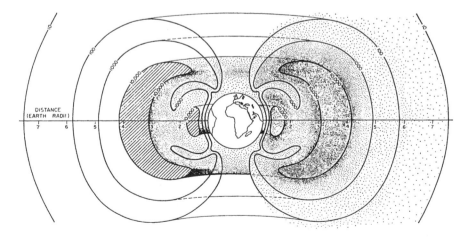

Figure 7.5 Earth's magnetic field and the Van Allen belts. Structure of the radiation belts shown by contours of radiation intensity (black lines) with shading (left). The dots (right) suggest a distribution of particles in the two belts. Contour numbers reveal the counts per second.

Southern. The aurorae are produced by the excitation of atoms and molecules of oxygen and nitrogen excited in collisions with electrons liberated in photo-ionization processes with high-energy solar radiation.

The region above 500 km marks the gradual transition from the terrestrial atmosphere to the interplanetary gas and is known as the *exosphere*. The pressures are now so low that collisions between molecules virtually cease and the molecules perform parabolic (ballistic) trajectories in the gravitational field such that neutral helium and hydrogen atoms, with low atomic weights, can escape into space. This is why there are low concentrations of atomic hydrogen and helium in the Earth's atmosphere.

The density of ionized particles increases through the exosphere and beyond to about 200 km in the *magnetosphere* where there are only electrons and protons derived from the solar wind. These charged particles are trapped by the Earth's magnetic field in two bands, the *Van Allen belts*, centred at about 3000 and 16 000 km (Figure 7.5).

Above about 8000 km (the heliosphere) the Earth's atmosphere merges imperceptibly with that of the Sun's.

7.3.2 Composition of the atmosphere

Table 7.1 lists the major chemical components of the terrestrial atmosphere. Nitrogen, oxygen, argon and carbon dioxide together make up more than 99.9997% of the total abundance of dry air.

In the troposphere, stratosphere and the mesosphere the major constituents, molecular nitrogen and molecular oxygen, comprise about 80 and 20% respectively of the atmospheric mass, such that the mean molecular mass of the air is essentially constant with altitude. This region of uniform chemical composition is collectively known as the *homosphere*. Within the thermosphere gas concentrations are extremely low and at altitudes above 500 km some of the gas molecules can overcome the gravitational pull of the Earth and escape into space. Thus, the atmosphere's chemical composition changes with altitude and for this reason is known as the *heterosphere*.

The concentration of carbon dioxide is variable within the lower troposphere, being affected by the type of local combustion, photosynthesis and exchange with the oceans (in which large amounts of global carbon dioxide emissions are stored), but at higher altitudes it is constant at around 360 ppm (parts per million by volume). It is common to describe the concentrations of trace compounds by ppm (or ppb, parts per billion) where parts means parts 'by volume' and this may have to be converted when considering concentrations by mass.

There is, however, one important gas that is variable in the lower layers of the atmosphere, and which has not yet been mentioned – water vapour. The condensation of water vapour in the troposphere forms the clouds, but occasionally clouds can be observed at

Table 7.1 Chemical composition of the terrestrial atmosphere.

Compound	Molar fraction
Nitrogen	0.7809
Oxygen	0.2095
Argon	0.0093
Carbon dioxide	0.00033

two higher levels. 'Mother of pearl' clouds, so-called because of their iridescent aspect, appear in the lower stratosphere at altitudes above 27 km and consist of frozen water droplets. At still higher altitudes 'noctilucent' clouds can be seen between 80 and 100 km. Visible after sunset, while still illuminated by the setting Sun, they consist of ice crystals, probably formed around the dust remnants of meteorites burnt up as they enter the Earth's atmosphere. Water vapour and its circulation through the lower regions of the atmosphere is an essential part of the Earth's weather system and discussion of this 'hydrological cycle' is made in Section 8.5.2.

Minor constituents of the terrestrial atmosphere are present to <0.003% or 30 ppm. Although these constituents might be described as 'trace compounds' they may dominate the local chemistry; for example, one such compound is ozone. Ozone is formed in a restricted layer of the Earth's atmosphere, the stratosphere, and plays a crucial role in main-taining life on Earth. Recently, global industrial pollution has led to drastic reductions in ozone concentrations over Antarctica and the Arctic. The environmental problem of ozone depletion and the development of the 'ozone hole' is one of the major environmental prob-lems facing the Earth today and will be discussed in detail later in this chapter.

Trace compounds are usually described by their abundance, variability, residence time and/or their origin. The four main chemical constituents of the terrestrial atmosphere account for 99.997% of all dry air and have concentrations >300 ppm. The second group in Table 7.2 are present in the dry air with concentrations from 1 ppb to 20 ppm. Variabil-ity is an important concept since it describes the behaviour of the compound and, in particular, its chemical reactivity. Thus, CO_2 is non-variable despite its localized source of production and diffuses throughout the atmosphere to produce at higher altitudes a glob-ally constant value. The more reactive SO_2, NO_2 and NO are however variable since they react quickly and are often produced at very specific sites. Hence, they can never diffuse throughout the atmosphere.

A parameter can therefore be introduced for each compound in the atmosphere which describes their mean lifetime or average *residence time* (τ) in the atmosphere:

$$\tau = M/F \tag{7.2}$$

Table 7.2 Trace compounds in the Earth's atmosphere.

Compound	Concentration (by volume)	Residence time
Neon	18 ppm	3×10^6 years
Helium	5 ppm	
Krypton	1 ppm	
Xenon	0.09 ppm	
Methane (CH_4)	1.5 ppm	3 years
Carbon monoxide (CO)	0.1 ppm	0.35 years
Nitrous oxide (N_2O)	0.25 ppm	<200 years
Ozone (O_3)	10 ppm in stratosphere 5–500 ppb in troposphere	
Hydrogen sulphide (H_2S)	0.2 ppb	10 days
Sulphur dioxide (SO_2)	0.2 ppb	5 days
Ammonia (NH_3)	6 ppb	1–4 days
Nitrogen dioxide (NO_2)	1–100 ppb	2–8 days

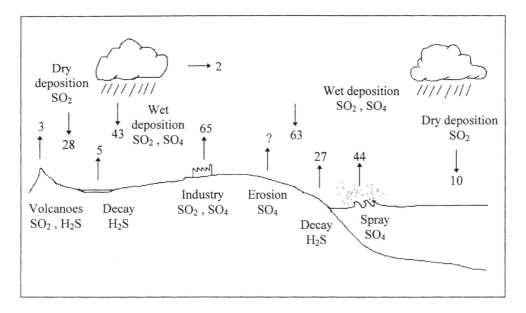

Figure 7.6 The sulphur cycle, numbers are in Tg (10^{12}g per year).

where M is the total average mass of the compound in the atmosphere and F the total average influx or outflux. $1/\tau$ is then defined as the *rate of turnover*. Hence, a very reactive gas will have a high rate of turnover while permanent gases (like nitrogen) will have a very slow rate of turnover.

Compounds can also be considered as part of elemental cycles, e.g. the carbon cycle (see Chapter 11), the sulphur cycle, the ammonia cycle, the NO_x cycle and the water (hydrological) cycle. Such cycles provide another convenient grouping for chemical substances and the role they play in the atmosphere. A complete understanding of the cycle requires a knowledge of the sources and sinks of the different compounds and the chemical transformations undergone within the cycle. Figure 7.6 illustrates the sulphur cycle and shows both the sources and sinks of sulphur compounds in the terrestrial atmosphere:

7.4   Atmospheric pressure

7.4.1   *Pressure and temperature as functions of altitude*

In Chapter 2 it was noted that in climbing mountains both pressure and temperature decreased with increasing height. If we journey further into the troposphere, it can be shown that the pressure decrease can be described by the following expression:

$$P = P_o.e^{-gh/RT}$$

or in the form

$$P = P_o \exp(-h/H) \tag{7.3}$$

where $H = k_B T/mg$ is known as the *scale height*, P_0 is the atmospheric pressure at the surface ($h = 0$), and P the pressure at a height h above the surface. The scale height is then the altitude above the Earth's surface at which the pressure is reduced by a factor of e^{-1} ($= 0.37$) or the height within which about two-thirds of the atmospheric mass would be contained.

The derivation of the pressure formula is given in Appendix 5.

Since the pressure falls exponentially with height, 90% of the mass of the atmosphere is contained within the first 21 km and 99.9% in the first 50 km; the pressure therefore drops from 10^5 Pa at the Earth's surface to 10^4 at 20 km to 10^2 at 50 km. At 100 km the pressure is only 0.1 Pa, so only one-millionth of the atmospheric mass will be above that level and 10^{-13} above 100 km. If these distances are compared with the radius of the Earth (about 6370 km) it can be seen just how thin the atmosphere is around the Earth.

The temperature also falls with altitude and it is called the *lapse rate* (dT/dz). Its derivation is shown in Appendix 6.

7.4.2 Escape velocity

Not all the Earth's atmosphere is retained by the Earth's gravitational field; some of it may 'leak' into space. Thus, gradually the composition of a planet's atmosphere may change or there may be a net loss of atmosphere. What limits such loss and why does the Earth retain its atmosphere?

Imagine a rocket of mass (m_R) is launched from the Earth's surface (S) so that it just escapes from the gravitational influence of the Earth.

$$\text{The work done} = m_R \times \text{potential difference between S and infinity}$$
$$= m_R(Gm_E/R_E) \tag{7.4}$$

where m_E and R_E are the mass and radius of the Earth respectively and G is the *gravitational constant*.

To escape, the kinetic energy of the rocket must be balanced by the work done in liberating the rocket from the Earth's gravitational potential such that

$$\frac{1}{2} m_R v^2 = m_R Gm_E/R_E$$

Thus,

$$v = (2Gm_E/R_E)^{1/2}$$

where v is the velocity of the rocket. However from Newton's Law of gravitation, since $m_R g = Gm_R m_E/R_E^2$, the velocity of the rocket is

$$v = (2gR_E)^{1/2} \tag{7.5}$$

As $g = 9.8$ m s^{-2} and $R_E = 6370$ km the escape velocity from the Earth is 1.1×10^4 m s^{-1} or 11 km s^{-1}. If a rocket is launched with a vertical velocity of 11 km s^{-1} it will completely escape the gravitational attraction of the Earth. The escape velocity is completely independent of the mass of the body. Hence, the escape velocity of a rocket is the same as

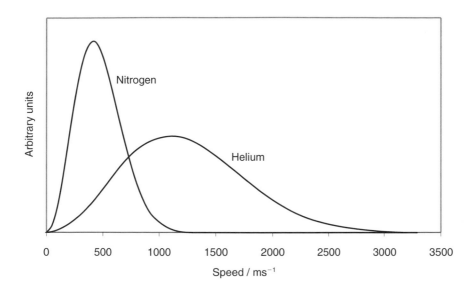

Figure 7.7 Maxwellian speed distribution for helium and nitrogen at standard temperature and pressure (STP).

for a molecule! If molecules reach a velocity of $11\,km\,s^{-1}$ they too can escape into space and leave the Earth's atmosphere.

Consider the molecules in the air at normal temperatures and pressures. Their speed is given by the *Maxwellian speed distribution* (Figure 7.7) and is dependent on their molecular mass (M) and temperature (T). The most probable speed, v_p, is given by

$$v_p = (2k_BT/M)^{1/2} \qquad (7.6)$$

For atmospheric gases, molecular oxygen and nitrogen, $v_p = 395$ and $420\,m\,s^{-1}$ respectively. This is much less than the Earth's escape velocity. Gravitational attraction, therefore, keeps the bulk of the Earth's atmosphere contained around the Earth.

However, there is a finite probability that atoms or molecules will have speeds $>11\,km\,s^{-1}$; the lighter the gas the larger the proportion of atoms or molecules having such speeds. Hence, since the creation of the Earth (over 10^9 years ago) much of the molecular hydrogen and helium formed at the creation has escaped into space such that the Earth's atmosphere is lacking in these molecules.

7.5 Solar radiation

7.5.1 *Solar spectrum*

The Sun emits electromagnetic radiation over a wide spectral region from the ultraviolet to the infrared (Figure 7.8). Almost all the solar radiation reaching the Earth arises from the solar *photosphere*, a relatively thin region in the outermost layers of the Sun and comprised mainly of hydrogen and helium gas at temperatures greater than 5500 K.

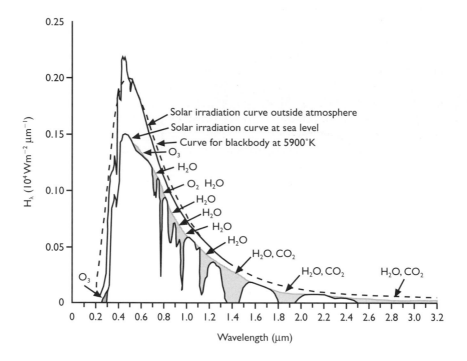

Figure 7.8 Solar flux [H_λ] at the top and bottom of the atmosphere.

Normal laboratory spectra arising from excited hydrogen and helium gas would show a distinct emission line spectrum but at the elevated temperatures and in the high pressure conditions existing in the photosphere the individual emission lines are broadened out to produce an almost continuous spectrum. Such a spectrum is indeed observed at the top of the Earth's atmosphere, but characteristic *Fraunhofer lines* can be seen if the Sun is viewed through a high-resolution spectrometer (Figure 7.9), a result that provided the first clear evidence that stars were in fact made up of burning hydrogen and helium.

Human vision has evolved such that our 'visible' spectrum is centred on wavelengths of about 500 nm, close to the wavelength of maximum solar emission. Viewed on their own, these wavelengths would appear green to the human eye, but the substantial range of wavelengths in visible sunlight coupled with the variation in sensitivity across the human eye produce the familiar golden appearance of the Sun.

Much of the radiant energy of the Sun is absorbed by the molecules in the atmosphere. As a result, the solar spectrum observed at the Earth's surface is therefore considerably

Figure 7.9 Fraunhofer lines seen through a solar spectroscope. Wavelengths presented in Ångstroms (Å) [$1\text{Å} = 10^{-10}\text{m}$].

different from that direct from the Sun. Figure 7.8 shows the solar spectrum at the top of the Earth's atmosphere and at the ground. At the ground the solar spectrum has been greatly altered. In contrast to the smooth continuous solar spectrum observed at the top of the Earth's atmosphere several maxima and minima can be observed. The minima are due to absorption bands in the atmosphere, while the maxima are due to transmission bands. The latter are due to there being no molecules able to absorb the solar radiation at these wavelengths such that the solar radiation can then reach the Earth's surface.

To understand how such radiation is absorbed remember how radiation may interact with atoms and molecules (see Case study 2.1). Solar radiation may excite, ionize or fragment atoms and molecules by the absorption of incident light. According to Planck's law, the energy of a photon can be written as $E = h\upsilon$, where h is Planck's constant. Thus, absorption of short-wavelength radiation transfers more energy to the atomic/molecular target than long-wavelength radiation. While infrared radiation ($\lambda < 800\,\text{nm}$) may only make the molecule rotate or vibrate, visible and ultraviolet radiation may excite atoms and molecules electronically. However, the molecule can only absorb the incident radiation if the photon energy is *resonant* with an excited state in the target atom or molecule. If no such state exists then light of that wavelength cannot be absorbed by that atom/molecule and the target is said to be *transparent* to that radiation.

Thus, if a wide wavelength range of light is shone on a gas of atoms/molecules only discrete wavelengths will be absorbed by the atoms/molecules corresponding to their characteristic excited states. Consequently, the parts of the spectrum corresponding to those wavelengths will appear dark after the light has passed through the gas. An *absorption spectrum* is then observed. Measurement of the wavelengths of these 'dark bands' will then indicate which atoms/molecules are present in the gas sample. This principle may be used not only to determine the concentrations of molecular species in our own atmosphere but has been used to determine the atmospheres of the other planets in the Solar System.

Thus, as the solar radiation passes through the Earth's atmosphere many solar wavelengths are absorbed by the constituent atoms and molecules producing the characteristic solar spectrum observed at the Earth's surface (Figure 7.8). Biologically harmful solar ultraviolet radiation ($\lambda < 300\,\text{nm}$) is filtered from the incident radiation by the presence of ozone in the Earth's stratosphere (see Section 7.6) while *greenhouse gases* (see Section 7.8) trap the re-emitted terrestrial infrared radiation warming our planet's surface and allowing life to evolve.

The Beer–Lambert law

The light flux intensity after passing through an absorbing gas may be determined from the Beer–Lambert law:

$$I_t = I_o \exp\left(-\sigma_{pa} N x\right) \tag{7.7}$$

where I_t is the transmitted light flux at a set wavelength, I_o the incident light flux, N the number density of the target gas and x the path length of the radiation through the gas. σ_{Pa} is known as the *photo-absorption cross-section* and is a measure of the efficiency with which any molecule absorbs light; the higher the cross-section the greater the absorption.

7.5.2 Earth's ionosphere

If the solar photon energy is greater than the ionization energy of the molecule then the molecule may be *photo-ionized* by the incident solar photons (energy $h\nu$) producing a positive ion and a free electron. This is the process by which the Earth's ionosphere is formed and absorbs much of the Sun's far ultraviolet radiation below 150 nm. For example:

$$O_2 + h\nu \ (\lambda < 102.6\,\text{nm}) \ \rightarrow O_2^+ + e$$
$$N_2 + h\nu \ (\lambda < 79.6\,\text{nm}) \ \rightarrow N_2^+ + e$$
$$O + h\nu \ (\lambda < 91.0\,\text{nm}) \ \rightarrow O^+ + e$$
$$N + h\nu \ (\lambda < 85.2\,\text{nm}) \ \rightarrow N^+ + e$$
$$NO + h\nu \ (\lambda < 134.1\,\text{nm}) \rightarrow NO^+ + e$$

The ionosphere can be studied from the ground using radiowaves. Indeed, it is the presence of the ionosphere that allows radio transmissions to be passed around the Earth. Using an instrument called an *ionosonde* pulses of radiowaves of between 1 and 20 MHz (300–15 m wavelength) are sent vertically upwards into the Earth's atmosphere. When the radiowaves reach the ionosphere the free electrons oscillate in response to the electromagnetic fields of the incident radiowaves and, if there is a high concentration of electrons, the radiowaves will be reflected backwards and might be detected by the ionosonde. The time interval between the transmission of the outward radiowave and detection of the reflected wave is typically 1 ms. The exact altitude at which the radiowaves are reflected is dependent on both the frequency and the free electron concentration, N, such that:

$$N = 4\pi^2 \epsilon_0 m \nu^2 / e^2 \tag{7.8}$$

where ϵ_0 is the permittivity of free space ($8.854 \times 10^{-12}\,\text{F m}^{-1}$), and m the mass of an electron ($9.1 \times 10^{-31}\,\text{kg}$) and e the electronic charge ($1.6 \times 10^{-19}\,\text{C}$). Hence, the concentration of electrons needed to reflect a specific wavelength can be simply stated as:

$$N = 1.24 \times 10^{-3} \nu^2\,\text{m}^{-3}$$

Ions will also reflect radiowaves but only at much higher concentrations due to their larger mass.

Using this technique it has been possible to probe the structure of the Earth's ionosphere and determine that it is in fact divided into four main regions defined as D, E and F1 and F2 (Table 7.3).

D is only observed during the daytime when the solar flux is highest and penetration deepest into the Earth's atmosphere; and F1 and F2 become a single region at night. This

Table 7.3 Classification of the regions of the Earth's ionosphere.

Region	Altitude (km)	Electron density N (m^{-3})
D	<90	10^9–10^{10}
E	90–140	10^{11}
F1, F2	>140	maximum of 10^{12} at 250–500 km

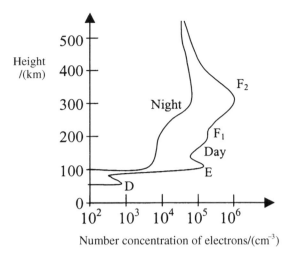

Figure 7.10 Electron density in the Earth's ionosphere.

is shown in Figure 7.10, where the electron densities are plotted (on a logarithmic scale) as a function of height. The D-layer has the highest molecular concentration and therefore electron collisions with constituent air molecules can occur leading to the formation not only of more positive ions, but also of negative ions by the process of electron attachment to a molecule AB: $e + AB \rightarrow AB^-$.

7.5.3 The aurorae

Observation of the Earth's aurorae is one of the most dramatic natural events anyone may experience (see Plate 5). Swirling curtains of red, green and purple light stretching across the night sky, they are usually observed in high latitudes; the Aurora Borealis (Northern Lights) in the Northern Hemisphere and Aurora Australis, in the Southern Hemisphere. A typical display may last half an hour with maximum intensity lasting only a few minutes.

The aurorae arise from the excitation of atoms and molecules in the Earth's upper atmosphere (at between 80 and 150 km altitude) by charged particle bombardment. This involves the recombination of free electrons with positive ions forming one excitation process while direct collisional electron impact provides another. The main auroral wavelengths arise from 557.7 nm green line emission from atomic oxygen, the 630.0 and 636.3 nm red emissions from atomic oxygen and strong ultraviolet emissions from molecular nitrogen.

Auroral displays are seen with greatest intensity and at lower (more inhabited) latitudes during the occurrence of a high solar activity leading to a solar storm. These conditions occur when there is a *solar flare* (Figure 5.4). Solar flares are extremely hot masses of gas ejected from the solar chromosphere in a parabolic path. Solar flares send intense pulses of ultraviolet radiation, electrons and protons across the Solar System some of which then impact with the Earth's atmosphere leading to increased ionization within the

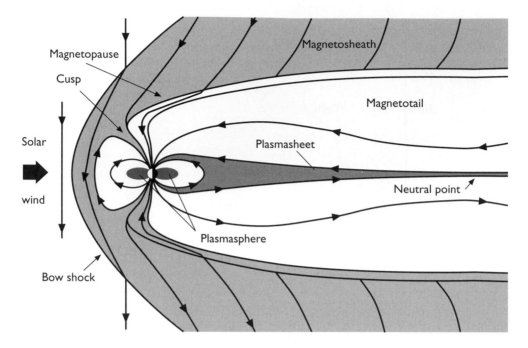

Figure 7.11 Impact of the solar wind on the Earth's upper atmosphere.

ionosphere, which in turn leads to greater molecular excitation and hence stronger auroral displays. The phenomena are concentrated at the geographic poles since the solar electrons and protons are funnelled into the Earth's atmosphere by the terrestrial magnetic field (Figure 7.11). Such solar (or magnetic storms) can also cause great disruption to telecommunication and radio signals and may even damage satellites. In 1989 a particular severe solar storm led to the loss of electricity across much of Canada for several days, when the electromagnetic fields induced by solar storm disturbances in the Earth's ionosphere led to the overload of the national grid. The cost of ensuring that such a catastrophic loss of power would not be repeated is estimated to have been over 500 million Canadian dollars.

7.5.4 *Solar photo-induced chemistry*

The solar photon energy may also be sufficient to break the chemical bond(s) dissociating the molecule into atomic/molecular fragments. For example, molecular oxygen can be dissociated into atomic oxygen:

$$O_2 + h\nu \rightarrow O + O$$

When dissociation occurs the products can carry away energy released by the rupture of the chemical bond. This energy is usually in the form of kinetic energy but one or other of the fragments may also be in an excited or even ionized state. Such dissociation processes require photons with a minimum (threshold) energy, but photons of *any* energy above the threshold may be absorbed and dissociate the molecule; the excess photon energy being transferred into internal or kinetic energy of the product fragments. Thus, the absorption spectrum for such a process shows continuous absorption above the threshold.

The formation of high kinetic energy or excited ('hot') atomic/molecular fragments has profound consequences on the local chemistry since the fragments are then 'reactive'. Such photon dissociation processes are the key to understanding much of the chemistry of the Earth's stratosphere and global ozone production and destruction mechanisms.

7.6 Ozone

7.6.1 The Earth's ultraviolet filter

Ozone is only a minor constituent of the Earth's atmosphere forming 0.2% of the terrestrial atmospheric mass, such that if all the atmospheric ozone was collected at the Earth's surface it would form a ring only 3 mm thick around the Earth. However, ozone is essential to the sustaining of both plant and mammalian life. It is the presence of ozone in the atmosphere that shields the Earth's surface from harmful solar UV radiation through its ability to absorb all solar radiation with wavelengths <295 nm. Figure 7.12 shows the photo-absorption 'cross-section' of ozone: the greater the cross-section the greater the efficiency of the molecules to absorb radiation. Uniquely among these molecules in the Earth's atmosphere ozone has a strong 'absorption band' (the so-called 'Hartley Band') between 210 and 300 nm. Hence, ozone filters out the Sun's ultraviolet radiation below 300 nm preventing these wavelengths from reaching the Earth's surface.

Biological molecules have evolved under conditions of such filtering and thus while they do not absorb visible light they can safely absorb ultraviolet light since such radiation

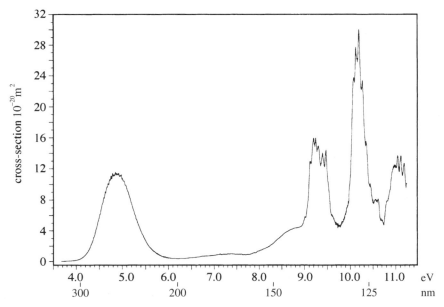

Figure 7.12 The photo-absorption spectrum of ozone, displayed as cross-section vs photon energy (ev) and wavelength to (nm).

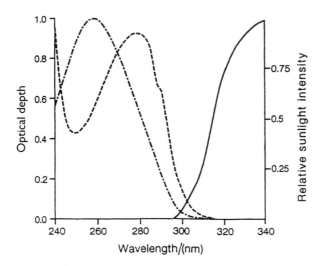

Figure 7.13 The absorption spectra of the biomolecules DNA [–·–·–] and α-crystallin [– – –] compared to solar flux at the Earth's surface [——]

does not reach the Earth's surface. Figure 7.13 shows the solar emission spectrum reaching the Earth's surface for wavelengths below 340 nm together with the absorption spectra of two important biomolecules, DNA, the carrier of the genetic code, and α-crystallin, the major protein of the mammalian eye lens. The absorption of light by both these biological molecules is essentially zero in the region $320 < \lambda < 400$ nm, the near-UV or UV-A region, but it is intense in the region $200 < \lambda < 290$ nm, the far UV or UV-C region. The presence of ozone in the terrestrial atmosphere ensures that the absorption spectrum of these biological molecules only overlaps the solar spectrum at the Earth's surface in the wavelength region $290 < \lambda < 320$ nm, the mid-UV or UV-B region.

DNA and other biomolecules are therefore extremely sensitive to changes in the radiant solar flux, an increase of which may therefore lead to severe damage of the genetic material (mutagenesis). It is in the mid-UV wavelength range that the reduction in the ozone shield will have the greatest influence on biological systems. In particular, it is now well-established that DNA is especially prone to UV damage, one consequence of which is the occurrence of erythema (sunburn) in human skin. Figure 7.14 shows the action spectrum (the damage caused by a unit of irradiation of a certain wavelength) for the production of erythema. The probability of recurring erythema increases by five orders of magnitude between 350 and 280 nm, precisely that region where ozone is absorbing solar radiation. Hence, even small changes in ozone density, and as a result changes in the solar UV flux reaching the Earth's surface, may lead to a dramatic increase in the number of cases of erythema. Similarly the mutagenesis of α-crystallin may lead to cataract formation in the human eye and, in the most severe cases, result in blindness. While humans may adapt to these changes (e.g. by wearing sun glasses and sun cream) other mammals are equally vulnerable and have less obvious methods of prevention. Nor are plants immune to the effects of changing solar UV levels. It is these potentially disastrous consequences of ozone depletion that have led to intense research, in the past two decades, on how increasing industrialization may influence global ozone levels.

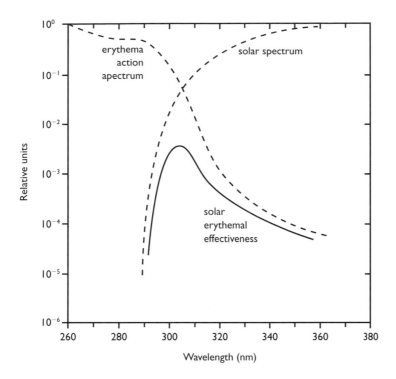

Figure 7.14 The erythema action spectrum compared to the solar spectrum at the Earth's surface.

7.6.2   Ozone chemistry

Ozone is restricted to a thin layer in the Earth's atmosphere, the ozonosphere, with a maximum concentration between 20 and 26 km above the Earth's surface. It is formed through the combination of atomic oxygen (O) and molecular oxygen (O_2). Atomic oxygen is formed by the photo-dissociation of O_2 at around 100 km by solar radiation with $\lambda < 175$ nm, the process for which may be summarized as:

$$h\nu(\lambda < 175\,\text{nm}) + O_2 \rightarrow O + O \tag{a}$$

The free oxygen atoms may then combine with oxygen molecules in a three-body collision (to conserve energy) to form ozone:

$$O + O_2 + M \rightarrow O_3 + M \tag{b}$$

where M is any atom or molecule (e.g. O_2, N_2) capable of absorbing the excess energy liberated in the exothermic chemical reaction.

Most of the ozone is produced in the equatorial regions where the amount of solar UV light is maximized. Ozone formed over these latitudes is then transported towards the Poles where it accumulates. Hence, ozone concentrations show significant seasonal fluctuations and tend to be at their highest in late winter and early spring.

Once formed there are two natural destruction mechanisms for ozone: photo-dissociation:

$$hv + O_3 \rightarrow O_2 + O \tag{c}$$

and collisional dissociation:

$$O + O_3 \rightarrow 2O_2 \tag{d}$$

the latter being a net result of complex catalytic chemical reactions, either natural or, more recently, man-made. Natural catalytic destruction occurs through the presence of the OH radical formed by photo-dissociation of water vapour in the atmosphere:

$$\begin{aligned} OH + O_3 &\rightarrow HO_2 + O_2 \\ HO_2 + O &\rightarrow OH + O_2 \\ \hline Net\ O + O_3 &\rightarrow 2O_2 \end{aligned} \tag{e}$$

The processes (a–e) explain why there is little ozone formed naturally in the troposphere where water is abundant and the flux of solar radiation ($<175\,nm$) is low, while lack of O_2 (restricting reaction (a)) prevents ozone formation in the atmosphere above the stratopause. As a result, the formation of ozone is restricted to altitudes between 10 and $50\,km$ where processes a–e have formed a stable equilibrium producing ozone concentrations of around $10\,ppm$. However, it is a fragile system and introduction of new chemicals (e.g. man-made pollutants transported from the troposphere) in the stratosphere can disturb this equilibrium and lead to rapid decrease in ozone concentrations.

7.6.3   'Ozone hole'

In 1985, Joe Farman, Brian Gardiner and Jonathan Shanklin of the British Antarctic Survey discovered a sharp reduction in ozone concentrations above Antarctica; this was soon termed the ozone 'hole' (see Plate 8). Satellite monitoring of the ozone 'hole' has shown that it is expanding and its depth increasing (Figure 7.15) and there is now evidence of similar but less marked reductions in ozone levels over the Arctic and the more densely populated northern latitudes.

An explanation for such dramatic ozone depletion was first proposed by Molina and Rowland in 1994. They suggested that the uncontrolled release of the chlorofluorocarbons (CFCs) used in refrigerators and aerosol cans into the terrestrial atmosphere would lead to the catalytic destruction of ozone. Although chemically stable within the troposphere, in the stratosphere the CFCs can be broken down by solar radiation and release chlorine atoms which in turn form ClO catalytic species such that:

$$\begin{aligned} ClO + O_3 &\rightarrow ClO_2 + O_2 \\ ClO_2 + O &\rightarrow ClO + O_2 \\ \hline Net\ O + O_3 &\rightarrow 2O_2 \end{aligned}$$

the ClO molecules being conserved.

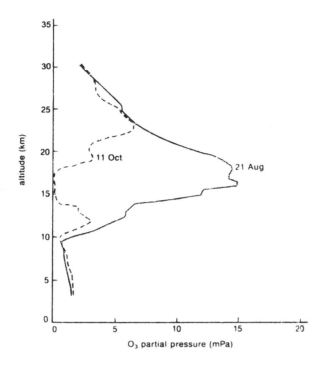

Figure 7.15 Ozone hole observed in October 1992 as function of altitude.

This mechanism is far more destructive than those naturally occurring mechanisms involving the OH radical as one ClO radical may destroy several hundred ozone molecules before it itself is removed from the ozone chemical cycle. Thus, Molina and Rowland suggested that even trace amounts (ppm) of any CFC in the stratosphere could have a dramatic effect on stratospheric ozone concentrations. Subsequently, their predictions were upheld and they were awarded the 1995 Nobel Prize for Chemistry, although the chemistry involved in ozone loss has in fact proved to be considerably more complex than this simple mechanism would suggest.

Mario Molina (1943–), F Sherwood Rowland (1927–) and Paul Creutzen (1933–).
Mario Molina, F Sherwood Rowland and Paul Creutzen shared the 1995 Nobel prize for Chemistry for their work in atmospheric chemistry, in particular for their work on global ozone depletion arising from industrial pollutants.

The possibility of nitrogen oxides destroying ozone in the Earth's stratosphere was first suggested by Paul Creutzen in 1970. In 1974 Rowland and Molina published a widely read article on the threat to the ozone layer posed by chlorofluorocarbons (CFCs) used in refrigerators and aerosol cans. In their article Rowland and Molina explained that CFCs could gradually be carried up into the ozone layer where, under the influence of intense ultraviolet light, they are decomposed into chlorine atoms which break down ozone by a catalytic process similar to that proposed for the nitrogen oxides by Paul Creutzen. Rowland and Molina proposed that unchecked use of CFCs would lead to a severe depletion of the Earth's ozone layer, a hypothesis that was confirmed by the discovery of the 'ozone hole' over Antarctica in 1985.

Rowland was born in Delaware, Ohio, USA and educated in Chemistry at the University of Chicago, USA, receiving his doctorate in 1952. He is currently at the University of California at Irvine, California, USA.

Molina was born in Mexico City and educated in Physical Chemistry at the University of California, Berkeley, USA. The first Mexican to receive a Nobel prize for science he is currently at the Massachusetts Institute of Technology, Cambridge, Massachusetts, USA.

Paul Creutzen was born in Amsterdam. Trained as a civil engineer, he worked in the bridge construction bureau of the city of Amsterdam before, in 1958, becoming a computer programmer in the Meteorology Department of the University of Stockholm. In 1963 he obtained his 'diplom' in mathematics, mathematical methods and meteorology before, in 1973, submitting his PhD thesis 'On the photochemistry of ozone in the stratosphere and troposphere and the contamination of the stratosphere by high-flying planes', this began his research career in the study of the effect on the Earth's climate by human activity. In the 1980s his pioneering theory of 'nuclear winter' showed how the cataclysmic fires ignited by a nuclear exchange would cost more human lives than the atomic explosions themselves. He is now director of the Atmospheric Chemistry Division of the Max Planck Institute for Chemistry.

7.6.4   Ozone loss in the Antarctic polar region

Continuous monitoring from satellites reveals that ozone concentrations in the stratosphere over Antarctica fall sharply in the local spring (September and October), when the atmospheric circulation established during the dark winter and the re-appearance of sun-

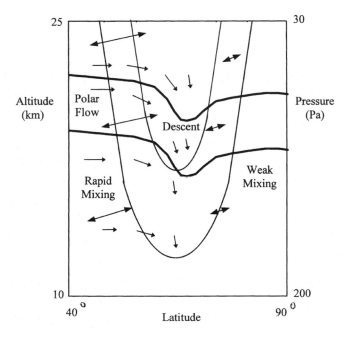

Figure 7.16 Schematic of the Antarctic polar vortex.

light combine to produce favourable conditions for the chemical reactions that lead to ozone destruction.

The increasingly stronger sunlight causes photochemical decomposition of the chemical compounds that release free chlorine, bromine and oxides of nitrogen in the stratosphere. During the winter a strong atmospheric vortex forms (Figure 7.16) over the polar continent in which westerly winds of speeds of up to $100\,\mathrm{m\,s}^{-1}$ or more effectively isolate the vortex air from the rest of the atmospheric circulation, trap the destructive chemical species and allow their concentrations to accumulate. The core of the vortex becomes very cold with temperatures as low as $-80°C$ such that water vapour, nitric acid [HNO_3] and sulphuric acid [H_2SO_4] gases condense on dust particles and freeze to form tenuous clouds composed of 'ice' particles in which the water combines with the acids to form solid hydrates. These are called *polar stratospheric clouds* (PSCs) (Figure 7.17). Those containing appreciable quantities of HNO_3 are called Type I, and those in which sulphuric acid hydrates predominate are termed Type II PSC.

The chemical reactions that take place on the surfaces of the cloud particles and lead to the destruction of ozone are called *heterogeneous reactions*. One important mechanism is thought to involve gaseous chlorine nitrate ($ClONO_2$) produced by:

$$ClO + NO_2 + M \rightarrow ClONO_2 + M \text{ [where M is any other atmospheric molecule (e.g. } O_2 \text{ or } N_2)]$$

This mechanism removes chlorine and nitrogen species from the cycles that destroy ozone. $ClONO_2$ may, therefore, be regarded as a reservoir for these species but only temporarily because chlorine nitrate is believed to be destroyed on the surfaces of the PSC particles by combining with hydrogen chloride (HCl) and water to release chlorine and retain the nitrogen as solid HNO_3, e.g.:

$$ClONO_2 + HCl \rightarrow Cl + Cl + HNO_3$$

Heterogeneous chemistry on polar stratospheric clouds (also known as "mother of pearl" clouds) has provided an explanation of the location and timing of polar ozone depletion

Figure 7.17 Polar stratospheric clouds.

The gaseous chlorine (Cl) can then re-enter the destructive ozone cycle, the reactions being very fast at low temperatures. Hence, PSCs play a vital role in ozone depletion in the cold Antarctic polar vortex.

Furthermore, a considerable enhancement in stratospheric ozone loss could occur if there was a large increase in the aerosol surface area by the injection of a cloud of sulphur from volcanic eruptions. The eruption of Mt Pinatubo in 1991 led to a massive increase in sulphate aerosols in the lower stratosphere. The deposits of huge plumes of SO_2 was estimated at about three times more as compared with the El Chichon eruption in 1982. The ozone loss due to the production of SO_2 has been explained by the following mechanism:

$$SO_2 + h\nu \rightarrow SO + O$$
$$SO + O_2 \rightarrow SO_2 + O$$
$$\underline{O + O_2 + M \rightarrow O_3 + M}$$
$$\text{Net } 3O_2 \rightarrow 2O_3$$

Therefore, the extreme ozone depletion in the Antarctic region measured in 1991–92 was caused by sulphur aerosols acting in addition to the heterogeneous chemical reactions on polar stratospheric clouds.

7.6.5   Ozone loss in the Arctic polar region

Ozone depletion is not as severe in the Arctic as in the Antarctic since the Arctic vortex is not as strong and does not become as cold as the one in the Antarctic due to the major difference in the meteorologies of the two polar regions. The South Pole is part of a very large land mass completely surrounded by the ocean, whereas the northern polar region lacks the land–ocean symmetry characteristic of the southern polar region area. As a consequence, the Arctic stratospheric circulation is less stable and the air is generally much warmer than in the Antarctic and gives rise to fewer PSCs that play a major role in ozone depletion in the Antarctic.

Such was the concern of the possible consequences of continued ozone depletion that global governmental meetings led to the agreement to phase out the production of all the CFCs (the Montreal Protocol 1987 and Rio Treaty 1992, see Appendix 8). However, due to long residence times of these compounds in the atmosphere it will be several decades before the negative influences of the CFCs are removed from the global atmospheric system. Thus, research into the physics and chemistry of the ozonosphere will continue, as will monitoring of global ozone levels.

7.7   Terrestrial radiation

7.7.1   Earth's energy balance

Not all the solar radiation incident over the Earth is absorbed by the atmosphere and the ground. A large fraction of it is reflected back into space from cloud tops and the ground (e.g. snow cover and the sea) The fraction of radiation reflected and hence lost to the Earth is called the *planetary albedo*, often denoted by the symbol *a*. An average planetary albedo from the Earth is estimated as 0.31 or 31%.

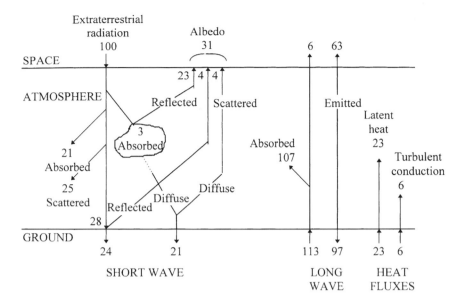

Figure 7.18 The Earth's energy balance.

Consider Figure 7.18. If 100 units of solar flux are incident per unit area on the top of the atmosphere, 31 are then reflected back into space mostly from cloud tops (23) with only eight from the ground, air molecules and particles (e.g. dust). Twenty-one units are absorbed by the atmosphere, 25 scattered by air molecules and clouds, 21 of which subsequently reach the ground, and only 28 reach the ground directly where 24 units are absorbed and four reflected back into space. Therefore, only 45 units of the 100 incident on the top of the atmosphere finally reaches the Earth's surface.

Should this energy be constantly absorbed the Earth's surface would rapidly heat up and become a molten surface unsustainable for life. Thus, the rate of absorption of solar energy must be balanced by the terrestrial output such that a steady-state is achieved.

As the Earth's surface is heated up, it emits radiation in the infrared – just like a domestic radiator. Ultimately, a radiative equilibrium is reached such that the re-emitted energy balances the incoming solar energy and a thermal equilibrium is reached at the Earth's surface.

However, much of this re-emitted radiation is absorbed by atmospheric molecules; only 6% escaping directly. The atmosphere subsequently itself radiates infrared radiation, 63 units into space and 97 units are absorbed by the ground.

Finally, there are two other heat fluxes in transporting energy from the ground to the atmosphere; the latent heat of evaporation of water (23 units) and the conduction by turbulent air motions (6 units).

All these gains and losses must compensate if the Earth is to be a steady-state system and result in the Earth's energy balance; this balance being met at every interface. Thus, at the space–atmosphere interface the 100 units of solar radiation are compensated by 31 units of albedo and 69 units lost to space as infrared radiation.

Table 7.4 depicts the nature of the gains and losses in the atmosphere.

Table 7.4 Gains and losses in the atmosphere.

Atmospheric gains		Atmospheric losses	
Solar absorption (by air and clouds)	24		
Infrared absorption	107		
Latent heat	23	infrared to space	60
Turbulent conduction	6	infrared to ground	100
	160		160

7.7.2   Earth as a black body

It has been shown how the absorption of solar radiation can excite a molecule or atom from one excited state to another of higher energy. Similarly, the transition to a state of lower energy is associated with the emission of a photon. Thus, any body can emit as well as absorb energy. If the body is in radiative equilibrium with its environment, it will emit as many photons as it absorbs per unit time, for each particular frequency interval.

The total emissive power, E, is then defined as the total radiant energy emitted per unit area per unit time. If E_λ is the radiant energy emitted per unit area per unit time, and between wavelengths λ and $\lambda + d\lambda$, then the total emissive power is:

$$E = \int_0^\infty E_\lambda d\lambda$$

Similarly, the absorbivity of a body, A_λ, can be defined as that fraction of the radiation incident on the body absorbed and the total or 'integral absorbivity':

$$A = \int_0^\infty A_\lambda d\lambda$$

A body that absorbs all the radiation incident upon it ($A_\lambda = 1$ for all wavelengths) is known as a '*black-body*' because it will appear black since it does not reflect any light.

Thus, one would also expect a black-body to be the best possible emitter at any given wavelength. The radiation emitted is then called 'black-body radiation' (see Section 2.4.3) and is independent of the nature of the body.

Assuming that the Earth emits terrestrial radiation as a spherical black-body with radius R_E and temperature T_E, Stefan–Boltzmann's law states (see Section 2.4.3) that the total power output is:

$$P = 4\pi R_E^2 \sigma T_E^4$$

The rate of absorption of solar energy is given by $S(1 - a)\pi R_E^2$, where a is the albedo, and S the solar constant. The terrestrial energy balance then requires that:

$$S(1 - a)\pi R_E^2 = 4\pi R_E^2 \sigma T_E^4$$

with the result that the Earth's 'effective temperature' is given by

$$T_{\mathrm{E}} = \left(\frac{S(1-a)}{4\sigma} \right)^{1/4} \tag{7.9}$$

Notice that T_{E} is *independent* of R_{E} showing that it is independent of the Earth's size and would apply equally to a satellite as to a planet the size of Jupiter. T_{E} is solely dependent on the albedo and the solar constant, which in turn is dependent upon the distance from the Sun.

7.7.3 Greenhouse effect

Given that the albedo of the Earth $a = 0.31$ and the solar constant $S = 1353\,\mathrm{W\,m^{-2}}$ using equation (7.9), the effective temperature of the Earth $T_{\mathrm{E}} = 255\,\mathrm{K}$ or $-18°\mathrm{C}$! However the average temperature of the Earth's surface is well above this ($\sim 288\,\mathrm{K}$ or $15°\mathrm{C}$). Therefore, this model for an irradiated and emitting Earth is too crude. We have neglected those atmospheric absorption/emission processes discussed above and in effect considered the space–atmosphere interface, not the atmosphere–ground interface. The 30°C elevation in surface temperature is due to the 'greenhouse effect', because it can be explained by considering the atmospheric mantle above the ground as if the atmosphere acts like the roof of a greenhouse (Figure 7.19).

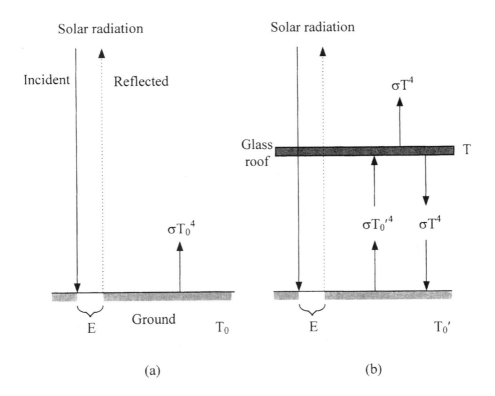

Figure 7.19 Simple model of the greenhouse effect.

Consider an area of the ground receiving and absorbing solar radiation. The surface will heat up due to absorbed energy (E) and the infrared radiation emitted will increase accordingly. As the ground can be considered a black-body the energy emitted will be σT_o^4, where T_o is the surface temperature. If there is no roof above the ground, T_o will increase until the emitted energy balances the incoming solar energy (Figure 7.19a)

Now consider a greenhouse with a roof above this same plot of ground. The glass lets the solar radiation through to the ground, but prevents the infrared radiation from being radiated into space; emitting this energy itself from its two surfaces according to σT^4, where T is the temperature of the glass (Figure 7.19b).

The ground is now receiving more energy than before and so its temperature will rise (T_o') until a new equilibrium is achieved, in which both the ground and glass emit as much as they absorb. In this new state the upward emission by the glass must again balance the incoming solar energy so:

$$E = \sigma T_o'^4 = \sigma T^4$$

or $T_o' = T$, and the glass now has the temperature that the ground had before the greenhouse was built over the plot. Balancing the energy per unit area of glass roof presents

$$2\sigma T_o^4 = \sigma T_o'^4$$

where T_f is the final temperature of the ground under the greenhouse or

$$T_o' = 2^{1/4} T_o = 1.19 T_o \tag{7.10}$$

So if $T_o = 255\,K$, $T_o' = 303\,K$ or 30°C, and where T_o' is the final temperature of the ground.

That the simple argument given above overestimates the actual observed surface temperature (288 K) is due to the atmosphere not acting as a perfect black body as was assumed for the glass roof.

Of course there is no glass roof over the Earth but molecules in the Earth's atmosphere act in a similar way. Molecular nitrogen and oxygen are poor absorbers of infrared radiation and play only a minor role in warming the Earth's surface but water vapour, carbon dioxide and ozone in the lower layers of the atmosphere exhibit the same 'selective absorption' effect as the glass roof and prevent infrared radiation emitted by the Earth from escaping. The absorption properties of these three *greenhouse gases* is illustrated in Figure 7.8.

Water vapour absorbs infrared radiation intensely at 6.3 μm, strongly for a band commencing at 9 μm (extending to longer wavelengths) and weakly for some bands below 4 μm. Carbon dioxide absorbs in the region 13–17 μm and ozone in an intense narrow band at 9.7 μm. This leads to the Earth's terrestrial infrared spectrum as observed from space (Figure 7.20) having regions of sharp minima which reflect the absorption properties of the 'greenhouse gases' in a similar manner to the solar flux spectrum having minima where atmospheric gases absorb the incident solar radiation as it passes through the atmosphere (Figure 7.8). Note however that between 9 and 11 μm there remains a region which is practically transparent; this is known as the *atmospheric window*. (see Section 6.2).

The absorbed radiation is ultimately re-radiated to space (or the atmosphere would continue to heat up) from levels somewhere near the top of the atmosphere (5–10 km altitude) and thus at lower temperatures. Since the molecules are colder at such altitudes,

Figure 7.20 Meteostat image of water vapour wavelengths for Africa and the Atlantic.

they will emit correspondingly less radiation. Hence, the net loss of energy from the Earth's surface is less than it would have been if the greenhouse gases were not present. They have therefore acted like a blanket over the Earth's surface and helped to keep it warmer than it would otherwise have been.

Increase in carbon dioxide and water vapour concentrations in the atmosphere or the introduction of molecules that absorb within the atmospheric window (e.g. methane and nitrous oxide (N_2O) by industrial emissions) will therefore lead to increased trapping of terrestrial radiation and hence an increase in the Earth's surface temperature. This generates the phenomenon of *global warming*.

7.8   Global warming

7.8.1   Enhanced greenhouse effect

The huge expansion of global industrialization in the twentieth century has led to the prediction that humanity is now altering the global radiation balance by enhancing the natural greenhouse effect, through increasing greenhouse gas concentrations in the Earth's atmosphere. The burning of fossil fuels, such as coal and oil, and natural combustibles such as wood during deforestation has led to a steady increase in CO_2 emissions since the Industrial Revolution. This has resulted in a 30% increase in CO_2 concentrations in the lower atmosphere since 1780 (Figure 7.21).

There have been similar changes in the concentrations of other greenhouse gases (Table 7.5) during the industrialized period; methane concentrations have more than doubled since the mid-nineteenth century, while nitrous oxide (N_2O) concentrations have increased by 8%. The introduction of the CFCs into the atmosphere has also provided a new source of greenhouse gases.

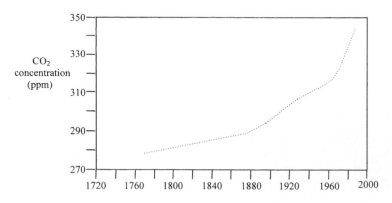

Figure 7.21 Global carbon dioxide concentrations since 1700.

Table 7.5 Concentrations of greenhouse gases.

Compound	Carbon dioxide	Methane	Nitrous oxide	CFC
Concentration	ppm	ppm	ppb	ppt
1800	280	0.80	288	0
1990	354	1.7	310	764
Annual increase (%)	0.5	0.9	0.25	4
Averaged residence times (years)	100	10	150	100

Increases in molecular concentrations alone are not a direct measure of the contribution to increased global warming by each of the greenhouse gases. To evaluate their contribution to possible global warming it is also necessary to know their absorption cross-section for infrared radiation. Table 7.6 shows the relative contributions of these gases to global warming since 1800 assuming current concentrations.

Of the CFCs, one in particular (CFC12) is estimated to provide 12% of the global warming. The ban on CFCs is therefore doubly effective in both reducing global ozone depletion and possible global warming.

Table 7.6 Relative contributions (%) of the greenhouse gases to global warming.

Carbon dioxide	55
Methane	15
CFC 12	21
Nitrous oxide	4
Ozone (tropospheric)	2
Others	3

7.8.2 Global warming: the evidence

Figure 7.22 shows the deviation from the Earth's mean surface temperature in 1960, during the period 1854 to present. Is the trend rising or is it a cyclic process following a little ice age in the mid-1800s? This remains the subject of much debate, since unlike the dramatic observation of the ozone hole the changes are small (±1°C).

Evidence for global warming is harder to obtain since there are many natural phenomena that can lead to variations in the Earth's global climate. Large volcanic eruptions (see Plate 9) can significantly affect the subsequent global weather since the transport of large amounts of ash and sulphur dioxide (capable of forming aerosols) into the upper troposphere can lead to significant cooling. The Mt Pinatubo eruption in June 1991 in the Philippines injected an estimated 20 million tonnes of sulphur dioxide into the atmosphere together with enormous amounts of dust. This dust caused spectacular sunsets all around the world for many months after the eruption, led to a 2% reduction in the amount of solar radiation reaching the Earth's surface and lowered the global average temperature by 0.25°C in the next 2 years. Hence, the base line on which global temperature changes are observed is complex. Climate extremes are not, in fact, unusual and every month somewhere in the world a climate record is broken. This is not evidence that our climate is changing, only that our data records are limited and that global climate patterns are complex entities. In examining the evidence for global warming one should look at the records at many sites and over long periods and not be mislead, by what are found to be short-term variations. In the late 1960s and early 1970s there was an actual drop in global surface temperatures leading to some popular speculation that we were about to enter an ice age!

Thus, although the late 1980s and 1990s have been the warmest since accurate measurements began a century ago and although the number of hurricanes, gales and droughts seems

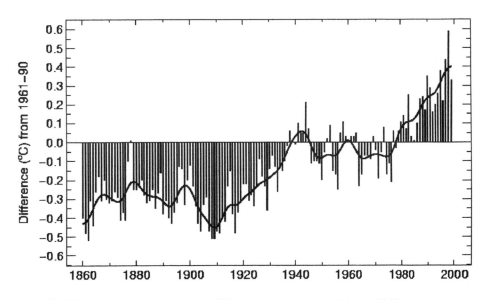

Figure 7. 22 Global surface temperatures 1854–present compared with the 1960 mean.

to have been higher, one should not conclude that this is direct evidence for global warming. There have been periods of extreme weather in the past at times when humanity was not producing any global warming gases, e.g. the little ice age in the Early Modern period, while in Roman times it was certainly warmer than it is now in northern Europe since grapes could be grown in northern England.

The study of *palaeoclimates* allows scientists to study climate patterns over several tens of thousands of years. Snow deposited at the polar regions is gradually compacted as further snow falls becoming solid ice, trapping within it tiny bubbles of air. Drilling out an ice core from such a polar ice cap and examining the composition of the air in the bubbles allows the scientist to determine the atmospheric content as a function of time. Samples at the top of the core are obviously more recent than those at the bottom. The results from one such core drawn from Vostock in Antarctica are shown in Figure 7.23. They show a strong correlation between the global surface temperature and the concentrations of carbon dioxide and methane, supporting the hypothesis that should the concentrations of either or both of these gases be raised then global surface temperatures will also rise. You may also like to compare these trends with the solar cycles as described in the Milankovitch hypothesis (Case Study 7.1).

7.8.3   Global warming: the predictions

Currently there is much scientific effort in setting up sophisticated computer models to determine whether greenhouse gas emissions will lead to major climatic change, not just globally, but on regional scales, since the latter is obviously of more concern to the local population and national politicians. Protective measures against climate change, e.g. rising sea levels, require regional action and often very expensive remedies may be needed, such as the building of sea walls. Hence, before embarking upon such projects governments require the most accurate forecasts that can be made both of the magnitude of the threat and the time scales over which such problems will become apparent.

Climate modelling is very complex, and the number of variables that need to be considered are extremely large. Therefore only a few research centres world-wide can undertake such modelling. Most of the models are based on those used for daily weather prediction (see Chapter 9) and incorporate all the atmospheric physics described here but in addition, since climate change acts over very much longer time scales than the weather, additional processes must be incorporated into the model. The atmosphere is coupled to the oceans, the land surface, the distribution of vegetation (the biosphere) and those parts of the Earth covered with ice (the cryosphere). These five components – atmosphere, ocean, land, ice, biosphere – must all be included in computer models of the global climate. Once these are established it is possible to predict how, for example, the doubling of carbon dioxide might increase global surface temperatures. If atmospheric carbon dioxide was to double in concentration, a temperature rise of about 2°C can be predicted.

However, this is too crude since such a simple model does not take into account *feedbacks*. Feedbacks may be positive or negative and show that when there is change in one parameter in the model others may be influenced. An example of positive feedback is the rise in surface temperature that will follow global warming, which in turn reduces the solubility of carbon dioxide in the oceans. This will release more carbon dioxide into the atmosphere. Similarly warming the oceans increases the evaporation of water from the sea

Plate 1 The components of Environmental Physics
This view of the English Channel illustrates the principal interacting components of Environmental Physics – the atmosphere, the lithosphere, the hydrosphere and the biosphere. Solar radiation is interacting with vegetation and the sea, and the cumulus clouds are a facet of the hydrological cycle. (Source: P. Hughes)

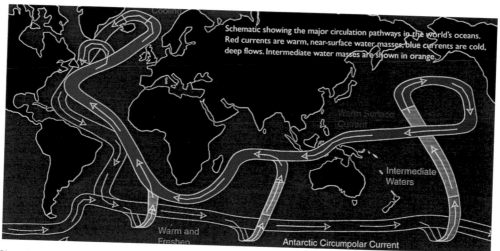

Schematic showing the major circulation pathways in the world's oceans. Red currents are warm, near-surface water masses, blue currents are cold, deep flows. Intermediate water masses are shown in orange.

Cooling

Warm Surface Current

Intermediate Waters

Warm and Freshen

Antarctic Circumpolar Current

Plate 2 The oceanic conveyor belt
The world's oceans play a profound role in the distribution of solar energy through the transport of mass and thermal energy through differences in density, temperature, and salinity. It has been suggested that global warming may result in the increasing melting of polar ice. Changes in the resulting salinity could disrupt the conveyor system. If this happened to the Gulf Stream the British Isles would experience a cooling effect. (Source: Natural Environmental Research Council)

Plate 12 Hurricane

A weather pattern normally found in the Caribbean and the North Atlantic, characterized by a very low pressure system with possible wind-speeds greater than 200 kph. The plate shows an infra-red satellite image of Hurricane Floyd moving rapidly over the eastern seaboard of North America, September 1999. This hurricane was rated category 4 on the Saffir-Simpson Hurricane Scale (which ranges from 1 to 5) with wind-speeds between 210 and 249 kph. (Source: NOAA)

Plate 13 Soil profile

The soil's vertical profile is comprised of four distinct horizons. At the top is the organic matter (humus) and then in descending order: silt, fine sands, coarse sand and the gravel. Larger rocks and boulders can be observed at the bottom.

In this plate British teachers and American students are engaged in an Earthwatch Millennium geophysics project. Data is being collected from gravel pits throughout the Stockholm Archipelago, in the form of the size and direction of depositions (such as sands, gravels and pebbles) made during the last deglaciation of the Scandinavian ice-sheet about 9000 years ago. Analysis of the data will provide information about the climate change that accompanied the retreat of the ice. (Source: P. Hughes)

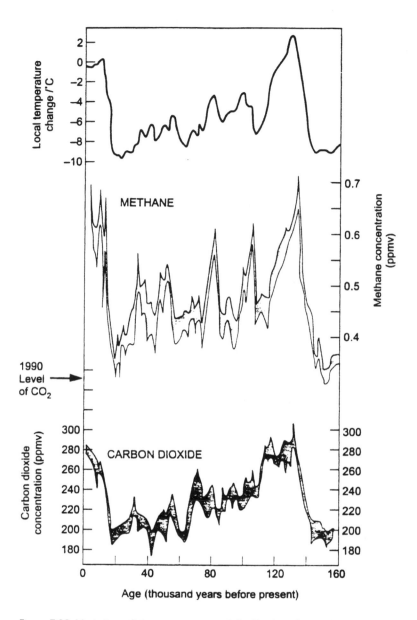

Figure 7.23 Variation of the temperature of the Earth and greenhouse gas concentrations from the Vostock ice core. Shading represents uncertainty in the measurements.

surface. This increases the water vapour in the atmosphere, which will lead to still further warming since water vapour is a good greenhouse gas. However, more water vapour may lead to increased cloud formation, resulting in a larger proportion of the solar flux being reflected back into space – an example of a negative feedback. But clouds can also act as a greenhouse roof and trap heat beneath them. Which effect dominates will depend on the

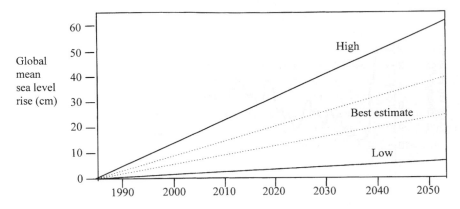

Figure 7.24 Global sea-level rise in the twenty-first century arising from global warming as predicted by various climate models, lowest and highest estimates are shown as well as the agreed best estimate.

type of cloud and its content of ice and water. There are many other examples of feedbacks all of which produce a different amount of global warming and hence make it difficult to forecast the climate in the twenty-first century. Nevertheless, there is now a convergence amongst the computer models which all predict that continued emission of greenhouse gases will lead to noticeable increases in global temperature in the twenty-first century and that the emissions of the past century will produce measurable effects in the next few decades. It is further thought that the land will warm more than the oceans and warming will be greater in the Northern rather than the Southern Hemisphere (see Plate 10). Further discussion of this problem will be made in Chapter 12 where the political and industrial response to global warming will be discussed.

7.8.4   Sea-level rise and global warming

One major consequence of rising global temperatures will be an increase in the temperature of the oceans and some melting of the ice at the poles. There is some evidence that sea levels have risen during the twentieth century. The sea level at Brest on the west coast of France, for example, has risen by 10 cm since the Industrial Revolution. Therefore there have been many cataclysmic predictions of the effects of sea-level rise (see Figure 7.24 and Plate 7) over the next 60 years. A rise of 15 cm by 2030 and 50 cm by 2100 may not seem much since most people live well above sea level and would not be directly affected. However, nearly half of the world's population lives near coastal regions and some countries (e.g. Bangladesh and the Pacific nations) are little more than 1 m above sea level. It is quite impractical to consider the construction of sea defences along the whole of their coast line, and even if this was possible the introduction of salt water into the fresh water table would make such defences ineffective as far as protection of agricultural land was concerned.

Most of the likely sea-level rise may be attributed to oceanic thermal expansion; however, the amount of expansion is dependent upon the temperature of the water. For cold water (e.g. polar) the expansion per °C surface temperature rise is negligible, while for water at 5°C a rise of 1°C causes an expansion of 1 part in 10000 and for water at 25°C a 1°C rise causes an expansion of 3 parts in 10000. Hence, global warming in the tropics may lead to higher sea-level changes than in the mid-latitude regions.

The remaining sea-level rise may be attributed to the melting of the glacial ice sheets. If all the glaciers outside Greenland and Antarctica were to melt, the rise in sea level would be <50 cm. However, as the sea warms there will be increased evaporation. This will lead to increased precipitation with higher snow falls over the glacial regions, while some land masses, such as Scandinavia, still rising from the loss of their last ice cap, will experience falls in the sea levels along the coasts for years to come. Hence, forecasting sea-level change on a regional basis is as difficult as forecasting temperature rises.

Case study 7.2   Greenhouse effect on other planets.

Our knowledge of the possible effects of global industrial emissions and the mechanisms of global warming have been enhanced by the ability to study the atmospheres of Earth's nearer planetary neighbours, Venus and Mars. Mars is smaller than the Earth and has only a thin atmosphere, but one comprised almost entirely of carbon dioxide and it thus experiences an accelerated greenhouse effect. Mars is 50% further away from the Sun than the Earth and thus has a lower solar constant. Applying equation (7.9) the surface temperature of Mars in the absence of its atmosphere would be 216 K, whereas it is actually 226 K; the greenhouse effect of its carbon dioxide atmosphere raises the surface temperature by 10 K.

Venus has a very different atmosphere to Mars. It consists mainly of carbon dioxide with dense clouds of sulphuric acid that almost completely cover the whole planetary surface, such that <2% of the solar radiation reaches the planet's surface. It might then be supposed that the planetary surface would be cool, but in fact probes suggest that the surface temperature is >800 K. The reason for these high temperatures is the very thick carbon dioxide atmosphere which prevents the infrared radiation from escaping. This dramatic rise in temperature (almost 500 K) was named the 'runaway greenhouse effect'.

Such a runaway greenhouse effect did not occur on Earth due to its greater distance from the Sun. The solar constant for Venus being twice that of the Earth, Venus would have a surface temperature of 325 K without any atmosphere and, hence, throughout the evolution of the Venusian atmosphere water (released from the planetary interior) would be continually boiling and evaporating leading to higher concentrations of this greenhouse gas in the atmosphere. This generated even higher surface temperatures until either the atmosphere became saturated with water or all the available water had evaporated. In the terrestrial atmosphere just such an equilibrium is reached between the rate of water evaporation and the amount of water in the atmosphere, but in Venus, due to the greater initial surface temperature such an equilibrium has not been reached. The cycle of water through the Earth's atmosphere is discussed in Chapter 8.

7.9   Summary

The Sun provides the energy that drives many physical processes within the Earth's atmosphere. In this chapter the mechanisms for the transfer of solar energy through the Earth's atmosphere on to the Earth's surface have been examined. The conversion of incident solar radiation into emitted terrestrial radiation from the Earth's surface has been described and the 'greenhouse effect' discussed. The environmental problems of ozone depletion and global warming have been reviewed. However, the Sun does far more than simply provide a heat source for the warming of the Earth, it drives the daily weather. Energy from the Sun leads to the formation of clouds and the generation of winds. In Chapters 8 and 9 cloud formation and the generation of the world's weather systems will be discussed before considering how the Sun supports the biosphere through, for example, photosynthesis.

References

Atmosphere

Barry, R. G. and Chorley, R. J., *Atmosphere, Weather and Climate*, 7th edn. London: Routledge, 1998.

Graedel, T. E. and Crutzen, P. J., *Atmospheric Change: An Earth System Perspective*. New York: Freeman, 1997.

Green, J., *Atmospheric Dynamics*. Cambridge: Cambridge University Press, 1999.

Houghton, J., *Global Warming: The Complete Briefing*. Cambridge: Cambridge University Press, 1997.

Jacobson, M. Z., *Fundamentals of Atmospheric Modelling*. Cambridge: Cambridge University Press, 1999.

Lutgens, F. K. and Tarbuck, E. J., *The Atmosphere*, 7th edn. Englewood Cliffs: Prentice-Hall, 1998.

McIlveen, R., *Fundamentals of Weather and Climate*.London: Chapman and Hall, 1992.

Schaefer, V. J. and Day, J. A., *A Field Guide to the Atmosphere*.Boston: Houghton Mifflin, 1981.

Thompson, R. D., *Atmospheric Processes and Systems*. London: Routledge, 1998.

Wallace, J. M. and Hobbs, P. V., *Atmospheric Science: An Introductory Survey*. San Diego: Academic, 1977.

Wayne, R. P., *Chemistry of Atmospheres*, 2nd edn. Oxford: Oxford University Press, 1991.

Wells, N., *The Atmosphere and Ocean: A Physical Introduction*, 2nd edn. Chichester: Wiley, 1998.

Web sites

A general web site that provides information on many of the topics discussed in this chapter is that of the Centre for Atmospheric science, Cambridge University: http://www/atm.ch.cam.ac.uk
Many web sites report the most recent news on ozone depletion, e.g.:
http://bsweb.nerc-bas.ac.uk/public/icd/jds/ozone/index.html provides an almost daily record of Antarctic ozone depletion
http://jwocky.gsfc.nasa.gov/and http://www.ozone-sec.ch.cam.ac.uk/provide a brief guide to global ozone depletion.
http://www.atm.ch.cam.ac.uk/tour/is a complete tour of the ozone depletion problem together with on-line movies of the ozone hole.
http://www.doc.mmu.ac.uk/aric/ace/ace_oz.html provides fact sheets that give a full discussion of ozone depletion, its causes and the possible consequences for human and plant life. It

also discusses the preventative measures being taken by the international community to mitigate the effects of ozone depletion and restrict global emissions of ozone-destroying gases.

Other ozone sites

NASA atmospheric chemistry and dynamics branch: http://geo.arc.nasa.gov/sgg.html
National Oceanic and Atmospheric Administration Climate Prediction Centre: http://nic.fb4.noaa.gov:80/products/stratosphere
National Oceanic and Atmospheric Nitrous Oxide and Holocarbons Group: http://www.cmdl.noaa.gov/noah-home/noah.html
Stratosphere ozone law, information and science: http://www.acd.ucar.edu/gpdf/ozone
US Environmental Protection Agency stratospheric ozone page: http://www.epa.gov/docs/ozone
World Meteorological Organization atmospheric research and environment programme: http://www.wmo.ch/web/arep/arep-home.html

Updates on global warming predictions and global treaties

Global climate information programme: http:/www.doc.mu.ac.uk/aric/gcciphm.html
NASA's Earth observation system project science centre: http://spso2.gsfc.nasa.gov/spso
National Oceanic and Atmospheric Administration Office Climatic Data Centre: http://www.ncdc.noaa.gov
National Oceanic and Atmospheric Administration Office of global programmes: http://www.noaa.gov/odp
US global change research programme: http://www.usgcrp.ov
The Intergovernmental Panel on Climate Change: http://www.ipcc.ch

Discussion questions

1   Different gases are more or less homogeneously distributed in the Earth's atmosphere. How do you expect the degree of homogeneity to be affected by the following: (a) their rate of turnover, (b) their origin and (c) their reactivity. Would you expect neon or sulphur dioxide to be the more homogeneous? Why?

2   Place the following in order of ascending height: troposphere, mesosphere, stratosphere, tropopause, ionosphere and stratopause.

3   Explain why atmospheric pressure decreases with height.

4   Discuss what happens to a balloon as it slowly rises through the atmosphere.

5   Why is the specific heat capacity at constant pressure (C_p) larger than the specific heat capacity at constant volume (C_v)? (Hint: apply the First Law of Thermodynamics to a rigid container and a plastic bag.)

6   If the ratio of ^{4}He to ^{40}Ar entering the Earth's atmosphere from radioactive decay during the past 10^9 years is close to unity, why is the present-day ^{4}He measured in the Earth's atmosphere much less than that of ^{40}Ar?

7   How does the amount of solar radiation received at a point on the Earth's surface depend upon (a) the time of day, (b) the time of year and (c) altitude?

8   What is the physical reason for the Earth's surface temperature being reduced under the following conditions: (a) when the Earth is covered with snow and (b) under the conditions of 'nuclear winter'.

9   Discuss the effect on the Earth's surface temperature under the conditions of an increase in the global production of rice.

10   It has been suggested that 65 million years ago an asteroid collided with the Earth. Discuss how such a bolide impact could induce the major climate change that could result in the mass extinction of whole species, such as the dinosaurs.

11   (a)   Discuss why ozone is produced mainly in the equatorial regions and hence how it is circulated around the Earth.

   (b)   Why is ozone restricted to such a thin layer in the Earth's atmosphere between 20 and 50 km?

   (c)   Discuss the problem of global ozone depletion and why the Antarctic ozone hole is deepest in the spring?

12   Explain why (a) the Sun is yellow and (b) the sky is blue.

13   Different molecules in the Earth's atmosphere absorb energy at different wavelengths. The approximate values of some of these wavelengths for certain molecules are listed below

Molecule	Water vapour	Carbon dioxide	Ozone
Absorbed wavelength (μm)	6	15	0.3

Discuss how the presence of water vapour, carbon dioxide and ozone in the Earth's atmosphere affects:

   (i)   the radiation received from the Sun at the surface of the Earth

   (ii)   the equilibrium temperature reached by the Earth

                    (London University: 'A' Level Physics: Earth and Atmosphere: January 1998)

14   (a)   In the Northern Hemisphere in winter the Earth is closer to the Sun by 5×10^6 km than it is in the summer. Explain why the average daytime maximum temperature in Britain is lower than that in the summer.

   (b)   In Northern latitudes the oceans are warmer in summer than they are in winter. In which season do the oceans lose energy most rapidly to the air by conduction? Give an explanation for your answer

   (c)   On which dates do all places on the Earth have 12 hours of daylight and 12 hours of darkness. Give an explanation of this phenomenon.
        (Oxford and Cambridge Examination Board: 'A' Level Physics: Physics in the Environment:
                                                                                June 1996)

15   Explain why it is imperative that there are greenhouse gases in the Earth's atmosphere. Cite two sources and sinks of atmospheric CO_2.

Quantitative questions

1   If the radius of the Earth is 6370 km and assuming that the pressure everywhere is equal to the atmosphere at the ground, what is the total mass of the Earth's atmosphere?

2   If the tropopause is at a pressure of 150 mb and the stratopause at 1 mb:

   (a)   Calculate the total mass per unit cross-section of the stratosphere.

   (b)   How thick would the stratosphere be if it was brought to ground level at standard temperature (273 K) and pressure (1 atm)?

3   Calculate the total number of molecules per cm^3:

   (a)   at the Earth's surface, $p = 10^5$ Pa, $T = 20°$C.

   (b)   at 100 km altitude, $p = 10^{-1}$ Pa, $T = 50°$C.

   (c)   at 300 km, $p = 3 \times 10^{-8}$ Pa, $T = 1500$ K.

4   Calculate the density of air at the summit of Mount Everest where the pressure and temperature are 3.13×10^4 Pa and $-38.5°$C respectively.

5   Given that the Martian atmosphere is mainly composed of CO_2, that its surface temperature is

210 K and that gravity is one-third of that on Earth, is the scale height of the Martian atmosphere more or less than that on Earth?

6   The root mean square speed of gas molecules is calculated from the relationship

$$\text{speed}_{rms} = \sqrt{3RT/M}$$

where R is the molar gas constant, T is the temperature of the gas in Kelvin and M is the molar mass of the gas. For oxygen $M = 32 \times 10^{-3}\,\text{kg mol}^{-1}$. Calculate the r.m.s. speed for (a) oxygen molecules and (b) hydrogen at 27 °C.

    The escape speed from the Earth is $1.1 \times 10^4\,\text{m s}^{-1}$. Hence or otherwise account for the lack of molecular hydrogen in the Earth's atmosphere.

                    (London University: 'A' Level Physics: Earth and Atmosphere: June 1999)

7   The radius of the Venusian orbit around the Sun is two thirds that of the radius of the Earth's orbit around the Sun. If the Earth's solar constant is $1350\,\text{W m}^{-2}$, what is the solar constant of Venus?

8   If sunlight consists of radiation of wavelength $\lambda = 250\,\text{nm}$:
    (a) Calculate the energy of one photon.
    (b) Is this sufficient to break the chemical bond of ozone, which has a dissociation energy of $2.5 \times 10^{-19}\,\text{J}$?

9   If the peak wavelength of the Sun's surface emission spectrum is 500 nm what is the temperature of the surface of the Sun?

10   If the maximum of E_λ occurs at $\lambda = 500\,\text{nm}$ for solar radiation and the temperature of the Sun is 6000 K, what is E_λ, the maximum of terrestrial radiation from the Earth's surface at 300 K?

11   The distance between the Earth and the Sun varies during the year, being a minimum in January and 3.5% larger at the maximum in July. What is the corresponding seasonal change in effective temperature, T_e?

12   More than two-thirds of the Earth's surface is covered by water. Solar radiation reaches the Earth as roughly parallel beams of mean intensity $1.4\,\text{kW/m}^2$; of this radiation, on average 50% reaches the surface. The oceans have a mean reflectivity of 7%, while they reradiate 35% of the energy they absorb.
    (i)   Using these data, estimate the mean annual rainfall over the whole of the Earth's surface, expressed as a depth in millimetres.
    (ii) Suggest two reasons why 50% of the Sun's radiation does not reach the surface of the Earth.

                            (British Physics Olympiad Paper: 1992–93)

13   (a) The value of the solar constant can be found using Stefan's law ($E = \sigma A T^4$). Use the data below to calculate its value.

The Stefan constant	$= 5.67 \times 10^{-8}\,\text{W.m}^{-2}.\text{K}^{-4}$
Surface temperature of the Sun	$= 5770\,\text{K}$
Radius of the Sun	$= 6.98 \times 10^8\,\text{m}$
Radius of Earth's orbit	$= 1496 \times 10^8\,\text{m}$

    (b) One consequence of a very large volcanic eruption would be that a great deal of dust and ash would be discharged into the atmosphere. As a result of this it has been predicted that land temperatures in the Northern Hemisphere could vary with time as shown below.

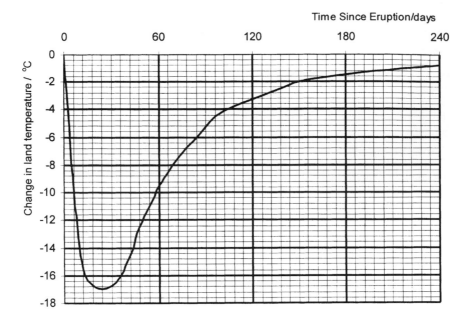

Discuss the effect that the dust and ash would have on the radiation balance at the surface of the Earth and hence explain the shape of the graph. Ash can cause ozone to break down. What additional problem would this cause if large quantities of ash penetrated to a height of 50km?

(London Examinations: 'A' Level Physics: Earth and Atmosphere: June 1998)

14   Given that the ozone concentration in the stratosphere is 10 ppm by volume, calculate the total mass of ozone available in the stratosphere. Assume the stratospheric pressure to be 10^4 Pa. Compare your answer with the total mass of the Earth's atmosphere calculated in question 1.

   If global tropospheric ozone concentrations were to rise by a uniform 50 ppb by volume, what percentage reduction in stratospheric ozone concentration would be compensated for?

15   Using the Beer–Lambert law, calculate the percentage increase in 260 nm UV radiation reaching the Earth's surface at the South Pole when the 'ozone hole' is 50% that of the normal concentration (3.2×10^{16} m^{-3}). Assume that the photo-absorption cross-section for 260 nm UV light is 10^{-21} m^2 and that the stratosphere is 40 km deep.

16   (a)   List the 4 principal layers of the atmosphere in order from the Earth's surface upwards. Within each of these layers, state how the temperature varies with height.

   (b)   (i)   The density of air is 1.2 kg/m^3 at the Earth's surface. Calculate the height of the column of air required to exert a pressure of 1 atmosphere (1×10^5 Pa) at its base.

       (ii)   At constant temperature the pressure of the atmosphere decreases exponentially with height according to the equation $p = p_o.e^{-kh}$ where p_o is the pressure at the Earth's surface. Given that p at a height of 5 km is approximately $0.5p_o$, estimate the height at which p will have fallen to $\left(\dfrac{1}{8}\right)p_o$.

       (iii)   Comment on your answers to parts (i) and (ii).

       (London Examinations Board: 'A' Level Physics: Earth and Atmosphere: January 1996)

17 Of all the water on the Earth, 97% is in the oceans and would, if evenly distributed across the Earth's surface, cover the Earth to a depth of 2.8 km. What would be the sea-level rise if the remaining 3% of the water (lying in the polar ice caps and glaciers) was to be released by global warming? Why, in fact, would the actual rise be less than this?

Chapter 8

Observing the Earth's weather

8.1   Introduction

Everyone scans the skies to estimate what the day's weather will be. Is it necessary to wear a raincoat and carry an umbrella or, in contrast, to put on sun cream and wear sunglasses? Humans have undoubtedly looked into the skies to assess the weather for thousands of years. Knowing what the weather is that day was, however, not sufficient for the development of civilization. The successful development of agriculture required the ability to predict how weather could affect the planting and harvesting of crops, while sailors and aviators needed to know of approaching storms. So how do we predict the weather? How is the information collected that allows the weather forecasts on the television to be made?

In discussing the 'weather', one tends to mean the instantaneous conditions within the atmosphere, while 'climate' in contrast monitors the mean conditions of the atmosphere over a period of time. In observing the weather it is necessary to measure several physical parameters within the Earth's atmosphere, more especially within the Earth's troposphere. Air temperature, atmospheric pressure, wind speed and direction, rainfall and humidity are traditionally the key parameters measured at any weather observation point. Figure 8.1 shows a typical weather station that can be used to monitor the weather at one specific site over a prolonged period. Collection of data from a network of such stations across the country and neighbouring countries and across the seas not only will allow today's weather to be monitored, but also will allow weather forecasters to predict what the weather may be tomorrow and several days ahead.

In this chapter, the instruments needed to monitor the Earth's weather will be described, and how measurements collected from these instruments is used to predict future weather patterns will be explained. It will then describe the characteristics and formation mechanisms of that most obvious feature of weather systems – clouds.

8.2   Observing the weather

In designing a weather monitoring station what parameters should be measured? Air temperature is an obvious choice as are the amounts of rainfall and sunshine, but it is also necessary to measure air pressure. Differences in air pressure lead to winds and a method must be developed for measuring both the wind speed and its direction. Less obvious is the need to measure the water content in the atmosphere; a property known as the *humidity* (see Chapter 3). The amount of cloud cover and the type of cloud should also be recorded

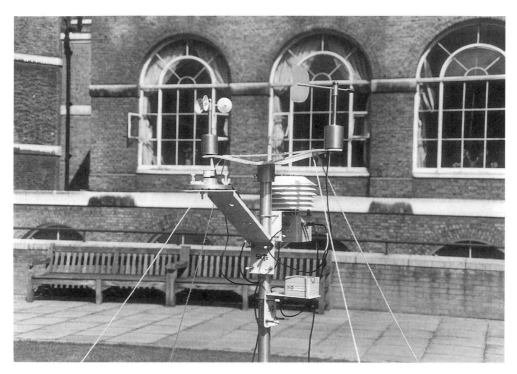

Figure 8.1 A mini weather station at University College, London.

since clouds reveal a good deal about the structure, movement and development of weather systems as fronts, depressions and storms. Another quantity that is recorded is *visibility* since this is a good indicator of atmospheric conditions and is obviously important for shipping and aircraft.

Having made these measurements they are compared with other weather stations such that the weather may be tracked across the country. Hence, it is necessary to ensure that all weather stations use standard apparatus and an agreed measurement technique. If this methodology is promoted across the whole country (and then across the whole globe) and all the data are recorded simultaneously a global picture of the Earth's weather can be established. Knowledge of the wind systems and hence the position of depressions and fronts will then be obtained allowing the movement of weather systems to be tracked and hence weather predictions to be made. This is the basis of the science of *meteorology* and the essential prerequisite for weather prediction (forecasting).

Let us first examine the instruments developed to monitor the Earth's weather.

8.2.1   *Air temperature*

Human bodies are very sensitive to the air temperature. They feel hot or cold depending on whether their bodies are having to waste or conserve heat to regulate the body temperature to about 37°C (see Chapter 2). Yet the human body does not provide a standard temperature sensor. The temperature one feels is dependent on the local environment. For

vice versa; the movement of the capsule being recorded to measure pressure changes. Aneroid barometers, however, must be calibrated against a mercury barometer to ensure absolute pressure values are recorded. The barometer found on the walls of many homes operates on the same principle; the capsule movement being displayed by a pointer moving over a graduated dial.

8.2.3 Wind measurement

Pressure differences in the Earth's atmosphere produce the winds (see Chapter 9), the direction of which is described in terms of compass bearings. A north-westerly wind means that the wind is blowing from the north-west or equivalently from 315° (measured clockwise from north at 0°). This is opposite to the definition in many other branches of science where the direction of any motion is considered to be the direction towards which the body is moving but is a natural adoption of the practices of sailors of measuring the direction from which the wind blew to fill the sails and the direction from which storm clouds approached the ships.

The strength of the wind is measured on the *Beaufort scale*, designed by Admiral Beaufort in 1805 (Table 8.1). On the scale global wind speeds are defined on a scale of 13 'Forces'. Force 0 is defined as *calm* where there is little or no wind, while Force 12 describes hurricanes where wind speeds are $>33\,\mathrm{m\,s^{-1}}$. Following the original definitions of Beaufort, each force might be described in terms of the observable effect of wind on the observer's surroundings. Thus, Force 5 described as a 'Fresh breeze' allows 'small trees in leaf to begin to sway and crested wavelets form on inland waters', while at sea ' moderate waves, taking a more pronounced long form are seen with many white horses being formed and the chance of some spray'.

Such observations are of course insufficient monitors of actual wind speed and many instruments have been developed to record local wind speed. The most common type of instrument is the *cup anemometer*, an example of which is shown in Figure 8.4. Three or more cups, each roughly conical in shape, are mounted symmetrically on arms placed at right angles to a vertical spindle placed at a standard height of 10 m above the surface. The cups are rotated by the wind, the rate of rotation being directly proportional to wind

Admiral Francis Beaufort (1774–1857). British admiral and hydrographer to the Royal Navy from 1829; the Beaufort scale and the Beaufort Sea in the Arctic Ocean are named after him. Born in County Meath, Ireland, in 1790 he enlisted in the Royal Navy. He conducted major surveying work, especially around the Turkish coast in 1812. As hydrographer to the Navy, he promoted voyages of discovery such as that of the botanist Joseph Hooker with the *Erebus*. Drawing up the wind scale named after him Beaufort specified the amount of sail a full-rigged ship should carry under various wind conditions. His recommendations were adopted by the Royal Navy in 1838.

Figure 8.4 A cup anemometer for the measurement of wind speed.

Table 8.1 The Beaufort scale.

Force	Mean wind speed (ms^{-1})	Description
0	0.0	calm
1	0.8	light air
2	2.4	light breeze
3	4.3	gentle breeze
4	6.7	moderate breeze
5	9.3	fresh breeze
6	12.3	strong breeze
7	15.5	near gale
8	18.9	gale
9	22.6	strong gale
10	26.4	storm
11	30.5	violent storm
12	>32.7	hurricane

speed. However, such instruments, even when properly calibrated in a wind tunnel, tend to overestimate the wind speed in gusty conditions.

8.2.4   Humidity measurement

Humidity is the measure of the amount of water vapour in the air. Humidity is usually recorded as relative humidity (see Section 3.6.2), which is the ratio of actual vapour density (or pressure) to the value which would produce saturation at that temperature.

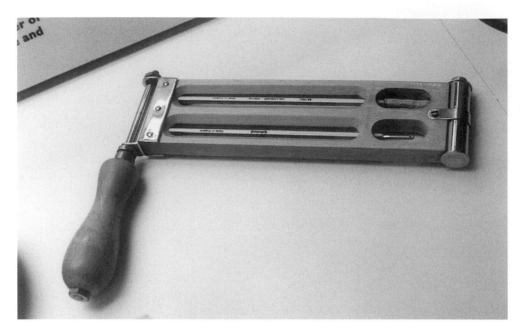

Figure 8.5 Wet and dry bulb thermometer used to measure humidity.

Animal tissue such as the hair and skin respond directly to humidity and are used in several measuring instruments known as *hygrometers*. In the common *hair hygrometer*, a hair is placed under tension such that the increase in its length with increasing humidity is recorded. Such a simple instrument provides surprisingly accurate and reproducible results but it needs to be calibrated against a standard if absolute measurements are to be made.

A more accurate measurement is made using the wet and dry bulb thermometer (also known as a psychrometer (Figure 8.5)). Two mercury thermometers are placed in a Stevenson screen. One records the air temperature directly, this is the dry bulb thermometer. The second thermometer has its bulb permanently wetted by a muslin wick. The wick is supplied with distilled water from a bottle. If the air surrounding the wet bulb is unsaturated, water will evaporate into it and cool the mercury by evaporative cooling. The rate of evaporation will be a function of the relative humidity of the surrounding air and hence the degree of cooling of the mercury will be a measure of the air's relative humidity. Thus, the wet bulb temperature is a measure of the air's water content and varies if water vapour is added (it will increase) or subtracted (it will decrease) from the air. Comparison of the temperature recorded with the wet bulb thermometer is then compared with the air temperature recorded by the dry bulb thermometer. The difference then allows the relative humidity to be determined (see Section 3.6.2).

8.2.5   Precipitation measurement

Much of the precipitation reaching the Earth's surface is in the form of rain, although the term 'precipitation' includes all forms in which water may reach the ground, i.e. snow and hail as well as rain.

Figure 8.6 A traditional rain gauge being set up at University College, London.

Measurement of the amount of rainfall is perhaps the easiest of all meteorological observations, precipitation being recorded using a *rain gauge*. The standard rain gauge is a metal cylinder of approximately 12 cm diameter, set in the ground with its upper rim one foot above the surface (Figure 8.6). Any precipitation is funnelled down into a collecting bottle, the observer then decants the water into a tapered measuring cylinder scaled in such a way that the reading represents the true depth of water (in millimetres) on the surface. However, even simple gauges need to be carefully sited if they are not to give spurious readings. More sophisticated gauges permit not only the amount of rainfall to be measured, but also its rate and duration. Such gauges are known as 'recording gauges'; including the tipping bucket type. Rainfall is also measured by precipitation radar. These use centimetre wavelengths to measure the power of the back-scattered radiation from the raindrops and, unlike rain gauges, provide spatial continuous instantaneous rates averaged over a square kilometre, each radar having a 'viewing' radius of about 120 km.

8.2.6 Sunshine

Measuring the amount of sunshine may seem unnecessary but the routine measurement of the recorded duration of bright sunshine is required for climatological purposes since annual records are required for agriculture and the tourist industry. Meteorologists have designed specific instruments to remotely and continuously record the sunshine in relation to the local time. Often positioned on a high building the *Campbell–Stokes sunshine*

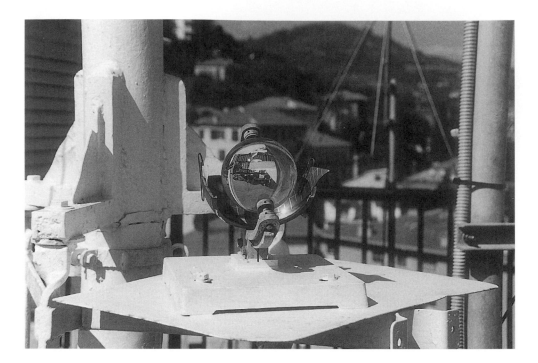

Figure 8.7 Campbell–Stokes sunshine recorder in Porto Maurizio, north-west Italy.

recorder (Figure 8.7) is a glass sphere which focuses the Sun's rays on to a specially manufactured card calibrated with hourly time markers and scorches a mark on it. The sphere is mounted concentrically within a portion of a spherical bowl. The sphere support is a semicircular brass bar attached symmetrically to the back of the bowl and concentric with it. The sphere is secured with two brass screws one fitted into a cup-shaped boss the other into a ball-ended boss, diametrically opposite on the sphere. Three overlapping grooves in the bowl then accept three different types of sunshine cards; the cards being secured with a clamping screw. The arc sphere is mounted in a grooved slide to allow correction for latitude and the slide is mounted on a T-shaped base, which is supported for levelling purposes on a fixed metal sub-base. The sub-base may then be securely fastened to the building.

Three types of sunshine card may be used; the type used depending upon the season (Figure 8.8). In the Northern Hemisphere the three cards used are as follows:

- Long curved cards during the summer, from 12 April to 2 September inclusive
- Short curved cards during the winter, from 15 October to the last day of February inclusive
- Straight cards are used around the equinoxes from 1 March to 1 April inclusive and again from 3 September to 14 October.

In the Southern Hemisphere, the short cards are used in the winter period between 12

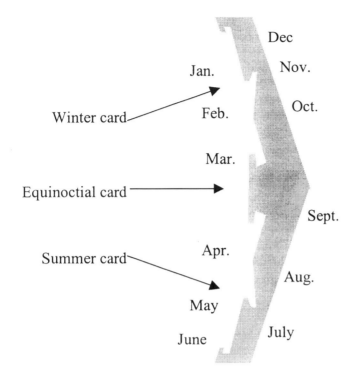

Figure 8.8 Sunshine cards mounted in the Campbell–Stokes sunshine meter.

April to 2 September and the long cards in the summer, from 15 October to the last day of February.

Each card is replaced each day ideally between sunset and sunrise, the cards being marked with the station name, year, month and date. Subsequent analysis of the burns on the card allows the daily sunshine record to be judged.

8.2.7   Visibility

Visibility is defined as the greatest distance at which an object can be seen and recognized in daylight. Clearly since it is not possible to achieve such conditions at night; lights may be used instead of objects with a relationship being established between the visual range of lights at night and the equivalent daylight meteorological visibility. For meteorological purposes it is necessary that visibility observations give a measure of the transparency of the atmosphere.

Clearly, the visibility recorded by any observer may be dependent upon the observer's own eyesight but in meteorological observations a good observer should attain an accuracy of reporting to within 10% of the actual visibility. Thus, objects chosen to gauge the local visibility should be chosen with care. The objects should be black or very dark coloured and stand well above the horizon. A tree at the edge of a wood, for example, would not be a suitable object nor would a white house particularly when the Sun is shining on it; but a group

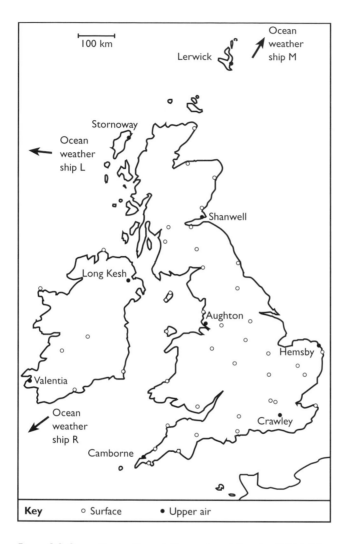

Figure 8.9 Synoptic weather station network for the British Isles.

09:00 GMT and are forwarded to central services monthly. The data are primarily used to provide long-term climatological data bases.

*   *Health resort stations* fulfil the same duties as climatological stations but observers make special reports at 18:00 hours clock time for inclusion in bulletins issued to the press.
*   *Agricultural meteorological stations* (agrometeorological) stations are maintained for research and operational decisions on field-work. They also record a daily observation at 09:00 GMT.

The majority of these stations provide, in addition, a monthly (or sometimes weekly) climatological summary which is collated by the central meteorological service and issued to the press. Such reports are often highlighted by the media, e.g. 'this month had the lowest/highest rainfall on record', or 'this summer had on average less sunshine'. It is these reports that we often remember most.

8.3.2   Upper atmosphere network

In addition to the ground-based network, the central meteorological service (in the UK it is the Meteorological Office in Bracknell) monitors weather conditions in the upper atmosphere. Upper-air stations record the temperature, pressure, humidity and wind in the upper air throughout the troposphere and lower stratosphere.

Observations of wind, temperature, pressure and humidity are recorded by *radiosondes* – on free flying balloons. The radiosondes are released at 00:00 and 12:00 hours and daily climb to between 20 and 30 km at 5 m s^{-1} whereupon they burst and the instrument package falls to Earth by parachute. During flight the radiosonde is tracked by radar, the position being recorded every minute from which the winds are calculated by the ground station. The temperature, pressure and humidity readings change the frequency of a radio-transmitter on the balloon and the signal is decoded by the ground station. At 06:00 and 18:00 hours additional balloons (*windsondes*) are launched to record only wind speed and wind direction.

Windsondes and the radiosonde data allow the weather systems to be mapped, but obviously they are restricted to those areas where governments have funds to support such a network. Although some sondes are launched from ships, most observations are made over the land. However, three-quarters of the Earth is water, so how is weather recorded over these regions?

The collection of meteorological data across the oceans was originally confined to instruments carried on ships. Thus, the accuracy of forecasting European weather was restricted to the availability of data on the weather fronts moving across the Atlantic. Special 'weather ships' were stationed in the Atlantic and simple instrumentation incorporated aboard many commercial vessels, but the former were very expensive to maintain and data from the latter were often conflictory and hence unreliable. Commercial airliners are also used to record upper atmosphere data and transmit them automatically to a satellite, which relays them to a major weather centre, such as at Bracknell.

The launch of satellites revolutionized the collection of meteorological data. For the first time, using satellite-borne instrumentation it was possible to cover the whole Earth's surface including areas where previously it had been impossible to establish a ground based network (e.g. the large desert areas and the polar regions). In addition, satellites can provide data on a local geographical scale on a time-dependent basis that no ground-based system can match. With the launch of the first meteorological satellites weather systems could be tracked across the Atlantic, and provided the ability to forecast the land fall and track of such storms across the Eurasian land mass. Today, much of the global weather monitoring and hence forecasting relies on satellites. The Meteosat weather satellite has already been discussed in Chapter 6 and provides the images so familiar to us when watching weather forecasts on television. Weather satellites traditionally measure air temperature, humidity and winds from the movement of clouds.

the Meteorological Office at Bracknell, UK (Figure 8.11)). All the data collected by all the weather stations over the world are put into the computer together with the equations describing an enormous amount of basic physics. The resulting computer model is used not only to provide the daily forecasts seen on television, but also by airlines to select routes of individual flights, by ships to chart their courses away from storm fronts and even to forecast how the global climate will evolve in tens or hundreds of years time (see Section 8.4.3).

L. F. Richardson was the first to design a simple numerical model to predict the weather, but modern modelling of the world's weather requires a mathematical and physical description of the way in which the Earth's atmosphere behaves. The equations in such models will be reviewed in Case study 8.2. Most of these equations are differential equations and describe the manner in which quantities like pressure and wind velocity change with time. If the rate of change of a quantity such as wind velocity and its value at a particular time are known (e.g. from observation) then its value at another time can be calculated. Repetition of this procedure many hundreds and thousands of times provides the model's predictive power. This allows the evolution of weather systems to be calculated and hence a forecast of the weather to be constructed.

8.4.3 Chaos in weather forecasting

It is not possible to predict the weather infinitely far ahead. Obviously the level of the prediction depends crucially on the initial observations since these 'initialize' the model with actual data. However, on longer time scales the data needed for the model would have to be much better and from a far wider geographic range. For instance, after a few days the weather patterns in the Southern Hemisphere affect those in the North. Nevertheless, even if one had a complete database of the world's weather at any one instant, one could still not forecast the weather over a particular place with 100% accuracy weeks later. Ultimately, the atmosphere is a *chaotic system* that is not totally predictable. This implies that very

Figure 8.11 The Meteorological Office at Bracknell, UK – one of the world's leading centres for weather forecasting and climate prediction.

small effects that cannot be accurately measured may have very large cumulative effects.

The American meteorologist and mathematician Edward Lorenz first mooted the idea that the *mathematics of chaos* could be used in forecasting in a seminal paper (1972) entitled 'Does the flap of a butterfly's wings in Brazil set off a tornado in Texas?' It was perceived that the atmosphere is a dynamical system which behaves chaotically and Lorenz's interests focused on the unpredictable nature of atmospheric convection currents. By reducing a multi-dimensional system to three differential equations he showed that their time-evolution, in three-dimensional (3D) space, generated a *strange attractor*. Within this context it is called the *Lorenz attractor* (Figure 8.12).

An attractor is the 3D mapping that results from the locus of a point as its position changes with respect to time. It involves dissipative situations, in which energy can be released, and forces, which may affect the system, being independent of time. The Lorenz attractor suggests that the predictability of atmospheric convection is extremely short term. The notion that such an attractor is 'strange' implies that it has fractal properties, which means that they have 'sensitive dependence on initial conditions'. Nevertheless, the idea of a strange attractor is currently being incorporated into weather forecasting, through such methods as ensemble forecasting in which finite-time integration of the Lorenz equations, derived from ensembles of initial data, are superimposed on the original Lorenz attractor.

Therefore, while there now may be confidence that reasonably accurate deterministic forecasts can be made up to six days ahead, forecasts for longer periods have to be increasingly probabalistic in nature. This statement should not however be read as meaning that models cannot be used to predict changes in *global climate*. The physics and mathematical laws are obviously rigid and from them we can predict how, for instance, the Earth's surface may warm under increased emissions of greenhouse gases (see Section 7.8). We shall return to the question of how computer models are addressing future global environmental risks in Chapter 12.

Lewis Fry Richardson (1881–1953). Richardson was a most unique British scientist. He was a Quaker who was concerned to know how conflicts start. He developed mathematical models to show how one country might wish to wage war with another, especially if it thought that it had the resources to win it. He also wrote treatises on Political Science. But he is most famous for his work on forecasting the weather. As a conscientious objector he volunteered as an ambulance driver during the First World War, and during lulls between battles he would develop simple climate models. Using only a slide rule, he solved the appropriate  mathematical and physics equations and produced a six hour forecast. It took him six months and was not very accurate. He was the first to apply non-linear partial differential equations, which he called primitive equations, in an attempt to predict the weather, and his ideas on computational weather forecasting were published in 1922 in the book 'Weather Prediction by Numerical Processes'. In a time when problem solving required an army of mathematicians, he estimated that he needed 64 000 of them to process the meteorological data. He was 40 years ahead of his time. During the 1960s his ideas were taken up by the Hungarian mathematician John von Neumann, who developed and applied the use of computers to predict weather. Increasingly sophisticated computerized mathematical models of the climate are now being developed to predict climate scenarios and monitor climate change; hence proving Richardson's hypothesis correct!

Figure 8.12 Lorenz attractor.

Case study 8.2   Predicting the weather: reducing the uncertainty.

Weather systems are notoriously hard to predict since the evolution and movement of such systems around the Earth are shrouded in uncertainty.

Science is a process of data collection, analysis and interpretation. An appreciation of mathematics and statistics is pivotal to this, and mathematics and physics, in particular, lie at the heart of environmental modelling. The evolution of the mathematical basis for weather forecasting provides a marvellous example of the development of computational modelling techniques. The Norwegian meteorologist Vilhelm Bjerknes (1862–1950) was one of the first to argue that the solution of the equations that describe the physical processes in the atmosphere provide a basis for forecasting. The British meteorologist L. F. Richardson was the first to use numerical modelling to forecast the weather, and this approach has developed over many decades into global circulation models (GCM), which require three dimensional models of the coupled atmosphere–oceanic interactions that describe the motion of momentum and energy through the solution of certain equations. This approach is underpinned by the notion that both mass and energy are conserved. Mathematics, therefore, has a central role in the computer modelling of the global climate. It requires the setting up of an algorithm that involves the parameters which have to be quantified. These include the geometry of the Earth–Sun system, the structure and the composition of the atmosphere and the role of the oceans in storing and transporting massive amounts of energy.

In predicting the weather, variables are required, and though beyond the scope of this book, in outline they are as follows. There are seven equations for the seven variables, which include the three components of wind velocity (u, v and w), density of the air (ρ), temperature (T), and pressure (P). There is also a mixing ratio, linked to the humidity of the atmosphere, i.e. which expresses the quantity of the water vapour present. The physical equations involve:

- *Conservation of energy*, as expressed through the First Law of Thermodynamics.
- *Ideal gas equation*, $P = \rho RT$ which is derived from the gas laws linking pressure (P), density (ρ) and temperature (T) of the gases in the atmosphere, as well as those governing change of states (e.g. latent heat).
- *Hydrostatic equation*, which incorporates the pressure at any point being given by the mass of the atmosphere above that point (see Appendix 5).
- *Conservation of momentum*, which through Newton's Laws of Motion can be applied to describe the motion of a parcel of air. This entails studying the horizontal acceleration of the parcel which is balanced by the horizontal pressure gradient and friction. Since the Earth is rotating this acceleration includes the so-called Coriolis acceleration (see Chapter 9).
- *Conservation of mass*, through the continuity equation.
- *Radiation laws*, which model the absorption and re-radiation of solar radiation. These include Wien's law and Stefan's law (see Sections 2.4.3 and 7.7.2).
- *Laws of energy transport* by conduction and convection.

In addition, equations are needed to describe the transport of water vapour, its condensation to form clouds and the development and fallout of precipitation (see Section 8.5).

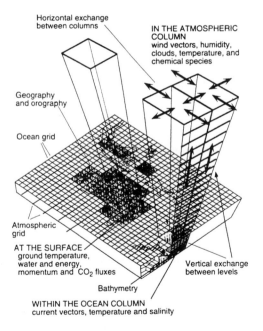

Horizontal exchange
between columns

IN THE ATMOSPHERIC
COLUMN
wind vectors, humidity,
clouds, temperature, and
chemical species

Geography
and orography

Ocean grid

Atmospheric
grid

AT THE SURFACE
ground temperature,
water and energy,
momentum and CO_2 fluxes

Vertical exchange
between levels

Bathymetry

WITHIN THE OCEAN COLUMN
current vectors, temperature and salinity

Figure 8.13 Global circulation model showing the breakdown of the atmosphere and the ocean into columns.

The wind equations, for example, have their own complicated beauty. Since winds are moving air masses and dynamical systems the principles of fluid mechanics can be applied. Some of the very great names in mathematics have contributed to the development of this branch of physics, including Euler, Lagrange, Poisson, Cauchy, Bernoulli and d'Alembert. The great Swiss mathematician Leonhard Euler developed differential expressions of Newton's Second Law ($F = ma$), and independently the French mathematician Claude Navier (1785–1836) and the Anglo-Irish mathematician and physicist George Stokes (1819–1903) developed the *Navier–Stokes equations*. These are a set of non-linear partial differential equations that can be applied to both the laminar and turbulent flow of air masses.

These equations are solved for time-step sequences to describe the prevailing atmospheric and oceanic scenarios for a given set of grid points defined by three-dimensional segments (Figure 8.13).

Non-linear equations are then set up that can then be solved numerically. There is thus a need for increasingly sophisticated computers capable of handling the massive datasets required for generating increasingly precise weather predictions. If one adds the coupling of oceans and atmosphere, and land and atmosphere with their accompanying fluxes, then it is possible, despite the added complexity, to apply ever-more advanced computers to make more accurate predictions.

8.5   Cloud physics

If we are to understand and forecast the world's weather, we must understand how weather systems (cyclones, anticyclones, fronts and depressions) are formed and transported around the world.

Clouds, by their presence or absence and by their type are the most obvious illustration of the local weather. They are also an important sink for the terrestrial radiation, reflect incoming solar radiation, adding to the Earth's albedo and, of course, provide the mechanism by which fresh water is transported from the seas across the continents. As an essential part of understanding weather systems and the Earth's climate it is necessary first to understand the physics which governs the formation of clouds.

8.5.1   Water: the unique molecule

Clouds are, of course, made of water. Of all the minor constituents of the atmosphere the most important to the existence of life is water. Life began in the seas and many animals and plants can still only live in water. The human body is composed of 70% water, while photosynthesis in plants can only function with water. Water's importance is due to the need of every living cell to transfer chemicals to and from itself. Water's ability to dissolve compounds thus makes cellular development possible.

Water is composed of two hydrogen atoms and one oxygen atom held together by covalent bonds. A covalent bond is an electron sharing bond. Each hydrogen atom shares its electron with the oxygen atom, and shares one of the oxygen's electrons, so it has two electrons to complete its outer shell. However, the electrons spend a longer time closer to the oxygen atom than the hydrogen atoms. This produces a slight positive charge on the

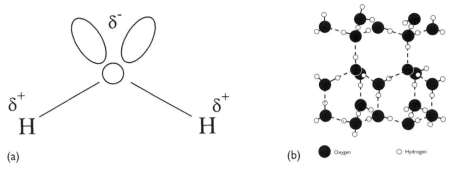

Figure 8.14 (a) Structure of the water molecule. (b) Structure of the ice lattice.

hydrogen end of the molecule and a slight negative charge on the oxygen end. Water is then said to be *polar* with a *dipole moment*.

The positively charged hydrogen ends of one molecule are then attracted to the negative ends of the oxygen atom in neighbouring water molecules, forming so-called *hydrogen bonds*. These hydrogen bonds give ice its structure and arrange the water molecules in a regular pattern or lattice (Figure 8.14).

As ice is heated the water molecules vibrate faster until at 0°C they move fast enough to break free from the lattice pattern, but the hydrogen bonds are strong enough to hold the molecules together such that, unlike CO_2, CH_4 and N_2O, water remains liquid at normal room temperature and is not a gas. Even when liquid water is heated to 100°C, the hydrogen bonds have an important binding effect and continue to hold molecules together in clusters of droplets (steam). Thus, without these attractive 'hydrogen bonds' water would be a 'normal' gas at normal temperatures, with the result that there would be no life on Earth.

None of the other planets in the solar system have liquid water on their surfaces; it is either too hot or too cold. Below 0°C water freezes; >100°C water boils and evaporates very quickly. Thus, if the Earth were a little closer to the Sun most of the water would have turned into vapour, a little further away, into ice. Thus, there is a 'habitable ring' around the Sun which can sustain life, as we know it. On Venus and Mars it is too hot, and on Uranus and Neptune it is too cold.

Water is at its most dense at 4°C (i.e. *above* freezing point) and so ice can float in water! As the temperature is lowered from 4°C water *expands*, at first gradually, but below 0°C rapidly, as it freezes into ice generating enough force to burst pipes and split rock!

Additional energy is absorbed in melting or evaporating a substance and emitted when freezing or condensing. This is called latent (hidden) heat (see Section 2.4.4) because it does not produce any change in the temperature of the body, only a change of state (e.g. water to ice, or water to steam). Thus, when water is boiling, its temperature remains steady at 100°C even though energy, called the *latent heat of vaporization*, is being supplied to it. Similarly, the temperature of water stays at 0°C while it is freezing into ice, with no fall in temperature until all the water has solidified, but energy, called the *latent heat of fusion*, is still being given out by the water. The latent heat of fusion for ice is $334 \, J \, g^{-1}$; the latent heat of vaporization of water is $2.3 \times 10^3 \, J \, g^{-1}$.

Latent heat plays an important role in controlling temperature of living organisms (see Section 2.4.4). You may have noticed that when you step out of your bath or the sea you feel cold? This is due to the water evaporating from the skin. Similarly, in hot climates cells can be easily damaged, so the human body temperature is regulated by sweat glands in the skin producing a watery liquid to lie on the skin. As the water evaporates so it takes heat energy (latent heat) from the skin and cools it down. Plants use a similar mechanism to cool their leaves. Water is allowed to evaporate through pores in the surface of the leaf, called stomata. As the water evaporates it takes heat energy from the leaf cells so cooling them down.

8.5.2 Hydrosphere

The total volume of water on the Earth has been estimated at 310 million miles³ (1384 million km³), 97% of which is in the oceans, such that if this was evenly distributed it would cover the Earth's surface to a depth of 2.8 km. The remaining 3% is at any one time in the atmosphere or on the land. Three-quarters of that 3% is retained in the polar ice caps and glaciers. Thus, the freshwater used in the soil, rivers and lakes is <1% of the total, and only an infinitesimal 0.035% is at any instant in the atmosphere. If all of the water vapour in the atmosphere was to be released at one instance, only 3 cm rainfall would be collected at the Earth's surface!

However, an average annual rainfall of 90–100 cm is found globally. How does one explain this discrepancy? There is, in fact, a constant exchange of water between the oceans and the atmosphere. This is known as the *hydrological cycle* (Figure 8.15).

Once again, it is the Sun that drives this vital mechanism. Most of the water vapour (84%) in the atmosphere comes from the oceans; transpiration from plant leaves accounts for most of the remaining 16%.

The Sun heats the water in the oceans and land surface leading to evaporation. The warm moist air rises, expands, cools and mixes with the surrounding cooler air so that the

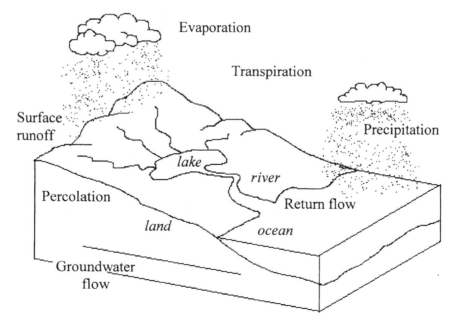

Figure 8.15 The hydrological cycle.

water vapour condenses to form clouds. The winds then carry the clouds across the Earth's surface until the water is released once more as precipitation (rain, hail or snow) to fall on the Earth's surface for further recycling. Most of the precipitation will fall back into the oceans (three-quarters of the Earth's surface being ocean) but that which falls on land replenishes our fresh water supply and sustains life.

The rate of circulation of water within the hydrologial cycle is very rapid. Since the total mass of water in the hydrosphere (i.e. water in all its forms close to the Earth's surface) is constant, precipitation must be balanced, on average, by evaporation. Comparing throughputs, it is found that the evaporated water remains in the atmosphere for only 10 days on average!

8.5.3   Types of clouds

The great variety in the form of clouds has necessitated a definitive classification for the purpose of global weather reporting (see Plate 11). The internationally adopted classification pattern is based on both the general shape, structure and vertical extent of the cloud and the height at which they form. Ten basic groups (also known as *genera*) are defined (Table 8.3); from the high *cirrus* clouds, composed of ice crystals, to the well-known dark, foreboding anvil-shaped *cumulonimbus*, the herald of thunderstorms.

8.6   Physics of cloud formation

Water evaporates by taking thermal energy (latent heat) from the surrounding air, which therefore cools. Solar energy evaporates water from the oceans into the atmosphere, at about 2.5 mm day^{-1}.

When the water evaporates, there is an increase in the air pressure due to the motion of

Table 8.3 Classification of clouds.

Cloud type (genus)	Height*	Composition	Description
Cirrus	high	ice	white bands or delicate filaments/patches with fibrous appearance
Cirrocumulus	high	ice	white patch regularly arranged in form of grains
Cirrostratus	high	ice	whitish veil covering large area of sky
Altocumulus	middle	water/ice	white/grey regular arrangement of small cloud elements
Altostratus	middle	water/ice	greyish or bluish layer fibrous appearance covering large area of sky
Nimbostratus	low	water/ice	grey thick layer often with snow and rain
Stratocumulus	low	water	grey or whitish layer with dark elements, regularly arranged
Stratus	low	water	grey layer with fairly uniform cloud base sometimes with snow and drizzle
Cumulus	usually low base extending several km	water	detached clouds, sharp outlines, 'cauliflower' clouds
Cumulonimbus	5–12 km	water/ice	anvil-shaped thundercloud

*Low is designated from the surface to 2 km, middle 2–7 km and high 7 km to the tropopause.

the evaporated water molecules. As more molecules evaporate, the vapour pressure increases steadily and forces some of the molecules to return to the liquid. Eventually, a dynamic equilibrium is established whereby the number of molecules returning to the water equals the number leaving the surface into the atmosphere. At this stage the air is said to be *saturated*.

The amount of water vapour that can be retained by a parcel of air (e.g. of unit volume) depends on both the pressure and the temperature of the air parcel and is defined by the *saturation vapour pressure* (3.6.1). It can be derived, thermodynamically, that the saturation vapour pressure, P_s, varies with the air temperature according to the Clausius–Clapeyron equation:

$$\mathrm{d}\ln P_s/\mathrm{d}T = l_v M_v/RT^2 \tag{8.3}$$

where l_v is the latent heat of vaporization, M_v the molecular weight of water and R the universal gas constant. Thus

$$P_s = \text{constant} \times \exp\left(-l_v M_v/RT\right) \tag{8.4}$$

which indicates a rapid increase of saturation vapour pressure with temperature (Figure 3.13) or, as the temperature cools (as in a parcel of air rising through the troposphere), there is a rapid reduction in the amount of water vapour the atmosphere can contain. Hence, as the warm air rises from the ocean surface into the troposphere it cools and water vapour condenses on to small particles suspended in the air once the vapour pressure approaches saturation.

These small particles are called *condensation* (mostly *hygroscopic*) *nuclei* and form part of what is known as the *atmospheric aerosol*. The aerosol includes dust blown up into the atmosphere from the Earth's land surface, ash from volcanoes, smoke particles, from both industrial and natural combustion processes, while a major component over the oceans arises from sea salt ejected into the air by bursting air bubbles in the foam. Such particles range from 0.001 μm to over 10 μm, though the larger particles do not remain airborne for long before falling to the surface. On average, there are 10^3 aerosol particles cm^{-3} over the ocean, $10^4\,\mathrm{cm}^{-3}$ over the countryside and $10^5\,\mathrm{cm}^{-3}$ in the larger polluted cities.

In the first stage of their growth water droplets grow quickly through the condensation of further water on to the surface of the droplet, but the rate of growth in the radius of the droplet decreases as the drop size increases since there is an increasingly larger surface area to add to for each increment in radius. The condensation rate is also limited by the rate at which the released latent heat can be lost from the drop by conduction to the surrounding air.

As the droplet grows larger a second process is introduced – *coalescence*. Coalescence becomes important when the water droplets are sufficiently large (radius >25 μm). They then collide with other droplets and combine to form an even larger droplet. This process depends upon the relative movement of the droplets and their size, and the probability of two droplets colliding, called the collision efficiency, increases with increasing size.

If a water droplet starts to fall in air it will accelerate under gravity but at the same time there will be opposing forces, due to friction from the surrounding air and the Archimedean upthrust, which tend to brake its fall. When the force acting downwards comes into equilibrium with the two forces acting upwards the velocity of the water droplet will be maximized and cannot be exceeded. This is known as the *terminal velocity*, which is summarized by *Stokes' law*, namely that the viscous drag on a falling droplet:

$$F_v = 6\pi\eta r v$$

where η is the coefficient of viscosity of the medium (in this case air), r is the radius of the rain drop and v is the terminal velocity. Larger droplets will have higher terminal velocities than smaller droplets, such that larger droplets will overtake and collide with smaller droplets and coalesce with them.

It is this combination of coalescence and condensation that forms the rain drops. Once the drops exceed a radius of $100\,\mu m$ they fall out of the shallow clouds as drizzle, but in deeper clouds they grow to between 1 and 3 mm. Raindrops as large as 6 mm diameter are observed in thunderstorms where strong updraughts in the atmosphere suspend the raindrops allowing them to grow larger. Above 6 mm the raindrops become unstable and break up during their fall.

8.7   Snow crystals

The remarkable beauty of snow crystals has long been recognized and recorded by the naturalist, the scientist and the artist. Composed of ice and formed by the condensation of water vapour around a minute nucleus, snow crystals appear in a wide variety of shapes and forms (Figure 8.16) that broadly fall into four main classes:

- Thin hexagonal plate
- Hexagonal prismatic column
- Long slender needle
- Six-pointed star.

Snowflakes are agglomerates of individual crystals in which the star-shaped crystals are generally prominent, but in which needles and plate forms may also appear. The largest snowflakes are several centimetres across and contain hundreds of individual crystals.

The outstanding common feature of the various forms of snow crystal is their hexagonal symmetry, a fact first recorded in ancient Chinese texts by Han Ying in 135 BC and by the poet Hsiao Thung (AD 501–31). The first European scientist to record such symmetry was Johannes Kepler in his 1611 essay *On the Six-Cornered Snowflake*, offered to his patron as a New Year's gift. 'Why six?', Kepler asked. He could give no satisfactory reason and it was not until three centuries later that the mechanisms for the formation of snowflakes and the explanation for their symmetry was solved.

Snow crystals grow from nuclei formed either by the freezing of supercooled water droplets or by the deposition of ice directly from the vapour on to a tiny fraction of the microcrystalline particles present in atmospheric aerosols. Droplets of very pure water, only a few microns (μm) in diameter, may be supercooled to $-40°C$, below which they freeze spontaneously. But at higher temperatures they freeze only if they contain foreign crystalline particles called ice nuclei. At temperatures above $-10°C$, natural clouds contain very few ice crystals because effective ice nuclei are only present in very low concentrations, sometimes $<10\,m^{-3}$. However, the number of ice nuclei increases by an order of magnitude for each 5°C fall in temperature to reach $1\,cm^{-3}$ at $-35°C$. Since these

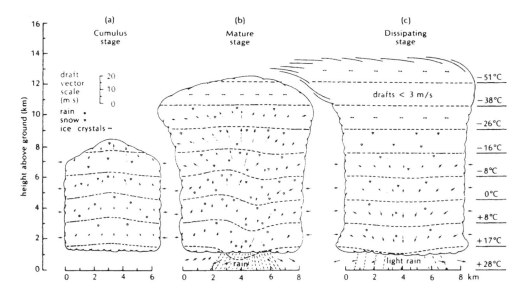

Figure 8.17 Mechanisms of a thunderstorm.

dercloud becomes charged. Thunderstorms (Figure 8.17) occur when moist, warm air near the ground becomes buoyant and rises producing many small cumulus clouds that at first form and dissipate without producing rain or thunder or lightning. However, under the special atmospheric conditions preceding a thunderstorm, as the day progresses, the clouds increase in size until several of them surge forward and combine to form the cumulonimbus anvil-shaped cloud. Such clouds may extend over 12 miles up into the stratosphere.

The upper parts of the thunderstorm clouds are positively charged, and the lower parts negatively charged, so that the storm cloud has an internal electric field that grows as the storm develops until the insulation of the air breaks down and a giant spark (lightning flash) passes and neutralizes some of the charge. This charge is then replenished to produce successive flashes.

How is this charge generated and separated into positive and negative regions? There is now general agreement that the charge is generated during the growth of small hailstones which become negatively charged and, falling under gravity, carry this negative charge towards the base of the cloud, while a compensating positive charge is attached to smaller particles (ice crystals and cloud droplets) that are carried to the top of the cloud in the strong updraughts.

The exact mechanism by which the hail pellets become negatively charged is still disputed, but the most convincing theory starts from the fact that a hail pellet falling in a vertical electrical field will have a fraction of its molecular dipoles orientated so that the bottom half of the hail pellet becomes positively charged and the top half negatively charged (Figure 8.18). Small ice crystals or cloud droplets will bounce off the lower half of the hail storm and carry away a positive charge, leaving the hail pellet with a net negative charge. As the hail pellets grow and make more frequent collisions with smaller pellets, they acquire increasing negative charge while falling towards the cloud base. The electric field increases and in turn increases the induced charges on the hail pellets. Hence, the whole process builds up rapidly to produce the first lightning flash. Calculations show that

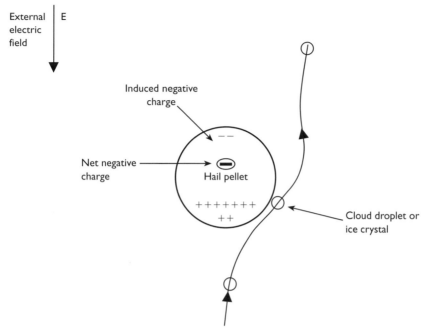

Figure 8.18 Mechanism for charging water droplets.

sufficient charge can be generated and separated within about 10 min and thereafter fast enough to produce lightning flashes at 20–30-second intervals.

Once charge separation has occurred an electric field (and hence potential difference) is established across the cloud and the air between may become ionized. As a result, a small region of the atmosphere is suddenly changed from being a good insulator to a good conductor. This discharge is lightning and is simply a high voltage spark. This discharge propagates through the cloud along a stepped and irregular path and is known as a *stepped leader*. After some 20 ms the tip of the leader approaches the ground (10–20 m) and the electric field under it is now so high that a new discharge is initiated from some pointed object (e.g. tall buildings or trees). This is called the *connecting discharge*. This goes to meet the tip of the stepped leader and continues back along the path already created by the downward leader stroke. Since a path already exists, the return stroke travels much more quickly into the cloud than the downward stepped leader. The return stroke carries currents of up to 10 000 amperes lasting for only about 100 μs and is the damaging component of the lightning flash.

The lightning may have finished with the return stroke but more often a multiple flash will occur with a new cloud-to-ground stroke being triggered, in turn sending a second return stroke back into the cloud.

Thunder is the sound wave produced by the lightning stroke. The sudden rise in pressure and temperature in the lightning channel produces an intense sound wave similar to that in an explosion. Since the sound waves travel at about 330 m s^{-1} while light travels a million times faster, the sound of the thunder arrives some time after the lightning flash causing it is seen. This time delay is often used as a method for determining the distance of the storm from you, counting 1 km for every 3 s.

Worked example 8.1   Lightning.

(a) If the potential difference between a cumulonimbus cloud and the ground is 10^5 V, what is the power developed at a peak current of $10\,000$ A during the return stroke in a cloud-to-ground lightning stroke?

(b) What is the average power generated if the lightning stroke lasts 0.2 s during which 20 Coulombs of charge is transferred?

Solution

(a) The power P developed $= VI = 10^9$ W or 1 thousand MW.

(b) Power $= VQ/t = 10^5.20/0.2 = 10^7$ W or 10 MW.

This is enough energy to keep a domestic air-conditioning unit running for several months.

8.9   Summary

The continual variability of the weather and the problems in forecasting it properly are of interest to everyone; indeed the daily weather forecasts regularly attract some of the largest viewing figures on national television. In this chapter, we have discussed how the weather is monitored and how clouds, the harbinger of rain, are formed. However, if one is to forecast tomorrow's weather accurately, one must understand how weather systems develop and are circulated around the Earth. This is the topic of Chapter 9.

References

Observing the weather

Coulson, K. L., *Solar and Terrestrial Radiation, Methods and Measurements*. New York: Academic Press, 1975.

Fritchen, L. J. and Gay, L. W., *Environmental Instrumentation*. New York: Springer Verlag, 1979.

Linacre, E. and Geerts, B., *Climates and Weather Explained*. London: Routledge, 1997.

Meteorological Office. *The Observers Handbook*. London: HMSO, 1982.

Cloud physics

Barry, R. G. and Chorley, R. J., *Atmosphere, Weather and Climate*, 7th edn. London: Methuen, 1998.

Houghton, J. T., *The Physics of Atmospheres*, 2nd edn. Cambridge: Cambridge University Press, 1995.

Iribane, J. V. and Cho, H. R., *Atmospheric Physics*. Dordrecht: D Reidel Publishing Co, 1990.

Mason, B. J., *Clouds, Rain and Rainmaking*, 2nd edn. Cambridge: Cambridge University Press, 1975.

Mason, B. J., *The Physics of Clouds*. 2nd edn. Oxford: Clarendon Press, 1971.

McIlveen, R., *Fundamentals of Weather and Climate*. London: Chapman and Hall, 1992.

Rogers, R. R. and Yau, M. K., *A Short Course in Cloud Physics*. Pergamon, Oxford 1989.

Salby, M. L., *Fundamentals of Atmospheric Physics*. San Diego: Academic Press, 1997.

Scorer, R. S., *Dynamics of Meteorology and Climate*. Chichester: Wiley, 1997.

Atmospheric electricity

Chalmers, J. A., *Atmospheric Electricity*, 2nd edn. Oxford: Pergamon Press, 1967.

MacGorman, D. R. and Rust, W. D., *The Electrical Nature of Storms*. Oxford: Oxford University Press, 1998.

Mason B. J., 'The generation of electric charges and fields in thunderstorms', *Proceedings of Royal Society* A415 303–15, 1988.

Uman, M. A., *Lightning*. New York: Dover, 1983.

Web sites

To observe the weather

http://weather.com/breaking_weather/encyclopedia/tropical/history.htm
http://www.yatcom.com/neworl/weather/whatis.html

Precipitation

From the 'Glossary', Internet Information Resource:
http://hyperion.advanced.org/17865/glossary/rain.html
http://hyperion.advanced.org/17865/glossary/snow.html

Discussion questions

1   What is the basic difference between (a) cirrus clouds and (b) cumulus and cumulonimbus and other clouds?

2   If air may contain many thousands of condensation nuclei m^{-3}, do clouds never contain more than a few hundred water droplets in the same volume?

3   In which case will rising air reach higher supersaturations – in a cumulus cloud, where the air is rising rapidly or in slowly developing layer clouds where the air rises at low velocities?

4   How does the terminal velocity of a drop depend upon its radius (a) in the range 10–30 μm and (b) when it is >3 mm?

5   Sodium iodide, a hygroscopic particle, has been suggested as a candidate for cloud seeding to produce more rain. On what idea is this based and what criteria would one apply to select an appropriate particle size?

6   A spherical water droplet moving in viscous flow through the air experiences a drag force given by Stokes' law $F_{drag} = 6\pi\eta r v$, where η is the coefficient of viscosity of the fluid, r the radius of the droplet and v the terminal velocity. Derive an expression $v = v(r)$ for the fall in the air of the droplets not exceeding about 30 μm, above which the air flow is no longer viscous.

7   Assume that an isolated thundercloud passes over an observation station. What changes will be observed in the vertical component of the electric field at the ground?

8   Why is the return stroke much faster than the stepped leader? Why does a connecting discharge start from a point on the ground only when the stepped leader arrives close to the ground?

9   A company states it will provide a weather centre with a low-price Stevenson screen. When shown the offered goods it is found that the screen has no floor and is made of transparent perspex but is otherwise identical to the traditional wooden Stevenson screen. Should the weather station buy a consignment of these screens and save money?

Quantitative questions

1   The horizontal acceleration on an air parcel of mass 1 tonne is $10^{-4}\,\mathrm{m\,s^{-2}}$:
    (a)   what is the net force on the air parcel?
    (b)   estimate the volume of the air parcel at sea level.

2   Given that a cumulus cloud is typically 2 km deep with a similar diameter and contains 5×10^7 water droplets $\mathrm{m^{-3}}$ each of 10 µm radius, calculate the depth of rainfall should the cloud release all its water in one instant.

3   Discuss (a) the hydrological cycle, and (b) the mechanisms of cloud formation. A spherical water droplet, radius r ($r < 30\mu\mathrm{m}$), moving in viscous flow with velocity v through the air experiences a drag force given by Stokes's law:

$$F_{\mathrm{drag}} = 6\pi\eta v r$$

where η is the viscosity of the fluid. Derive an expression for the terminal velocity of the droplet as a function of droplet radius. Sketch your results.
How does the terminal velocity of a droplet depend upon the radius for $r > 30\,\mu\mathrm{m}$?
Describe the possible charge processes that produce charge separation in a cumulonimbus cloud. Why does lightning not occur in shallow layer clouds?
    (University College London, London University: Environmental Physics: May 1999)

4   If the total pressure (p) of moist air changes, how does its partial pressure (e) of water vapour change, assuming that there is no mixing with the surrounding environment?
    Given that the Clausius-Clapeyron equation shows that $de/dT = L_v.e/RT^2$ where T is the air temperature and L_v is the latent heat of vaporization, show that the heat lost by the air (δQ) in forming a fog is given by:

$$\delta Q = (C_p + L_v.e/RT^2.p).dT$$

where C_p is the specific heat capacity of air at constant pressure.
    (University College London, London University: Environmental Physics: May 1999)

5   Two airports, 2500 km apart are served by an airline whose jets fly at an average $250\,\mathrm{m\,s^{-1}}$. At the normal cruising altitude for airline jets a jet stream flows from A to B at $100\,\mathrm{m\,s^{-1}}$. Calculate the journey times for the jets in each direction.

6   Calculate the space charge, assuming that there are 1000 positive ions and 900 negative ions $\mathrm{cm^{-3}}$ in the air.

7   0.50 kg of water vapour condenses to make a cloud the size of an average room.
    (a)   Assuming that the latent heat of condensation is 2500 $\mathrm{kJkg^{-1}}$, how much energy would be released when this condensation takes place?
    (b)   When the total mass of air before condensation is 100 kg, how much warmer would the air be after condensation has taken place? Assume that the air is not undergoing any pressure changes. Assume specific heat capacity of air is 1000 $\mathrm{J\,kg^{-1}\,K^{-1}}$.
    (Oxford and Cambridge Examination Board: 'A' Level Physics: Physics in the Environment: June 1996)

Chapter 9

Global weather patterns and climate

9.1   Introduction: atmospheric motion

Understanding the prevailing weather conditions in any specific geographical region requires a knowledge of the global climate. To understand the Earth's climate, it is necessary to investigate the atmospheric dynamics that lead to the formation of global circulation patterns, global wind patterns, and the formation and transport of weather systems around the Earth. This chapter describes the dynamics of air motion in the Earth's atmosphere and how it produces characteristic weather patterns and climatic regimes around the world.

The atmosphere acts rather like a gigantic heat engine (see Section 2.2.2) in which the temperature difference between the polar and equatorial regions provides the energy supply necessary to drive atmospheric circulation. The conversion of heat energy into kinetic energy leads to motions within the atmosphere which occurs as the winds. Winds are literally millions of tonnes of air in motion – a huge flow of mass that transports warm and cold air, and dry and moist air about the Earth's surface and throughout the depth of the atmosphere.

The wind is a basic aspect of life outdoors. In the wintertime it tends to blow more strongly than in the summer, and can be harnessed as a source of power when it blows strongly with a high velocity (see Chapter 5). It provides the motive force for those who sail and, when very strong, can be a major cause of damage in many parts of the world. To appreciate what makes the air move in the first place is, therefore, a crucial ingredient in the understanding of the weather and climate. First, it is useful to discuss the concept of air masses.

9.1.1   Air masses and weather fronts

An air mass is a large volume of air, extensive enough to cover an area of several million square kilometres across which the temperature and humidity remain reasonably constant. The prevailing air mass therefore has an overriding influence on the type of weather to be expected over a broad area, although the actual weather at a local region may be modified by topographical features such as hills, towns and lakes. Air masses are classified mainly according to two factors: their source region and the nature of the surface over which they travel.

Air masses originate within the extensive regions of high pressure that characterize certain areas of the Earth's surface. Highs (or anticyclones) reside typically over the subtropical oceans throughout the year and over mid- and high-latitude continents principally in the winter. The highs are the source of air masses and at the surface air spirals out of these as characteristic wind systems. For example:

- From the subtropical highs, e.g. the Azores anticyclone (so-called as it lies over the Azores islands in the Atlantic), the south westerlies blow towards the North Pole and the North-East Trades towards the Equator. This warm and humid air is strongly influenced by the underlying surface, so it is classified in mid-latitudes as 'tropical maritime' (mT).
- In winter continental highs (e.g. over North America and Eurasia) are source regions for cold and dry air and are termed 'polar continental' (cP).
- Air that spirals out from the anticyclones can travel long distances across oceans and, hence, be modified before it reaches new land masses. For example, the 'polar maritime' air mass reaches Britain having started as cP air over North America and Greenland but is warmed and moistened in its 1000 km journey over the Atlantic Ocean.
- The Arctic Basin is frozen in winter and acts as if it was a continental land mass, but in summer it is similar to a shallow cool sea. In the winter it provides a very cold, dry Arctic continental (cA) air mass, while in the summer it is the source of moisture, and cool maritime (mA) air.
- Tropical continental air occasionally influences mid-latitude regions as, for example, outbreaks of hot, dry air from the Sahara Desert.

The British Isles and other parts of Western Europe are influenced by day-to-day changes in air mass types since over these regions air masses arrive from different directions depending on weather patterns. The elongated boundaries that separate air masses from different sources and therefore different properties are known as *fronts* and divide air masses with regions of contrasting weather. In mid-latitudes, for example, a cold front is the leading edge of cool, shower-laden polar maritime air. Ahead of the cold front, layer cloud, light widespread precipitation and mild conditions typify the tropical maritime air.

9.2 Principal forces acting on a parcel of air in the atmosphere

To understand fully the wind patterns across the Earth's surface, it is necessary to understand the nature of the four forces experienced by a parcel of air in the atmosphere. They are:

- Gravitational force
- Pressure gradient force
- Coriolis force
- Frictional force

9.2.1 Gravitational force

The gravitational force is the force that retains the atmosphere around the Earth and is directed towards the centre of the Earth. For an air parcel of volume, dV, and density, ρ, the gravitational force F_g is:

$$F_g = dV.\rho g \qquad (9.1)$$

where g is the acceleration due to gravity. Since the atmosphere is thin compared with the radius of the Earth, g may be assumed to be constant ($9.8\,\mathrm{m\,s^{-2}}$).

9.2.2   Pressure gradient force

Consider a slab of air with cross-sectional area, dA, and length, dx, (Figure 9.1). The pressure at one end of the column is p and at the other end $p + \mathrm{d}p$. Since pressure is defined as the force per unit area the total force exerted in the x-direction (F_x) by the air column is then

$$F_x = p\mathrm{d}A - (p + \mathrm{d}p).\mathrm{d}A = -\mathrm{d}p.\mathrm{d}A. \tag{9.2}$$

If ρ is the density of air then the mass of the air parcel is given by $\rho.\mathrm{d}A.\mathrm{d}x$ and the force per unit mass is

$$F_x/(\rho.\mathrm{d}A.\mathrm{d}x) = -(1/\rho)\mathrm{d}p/\mathrm{d}x. \tag{9.3}$$

The minus sign indicates that the direction of the force points from high towards low pressure.

   The pressure gradient force has both vertical and horizontal components but, as was seen in Chapter 7, the vertical component is more or less in balance with the gravitational force. Thus, equation (9.3) can be generalized such that the pressure gradient force per unit mass, F_p, can be expressed as a function of the horizontal pressure gradient (dp/dn). When dn is the isobaric separation

$$F_p = \frac{-1}{\rho}.\frac{\mathrm{d}p}{\mathrm{d}n} \tag{9.4}$$

Hence, if a pressure gradient exists across a parcel of air the resultant force on it will cause it to move towards the region of lower pressure. The pressure gradient force is therefore the primary one in the movement of air in the Earth's atmosphere and in the formation of winds.

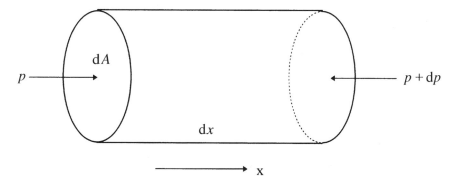

Figure 9.1 The pressure gradient force.

9.2.3   Coriolis force

The Earth rotates on its own axis once every 24 hours. A coordinate system fixed to the Earth's surface therefore rotates with the Earth. This means that a force must therefore be introduced to represent the effect of this rotational motion. This force is called the *Coriolis force*.

The simplest way to understand this deflecting force is to picture a rotating disc with angular frequency ω on which moving particles are deflected (Figure 9.2). A particle starts to move horizontally away from the centre O towards a point A. If no forces act on the particle, by Newton's First law it will follow a direct undeflected path to A. However, as the disc rotates the projection of the particle on to the disc surface below it will follow the curved path OA′

Gaspard Gustave Coriolis (1792–1843). Born and educated in Paris. He graduated in highway engineering and was Professor of Mechanics at the Ecole Centrale des Arts et Manufactures 1829–36, and at the Ecoles des Ponts et Chaussees. In 1838 he became Director of Studies at the Ecole Polytechnique.

From 1829, he was concerned that proper terms and definitions should be introduced into mechanics. He succeeded in establishing the use of the word 'work' as a technical term in mechanics, defining it as the displacement of force through a certain distance.

Investigating the movements of moving parts in machines and other systems relative to the fixed parts, Coriolis explained how the rotation of the Earth causes objects moving freely over the surface to follow a curved path relative to the surface – the Coriolis effect.

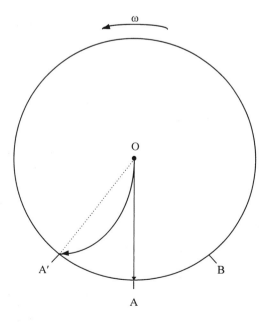

Figure 9.2 The Coriolis force.

rather than the straight line OA. In fact, the particle has travelled in a straight line but by the time it reaches the edge of the disc, A′ has rotated to A and A has rotated to B. To an observer rotating with the disc it will then appear that a deflecting force is acting on the particle deflecting the particle from A towards A′. This fictitious force is known as the Coriolis force.

In the analogous case of the rotating Earth (with rotating reference coordinates of latitude and longitude), there is an apparent deflection of moving objects to the right of their line of motion in the Northern Hemisphere and to the left in the Southern Hemisphere, as viewed by observers on the Earth's surface.

Imagine you wish to send a letter from London to Accra in Ghana using a small rocket. Ghana is on the same longitude as London and is 5100 km away. If the rocket flies at 308 m s^{-1} – or some 1110 km h^{-1}, it should take just over 4.5 h to get to Accra. So, you launch the rocket due south at 12:00 hours and telephone Accra to look out for it at 16:30 hours. After an hour you decide to check the *Global Positioning System* (GPS) to make sure that the rocket is travelling safely on its way. To your surprise you find that instead of flying over north-east Spain the rocket is over the Atlantic to the north-west of Spain. It has been flying to the right of its 'proper' track because the Earth has rotated from west to east through 15° of longitude during the hour in question. After 4.5 h as this movement to the right persists, your rocket and letter lands not in Accra but falls into the North Atlantic. It would seem to you that throughout its flight a force was acting on the rocket forcing it out into the Atlantic; this is the Coriolis force, F_c. The force acts in the same way on any parcel of air moving across the Earth's surface and influences large-scale ocean circulation too.

The deflective force F_c (per unit mass) is expressed by:

$$F_c = 2\omega V \sin \phi \qquad (9.5)$$

where ω is the angular velocity of the Earth (i.e. $2\pi/24$ radians h^{-1} = 7.29×10^{-5} radians s^{-1}); ϕ is the latitude and V the speed of the rocket. The effect is, therefore, a maximum at the Poles where $\phi = 90°$ and zero at the Equator where $\phi = 0°$. $2\omega \sin \phi$ is often written as f_c and termed the *Coriolis parameter*. Values of f as a function of latitude are given in Table 9.1.

9.2.4   Frictional forces

Irregularities in the topography of the Earth's surface naturally influence the flow of the wind. For example, lines of tall trees may be planted to act as 'wind breaks' and decrease the damage of wind to crops. Wind may be channelled down streets or through mountain passes. These effects may be described by introducing a frictional force between the atmosphere and the Earth's surface. The nature and magnitude of this force is difficult to describe and estimate but near the Earth's surface friction reduces the wind speed. The layer in which frictional forces play a role is known as the *planetary boundary layer*. Its depth may vary from a few hundred metres in still air at night to 4–5 km in altitude over hot surfaces during strong convections. Quantifying the magnitude of the frictional force is a rather difficult subject since it is strongly dependent on the local geography.

Table 9.1 Coriolis parameter f_c as a function of latitude.

Latitude	0°	10°	20°	45°	90°
$f_c(10^{-4}\,s^{-1})$	0	0.25	0.50	1.00	1.46

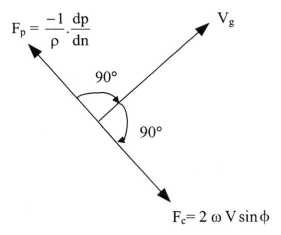

Figure 9.3 Geostrophic balance.

9.3   Pressure gradients and winds

Having reviewed the four forces acting on a parcel of air in the Earth's atmosphere consider the motion of such a parcel of air at an altitude sufficiently large so as to be able to ignore frictional effects. Since the gravitational force acts normal to the horizontal plane, if a steady state is to be obtained the pressure gradient force must be balanced by the Coriolis force (Figure 9.3), a condition known as *geostrophic balance*. Hence:

$$2\omega V \sin \phi = -(1/\rho)dp/dn \tag{9.6}$$

The speed of the wind is then given by

$$V_g = -(1/(2\rho\omega \sin \phi)) \times dp/dn \tag{9.7}$$

where V_g is known as the velocity of the *geostrophic wind*. Except in very low latitudes, where the Coriolis force approaches zero, the geostrophic wind speed is a good approximation to the measured wind velocity.

9.3.1   Cyclonic motion

In geostrophic balance the Coriolis force equals the pressure gradient force. The direction of the geostrophic wind may be found by rotating the pressure gradient force by 90° clockwise in the Northern Hemisphere (90° anticlockwise in the Southern Hemisphere) (Figure 9.3). Hence, the geostrophic wind blows parallel to the isobars (lines of equal pressure). The wind is in the direction with the low pressure to the left hand side of the wind vector in the Northern Hemisphere and to the right in the Southern (Figure 9.4).

This means that around a low-pressure area in the Northern Hemisphere (Figure 9.5a) the wind circulates in an anticlockwise direction (clockwise in the Southern Hemisphere). Such motion is known as *cyclonic*. A low pressure weather system is, therefore, known as a cyclone. Around high-pressure areas in the Northern Hemisphere (Figure

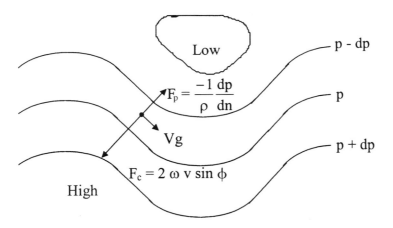

Figure 9.4 Geostrophic wind.

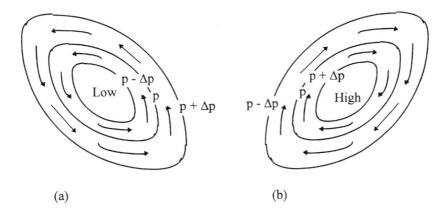

(a)   (b)

Figure 9.5 (a) A cyclone and (b) an anticylone in the Northern Hemisphere.

9.5b) the geostrophic wind circulates in a clockwise direction (anticlockwise in the Southern Hemisphere) and this type of motion is known as *anticyclonic motion*.

Cyclonic systems are found in low, middle and high latitudes. Virtually all cyclones (depressions) are associated with 'bad weather' with cloudy skies and often widespread rainfall. Low-latitude cyclones can develop into hurricanes or typhoons, systems of extreme intensity with surface wind speeds as high as $100\,\mathrm{m\,s^{-1}}$ (see Plate 12).

Anticyclones are generally associated with settled weather but are, none the less, active weather systems. They are characterized by mainly dry conditions but can have extensive low-level layer cloud 'trapped' within their circulation. The dullest conditions in the British Isles are often when highs dominate the weather with their 'anticyclonic gloom'. Since winds are generally light, under these conditions, pollution in the form of smoke or smog can build up, with subsequent detrimental effects to human and plant life.

Such high-pressure systems are often short-lived, typically 4–5 days, but occasionally they are of longer duration and cover an area of >1500 km across. Such large systems are known as

Worked example 9.1   Calculating the balance wind velocity.

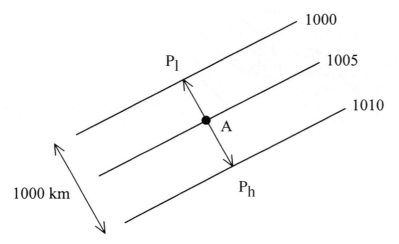

Assuming a geostrophic balance the speed of the geostrophic wind (V_g) can be calculated at the point A. Using equation (9.6):

$$V_g = -\frac{1}{\rho f} \cdot \frac{dp}{dn}$$

where $f = 2\,\omega \sin \phi$.

Assuming that the air density $\rho = 1.2\,\text{kg m}^{-3}$ and the Coriolis parameter $f = 2 \times 7.2 \times 10^{-5} \times 0.7071 = 1.02 \times 10^{-4}\,\text{s}^{-1}$ (at 45° latitude). The higher pressure $p_h = 1010\,\text{mb}$ and the lower pressure $p_l = 1000\,\text{mb}$; and the isobaric spacing is 1000 km. Determine the geostrophic wind velocity.

Solution
The pressure gradient,

$$\frac{dp}{dn} = \frac{(1010 - 1000)\,\text{mb}}{1000\,\text{km}} = \frac{10\,\text{hPa}}{10^6} = \frac{10 \times 10^2}{10^6} = 10^{-3}\,\text{kg m}^{-2}\,\text{s}^{-2}$$

where $1.01 \times 10^5\,\text{Pa} = 1010\,\text{mbar}$.

Therefore, the geostrophic wind velocity,

$$Vg = \frac{1}{1.2 \times 1.02 \times 10^{-4}} \times 10^{-3}\,\text{kg m}^{-2}\,\text{s}^{-2} = 8.17\,\text{m s}^{-1}$$

The formula for the magnitude of the speed of the geostrophic wind shows that it is directly proportional to the horizontal pressure gradient and inversely related to the air density, the rotation rate of the planet and the latitude.

blocking highs, since they block the normal west to east progression of cyclones. Such blocking highs also lead to periods of little or no rain and, hence, to local droughts. One such anticyclone led to the long period of hot and dry weather across Britain and western France in 1976.

9.3.2 Depressions and fronts

Depressions are low-pressure regions usually generated at the Polar front. Mid-latitude depressions generally travel eastward and poleward, except when steered around blocking highs. A depression forms at the boundary between two air masses, one of cold and one of warm air (Figure 9.6a). At first the winds flow parallel to the isobars and continue to do so until a wave develops (Figure 9.6b). Air then starts to flow across the isobars and a distinct low-pressure area forms with cyclonic motion. The whole depression moves in accordance with the winds in the warm sector during its early life (Figure 9.6c). Since the cold air

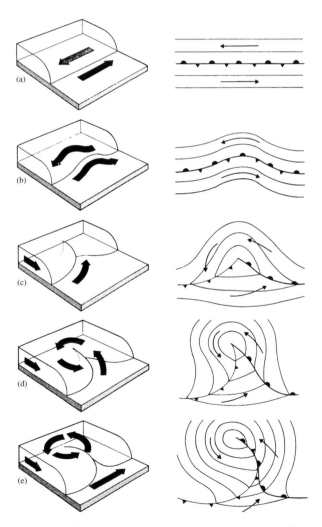

Figure 9.6 The formation and development of a depression.

tends to move faster than the warm, the cold front begins to catch up with the warm front as the warm air rises above the cold (Figure 9.6d). This process, in which the warm air is raised away from the ground, leads to an *occlusion*. An occluded front, therefore, has two cold air masses in contact at the surface and warm air aloft (Figure 9.6e). The cold continues to mix with the warm until they reach equilibrium whereupon the weather system has dissipated. Typically, such a process takes five days but a whole family of depressions may develop each forming on the trailing cold front of its predecessor. Mid-latitude areas like western Europe suffer prolonged wet and stormy weather under these conditions.

9.4   Thermal gradients and winds

Winds occur if there are pressure differences between one region of the atmosphere and neighbouring areas. Such pressure differences may be induced by uneven heating or cooling within the atmosphere. Consider a region that has an even temperature distribution (Figure 9.7a). Suppose that thermal energy is added to one end of the region and is extracted at the other. The air will be heated at one end of the region and cooled at the other. This will increase the internal energy of the air at the former and reduce it at the other. This internal energy is proportional to the region's total potential energy. When an air column is heated, so increasing its internal energy, the air column must expand vertically so as to raise its potential energy proportionally. Similarly, as an air column is cooled it will contract in the vertical. Thus, a horizontal pressure gradient is produced between the warm and cold regions (Figure 9.7b).

As soon as such a gradient is formed the air will begin to flow from the warm area to the cold. This air flow decreases the pressure in the warm area and increases that in the cold (shown by the arrows), because the pressure at any point is equal to the weight of air above it (see Chapter 7). The rate of decrease of pressure at a fixed height in the warm area is equal to that mass of air flowing out above it. Therefore, the rate of pressure decrease due to the outflow of air in the warm region should be larger in the lower levels than the upper levels. Similarly, the rate of increase of pressure in the cold region will be larger in the lower levels than in the higher levels.

Eventually equilibrium is restored and a steady state is reached. In the lower regions the pressure of the air column will be lower in the warm area and higher in the cold areas, whereas in the upper regions of the column the reverse is true; the pressure is higher in the warm areas and lower in the cold areas (Figure 9.7c). The direction of air flow in the lower levels will, therefore, be reversed from that in the higher.

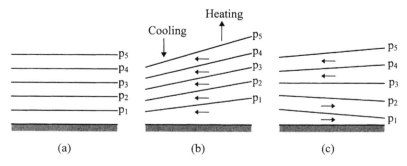

Figure 9.7 Schematic diagram of the thermal circulation process.

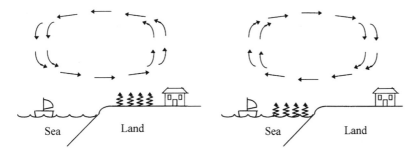

Figure 9.8 Sea and land breeze.

This is one basic way in which air convection is organized on a large scale and is related to the thermal circulation patterns in the atmosphere. It is one which is experienced at the seaside. Consider Figure 9.8. During the day the land surface is heated by the Sun and the temperature rises above that of the sea (the specific heat capacity of the land being lower than that of water). The air over the land is then warmer than that over the sea which results in the establishment of a thermal circulation pattern where the low-level air blows from the sea towards the land. This surface wind is called the 'sea breeze'. In contrast, at night, the land surface cools below that of the sea, and the thermal circulation cell is now in the reverse direction with the wind blowing from the land towards the sea as the 'land breeze'.

9.5 Global convection

The first model to describe larger-scale global convection was proposed by George Hadley in 1735. He noted that due to the uneven distribution of the solar flux reaching the Earth's surface, air in lower (equatorial) latitudes is warmer than in higher (polar) latitudes. Tropical air, he thought, should therefore rise vertically from the equatorial regions and, in the Northern Hemisphere, move northward at higher altitudes. This would be linked to a low-pressure region at the Equator due to the outflow of air mass while the colder, polar air should move southward at low levels (Figure 9.9). As the tropical air moves north, it will lose energy by radiation before descending to replace the southward moving cold air. This would produce a high pressure region. Similarly, the cold air will gain thermal energy from the surface on its

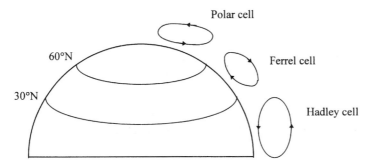

Figure 9.9 Tricellular model of atmospheric circulation.

journey to the equatorial regions (itself radiatively heated) and so rise. Thus, a thermal circulation cell will be established in the Earth's atmosphere capable of transporting thermal energy from the equator to the polar regions. This is called the *'Hadley Cell'*.

There are, however, significant differences between Hadley's model and the observed global air circulation patterns. Over the equator there is a low pressure belt (*the equatorial low*), while at the Poles there are high-pressure regions just as the Hadley model predicts. However, at about 30°N and 30°S there are two *subtropical high pressure belts* (Figure 9.10). A circulation pattern exists that 'connects' the subtropical high-pressure and the equatorial low-pressure belts, which results in the wind patterns known as *'Trade winds'* (so-called because of their importance to the conduct of trade during the age of sailing ships). The Trade winds in the Northern Hemisphere are north-easterly (i.e. they blow from the north-east) while in the Southern Hemisphere they are south-easterly.

Another circulating cell of air exists between about 30°N and 60°N in which air rises in the colder regions around 60°N and descends in the warmer regions around 30°N, the motion being in the *opposite* direction to the motion in the Hadley cell. This cell is known as the *Ferrel cell* named after William Ferrel, a nineteenth-century American meteorologist. A similar cell exists in the Southern Hemisphere between about 30°S and 60°S, as does a matching Hadley cell between about 0° and 30°S.

George Hadley (1685–1768). An English physicist and meteorologist who was the first to formulate an accurate theory describing the Trade Winds and the circulation patterns in the Earth's atmosphere.

Educated as a lawyer Hadley preferred physics to legal work. For seven years he was responsible for meteorological observations for the Royal Society of London. Having made the first really accurate study of the trade-wind currents, he explained their relation to the Earth's daily rotation and discussed the relevant atmospheric motions and their causes due to convection. He presented his ideas in a paper to the Royal Society in 1735 entitled 'Concerning the Cause of the General Trade Winds'. His arguments were not however appreciated until recognized by the famed British scientist John Dalton in 1793.

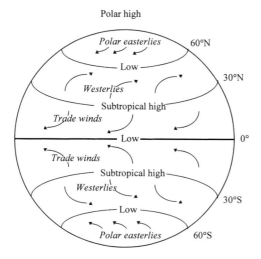

Figure 9.10 Global circulation patterns.

A third circulation cell occurs at high latitudes which, like the Hadley cell, is due to thermal convection but which is much weaker than the Hadley cell. Air rises above 60°N and 60°S and descends in the cold regions around the Pole. These are the *Polar cells*.

Seasonal variations in these cells are observed and have important consequences for regional climates. In the Northern Hemisphere the Hadley cell is stronger in the winter than in the summer; extending north to lie between 10°N and 45°N while the Ferrel cell lies between 45°N and 65°N in the summer, and between 35°N and 60°N in the winter, which in turn leads to changes in wind patterns in some geographical areas.

Worked example 9.2   Wind as mass in action.

In the Northern winter surface winds around the British Isles have a mean value of $4\,\mathrm{m\,s}^{-1}$. Estimate the mass of air, on average, flowing through the lowest 100 m of the atmosphere from the Western Isles of Scotland to the English Channel every second.

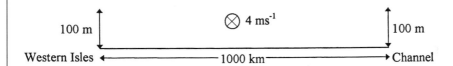

Solution

Assume that the distance across the ground is 1000 km and that the $4\,\mathrm{m\,s}^{-1}$ is the mean wind speed throughout the 100 m layer, and that the layer mean air density is $1.2\,\mathrm{kg\,m}^{-3}$.

The flow of mass across this vertical plane = wind speed × area × density each second

$$= V \times A \times \rho$$
$$= 4\,\mathrm{m\,s}^{-1} \times 10^6\,\mathrm{m} \times 10^2\,\mathrm{m} \times 1.2\,\mathrm{kg\,m}^{-3}$$
$$= 4.8 \times 10^8\,\mathrm{kg\,s}^{-1} = 480\,000\ \mathrm{tonnes\,s}^{-1}.$$

9.6   Global weather and climate patterns

Having discussed the global convection patterns let us now look at the global weather patterns by drawing maps of average global weather in a similar manner to the local weather charts described in Chapter 8. Global charts or, as meteorologists call them, 'fields' may be drawn for many variables including pressure, temperature, humidity, cloud cover and precipitation.

9.6.1   Global pressure field

Northern winter/southern summer

The time-averaged pressure field for December to February is shown in Figure 9.11a and illustrates the characteristic regional weather. The principal features are:

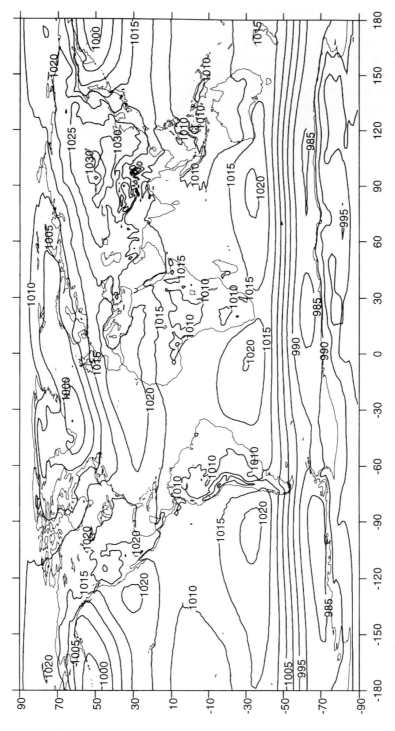

Figure 9.11 (a) Global pressure fields (hPa) in northern winter/southern summer.

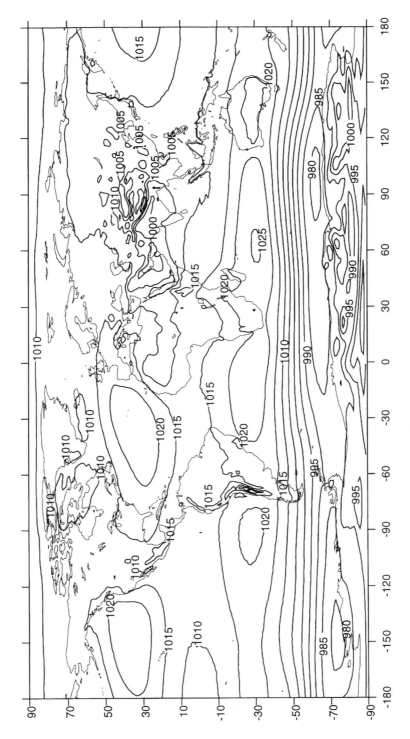

Figure 9.11 (b) Global pressure fields (hPa) in northern summer/southern winter.

- Weak high pressure across part of the Arctic Ocean.
- Low pressure centres south-east of Greenland and in the central North Pacific (commonly known as the 'Iceland' and 'Aleutian low' respectively).
- Extensive region of high pressure stretching from the subtropical eastern Pacific across much of North America and the North Atlantic between about 15° and 45°N, and on to North Africa through to much of Asia. Note the relatively low pressure lying over the Mediterranean within this otherwise extensive high pressure.
- Belt of weak pressure gradient circumventing the globe at low latitudes, a broad zone known as the Doldrums; since it is a region of light, variable winds associated with the weak pressure gradient.
- Weaker low pressure centres over the warmer southern continents.
- Subtropical anticyclones that display distinct centres over the South-east Pacific, South Atlantic and Southern Indian Oceans.
- Circumglobal belt of rapidly changing pressure gradients at mid-latitude in the Southern Hemisphere resulting in strong wind patterns known as the 'Roaring Forties', so named because they are persistently strong all year-round with mean speeds often exceeding 40 knots.
- Two low-pressure centres that flank the Antarctic with an elongated west–east low.
- High pressure across the Antarctic continent.

Northern summer/southern winter

Figure 9.11b shows the average mean sea-level pressure field for the June to August season. There are substantial differences in comparison with the December to February weather patterns. Some of the most notable are:

- Weakening of the mid-latitude oceanic low centres in the North Atlantic and North Pacific.
- The 'flip' from extensive high pressure to extensive low pressure across much of central and southern Asia in association with the evolution of the monsoon.
- Change to low pressure (from winter highs) over the subtropical/lower mid-latitude reaches of Africa and North America.
- Change from summer low pressures to winter high pressures over the subtropical/lower mid-latitude tracts of South America, Southern Africa and Australia.
- Intensification and shift towards the Poles of the subtropical anticyclones in the North Atlantic and North Pacific.
- Deepening and extension of the low-pressure belts around the Antarctic.

9.6.2 Global wind patterns

Seasonal changes in the intensity and location of the major pressure systems are of course reflected in changes in the wind strength. If the horizontal pressure gradient is steep then the air speed will be high and the winds will be strong.

Figure 9.12 illustrates clearly the principal surface wind patterns. Comparison with Figure 9.11 shows that there is a great similarity between the mean wind direction and the orientation of the mean isobars. The wind does not simply blow at right angles to the isobars but circulates out of the high pressure regions to flow across the isobars into the low pressure regions. This is a consequence of the modification of the geostrophic balance by frictional forces that produce a deflection of the wind across the isobars into lows and out of highs.

Northern winter/southern summer

The strongest wind patterns stand out clearly in Figure 9.12a, with the marked westerlies over the mid-latitude oceans being more 'confined' by the northern ocean basin shape compared with the comparable winds in the Southern Hemisphere. These westerlies blow on the equatorward flank of the mid-latitude depressions and transport more warm air into higher latitudes in the Northern Hemisphere than the Southern. Within the Northern Hemisphere, the south-westerlies penetrate further from the subtropical regions in the North Atlantic than in the Pacific. Note also the belt of strong north-easterlies that stretch from the Arabian Sea to the South China Sea, flowing out from the huge winter-time Asian anticyclone. The northerlies and north-easterlies across North Africa converge with south-westerlies near the Gulf of Guinea coast of West Africa.

Another major wind pattern are the strong Trade winds that blow most markedly across the tropical oceans. The *Inter-Tropical Convergence Zone* (ITCZ) is the elongated mainly west-east region into which the huge fluxes of air and moisture carried on these winds converge. Careful inspection of Figure 9.12a shows that the ITCZ is essentially west–east across all tropical oceans and 'dips' into the Southern Hemisphere over the heated southern continents.

It is clear that the strongest air flow is associated with the steepest horizontal pressure gradients (Figure 9.11), while the weaker air flow tends to characterize weaker pressure gradients across the centre of anticyclones. Strong winds on the eastern flanks of the oceanic subtropical anticyclones drive cool equatorward currents in the upper layers of the

---

Worked example 9.3 Mass transport into the Inter-Tropical Convergence Zone.

Assuming that the average air density (ρ) in the lowest kilometre of the troposphere is $1.0\,\mathrm{kg\,m^{-3}}$, estimate the mass flux that streams towards the ITCZ within the NE Trades around 30°W, 15°N.

Solution

Take the 1 km layer mean wind speed to be the same as that at the surface, which is $8\,\mathrm{m\,s^{-1}}$, and consider a vertical wall at right-angles to the NE Trades (i.e. aligned north-west to south-east) that stretches 5° of latitude either side of the central location.

The wall will then be 1 km high and 1110 km wide.

The seasonally-averaged mass flux F is, therefore, $F = V\rho LH$, where $V =$ Trade wind speed normal to the 'wall', L and H are the two dimensions of the 'wall' through which the air is flowing, and ρ is the mean air density.

The flux, F, is then $8.88 \times 10^9\,\mathrm{kg\,s^{-1}}$ or about 9 million tonnes s^{-1}! This is only a limited section of the atmosphere. The mass flux will diminish rapidly with height, for the same speed, because density falls off rapidly with height.

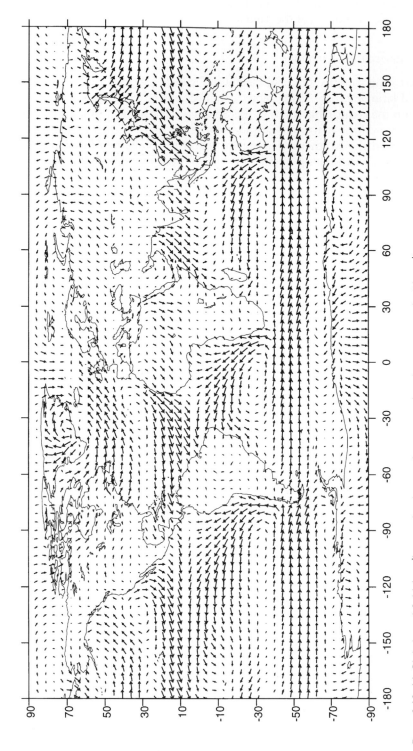

Figure 9.12 (a) Global wind field (ms^{-1}) at sea level in northern winter/southern summer ($\rightarrow$ 10ms^{-1}).

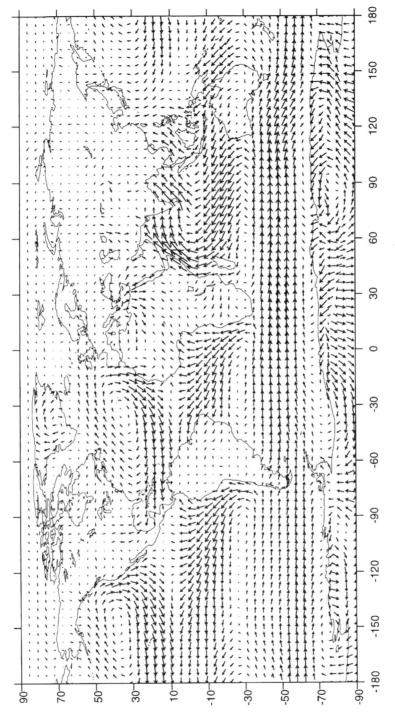

Figure 9.12 (b) Global wind field (m s^{-1}) at sea level in northern summer/southern winter ($\rightarrow$10m s^{-1}).

ocean. These occur off western South America (Peru or Humboldt Current), Southern Africa (Benguela Current), Australia (West Australian Current), North Africa (Canaries Current) and North America (California Current). The North Atlantic Drift and the North Pacific Current are similarly wind-driven, though they are important contributors to the poleward flux of heat from lower latitudes.

Northern summer/southern winter

There are major changes in the wind patterns compared with the December–February period (Figure 9.12b). The westerlies over the northern mid-latitude oceans are weaker. The wind speed near 50°N, 30°W has decreased from its wintertime value to about $5\,\mathrm{m\,s}^{-1}$ so that the low-level mass flux towards the north-east is some 65% of the wintertime's.

The oceanic ITCZ has moved northwards compared to its December–February location but not as far as its continental equivalent which has penetrated deeply into sub-Saharan Africa and into a broad region of southern Asia. There is an amazing reversal of the wind direction across the area stretching from the Arabian Sea to the South China Sea – from north-easterlies in the northern winter to powerful south-westerlies in the summer. This is part of the 'monsoon'. Indeed, the word 'monsoon' is believed to originate from the Arabic word 'mausam' meaning reversal. A similar reversal of the surface winds is clear across West Africa, where another monsoon occurs. The 'Roaring Forties' are well established as the circumpolar westerlies are stretched around the Southern Ocean.

Mid-tropospheric wind patterns

The wind patterns at higher altitudes in the troposphere are much simpler than at the surface. At lower levels, the continents, oceans and extensive mountain ranges all have a strong impact on wind flow patterns. However, their influence diminishes with height. The wind field mapped in Figure 9.13a is for the 500 mbar surface (about 5 km above the mean sea level) in the northern winter/southern summer. The strongest seasonally-averaged winds occur between 30° and 50°N. This fast-flowing current is known as a *jetstream*, and it reaches its maximum intensity somewhat higher in the upper troposphere at around 10 km. Note how the strongest flow occurs downstream of the continents off the eastern USA and China, where mean speeds in excess of $30\,\mathrm{m\,s}^{-1}$ are widespread.

These winds are related to the frontal storm tracks over the North Atlantic and North Pacific. The winds above the tropics are generally light and easterly. They intensify again above the southern middle latitudes where they are clearly much more 'zonal' (or parallel to the latitude lines) than in the Northern Hemisphere because their course is not influenced so much by broad continents and extensive mountain chains.

In the northern summer/southern winter the most significant change in the higher altitude wind patterns occurs within the tropics in association with the evolution of the Asian and West African monsoons. Deep, warm anticyclones form over North Africa and the Arabian Peninsula (and over northern Mexico) with strong westerlies on their northern, and strong easterlies on their southern flanks (Figure 9.13b). The moist monsoon flow is quite a shallow low-level layer that is capped by drier air – for example, over the Sahara. It is within this zone of vertical wind 'shear' that disturbances called 'easterly waves' form over West Africa and run westwards – some of these eventually become hurricanes a week or more later, battering the Caribbean. The jet maxima shift seasonally in unison with the

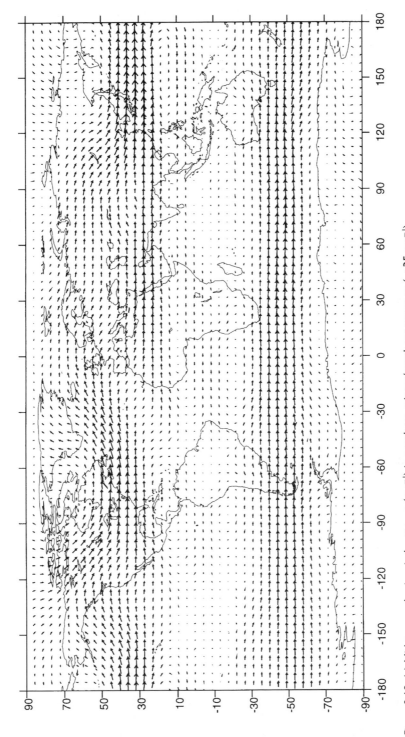

Figure 9.13 (a) Mid-tropospheric wind patterns (ms^{-1}) in northern winter/southern summer (→35 ms^{-1}).

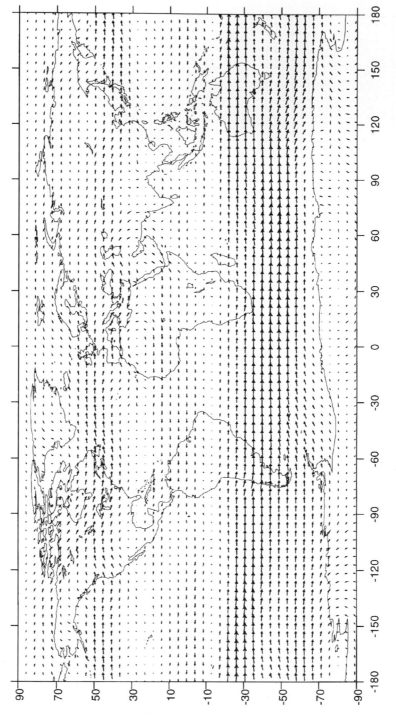

Figure 9.13 (b) Mid-tropospheric wind patterns (m s^{-1}) in northern summer/southern winter ($\rightarrow$35 ms^{-1}).

seasonal movement of the Hadley cells. Thus, they are further poleward in the summer and equatorward in the winter.

The high latitude middle and upper tropospheric circulation in the depth of the southern winter is significantly different from the cold season above the Arctic. The marked westerly winds high over the flanks of the Antarctic flow, more-or-less, in a circular pattern due to the 'quasi-circular' Antarctic continent lying beneath them, surrounded as it is by a circumpolar ocean. This flow has the effect of preventing any heat transport into its centre by depressions from lower latitudes – implying that stratospheric wintertime temperatures over the Antarctic are significantly colder than those in the Arctic winter stratosphere. This intense frigidity plays a crucially important role in the artificial depletion of ozone above Antarctica (see Chapter 7).

9.6.3   Temperature fields

Figure 9.7 illustrated the way in which horizontal thermal gradients produce winds in the atmosphere. There are of course regions where thermal gradients are very strong – in the wintertime especially these occur off the eastern coasts of the strongly cooled mid-latitude continents. These are often linked to frontal systems where the cold air flowing off the eastern coasts of North America or North-East Asia meets much warmer, moister air that circulates around the Azores and Hawaiian anticyclones. This culminates in the jetstream in the upper troposphere. In general, the steepest horizontal gradients of temperature flank the colder continent or mountainous regions such as those south of the Himalayas.

Northern winter/southern summer

Figure 9.14 illustrates that within the extensive wintertime anticyclones (Figure 9.11a) strong radiative cooling promotes low temperatures. The lowest mean temperatures during this season over the Northern Hemisphere are located around Verkhoyansk in north-east Russia, where the average is a cold −42°C.

The hottest conditions during a typical southern summer occur across the strongly heated continents, with mean temperatures of >30°C over the sunny northern Australian desert and in the interior of southern Africa. However, even though it is summertime, the high elevations of the Antarctic continent still record averages below −35°C.

Cold, dry air that flows across the very much warmer ocean off the east coasts of North America and North-East Asia is very strongly warmed by fluxes of latent heat from the ocean surface. The cool ocean currents that are driven equatorward along the eastern flanks of ocean basins are quite influential in determining the nature of coastal climates in these regions. Note how the isotherms (lines of constant temperature) bend towards the equator in these zones.

Northern summer/southern winter

The most obvious seasonal changes in temperature are those that occur outside the tropics; the annual variation within much of the tropics being only a few degrees Celsius. The greatest variation is observed in the extratropics where the highest of latitudes change from polar day to polar night, while the mid-latitudes generally experience only 'intermediate' seasonal variation (Figure 9.14b).

Figure 9.14 (a) Global temperature field (°C) in northern winter/southern summer.

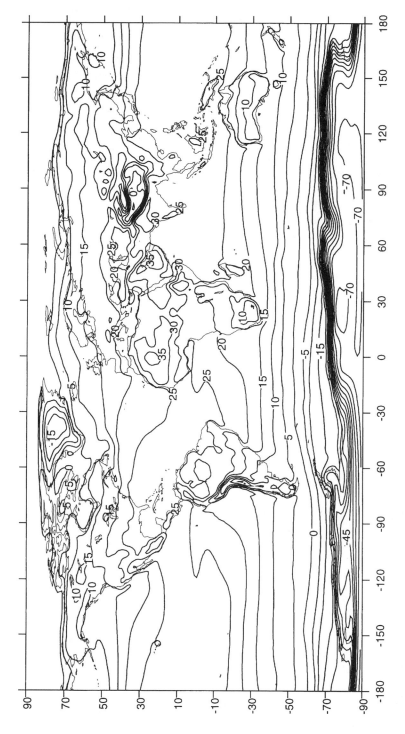

Figure 9.14 (b) Global temperature field (°C) in northern summer/southern winter.

The hottest conditions in the northern summer are not found at the equator but in the Saharan interior where the mean reaches more than 35°C. Indeed, the equator is hardly ever the hottest place on Earth, that record belongs to these strongly heated subtropical deserts in their summer season.

Generally, the tropical oceanic thermal gradients are relatively gentle and tend to be steepest over the higher latitude oceans. The summer season sees the continents acting as the major heat source for the Earth's atmosphere, while in the winter it is the warmer oceans (principally outside the tropics) that fulfil this role. Extratropical convection therefore tends to be more common over land in the summer and over the sea in the winter. This heating is also associated with the generation of sea breezes (see Section 9.4).

The influence of the cool coastal currents in the subtropics is also noticeable in the mean temperature field. These areas are sometimes marked by the presence of extensive low stratiform cloud as the mild, damp air flowing around the subtropical highs is cooled to its dewpoint temperature by the ocean. An example of this is seen off the coasts of Namibia and Angola, stretching equatorwards over many thousands of square kilometres across the eastern South Atlantic.

Mid-tropospheric temperatures

Like the wind patterns, the thermal patterns are much simpler at higher levels than at the surface, where the local effects of all the continents and oceans may dominate. The warmest region at an altitude of some 5 km in the northern winter/southern summer (Figure 9.15a) exists in a broad band around the tropics. Not surprisingly, there is a poleward decline in temperature with the coldest conditions over the winter high latitudes. The warmer regions above both the eastern North Atlantic and North Pacific are expressions of the thermal energy transported by the north-eastward-tracking frontal depressions. The poleward gradient of temperature is steeper in the winter hemisphere and is linked to stronger air flow. The poleward temperature gradient is generally much steeper in the southern, winter hemisphere (Figure 9.15b) – partly because the Antarctic is quite different from the Arctic. Since much of the Antarctic continent lies at high altitudes and stretches up into the lower troposphere, it influences temperatures at higher altitudes more than the lower lying Arctic. This intense frigidity plays a crucially important role in the artificial depletion of ozone above Antarctica since it is one factor in the formation of a stable polar vortex throughout most of the winter (see Chapter 7). The warmest region – where temperatures are actually above 0°C even at an altitude of close to 5 km – lies over the Himalayan plateau, which acts as an elevated heat source in the northern summer. This is extremely significant for the development of the Asian summer monsoon.

9.6.4   Global humidity patterns

Humidity may be measured in many ways. The most common is by using the wet bulb thermometer discussed in Chapter 8. If the air surrounding the wet bulb thermometer is unsaturated, water will evaporate into it leading to a decrease in the temperature measured by the 'wetted' bulb. The difference between the dry bulb and wet bulb temperatures is inversely related to the relative humidity.

The *dew point temperature* is another measurement of the humidity. It is the temperature to which a sample of air must cool, at constant pressure and constant water vapour mixing ratio, to become vapour saturated. Thus, the dew point temperature is a measure of the actual water vapour concentration of an air sample. It is the value plotted routinely on weather maps. From the dew point temperature and the corresponding value of pressure it is possible to calculate the number of grams of water per kilogram of air.

Northern winter/southern summer

The highest dew points are, not surprisingly, found at low latitudes and, generally, at sea level. The combination of high evaporation over windy tracts of warm tropical ocean and the convergent transport of large water vapour concentrations towards the ITCZ on the Trade winds are responsible for the high dew point values from the western equatorial Pacific across to the 'Island Continent' of Malaysia and Indonesia and into northern Australia, and for the zone that stretches from Madagascar to the Gulf of Guinea coast, and for the large values over Amazonia (Figure 9.16).

Dew point values, typically in the mid-20s°C, are recorded in these regions, equating to a 'water vapour mixing ratio' of approximately $18.0 \, g \, kg^{-1}$ (mass of water vapour in a unit mass of dry air within which it is contained) – a very high concentration. In Worked example 9.3 the mass flux in the NE Trades was $8.88 \times 10^9 \, kg \, s^{-1}$. Hence, $1.60 \times 10^8 \, kg$ of water is similarly transported across the NE trades and into the ITCZ, explaining the torrential rainfall seen in these regions during the summer months.

Worked example 9.4   Water vapour transport into the ITCZ.

Worked example 9.3 indicated a mass flux of about 9 million tonnes s^{-1} into the lowest kilometre across a stretch of subtropical North Atlantic. Estimate the associated flux of water vapour across this area towards the ITCZ.

Solution

Take the water vapour mixing ratio of $18 \, g \, kg^{-1}$ to be a mean value in the lowest kilometre of the atmosphere. It is then possible to calculate the flux of water vapour driven by the mean wind speed towards the ITCZ.

This is the layer mean mass flux, F, multiplied by the layer mean water vapour concentration, q. So

$$\begin{aligned} Fq &= 8.88 \times 10^9 \, kg \, s^{-1} \times 18 \, g \, kg^{-1} \\ &= 159.8 \times 10^9 \, g \, s^{-1} \\ &= 159.8 \times 10^3 \, tonnes \, s^{-1}. \end{aligned}$$

This approximation means that there is a flux of about 160 000 tonnes of water vapour flowing into the ITCZ across this area every second during the northern summer.

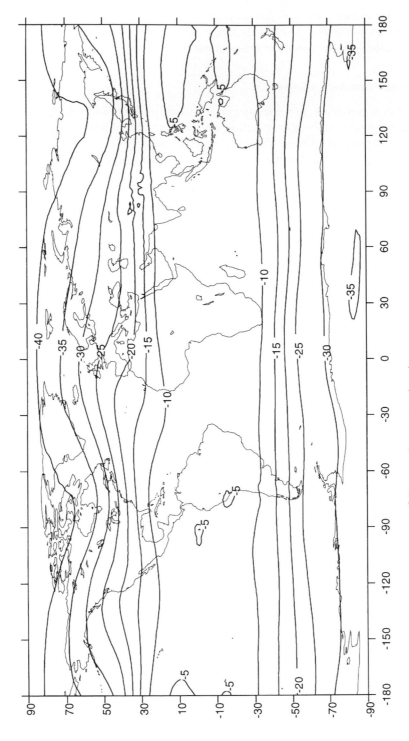

Figure 9.15 (a) Mid-tropospheric temperatures (°C) in northern winter/southern summer.

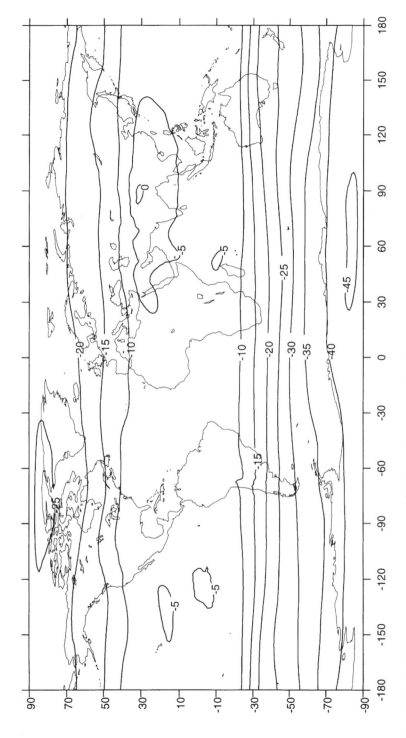

Figure 9.15 (b) Mid-tropospheric temperatures (°C) in northern summer/southern winter.

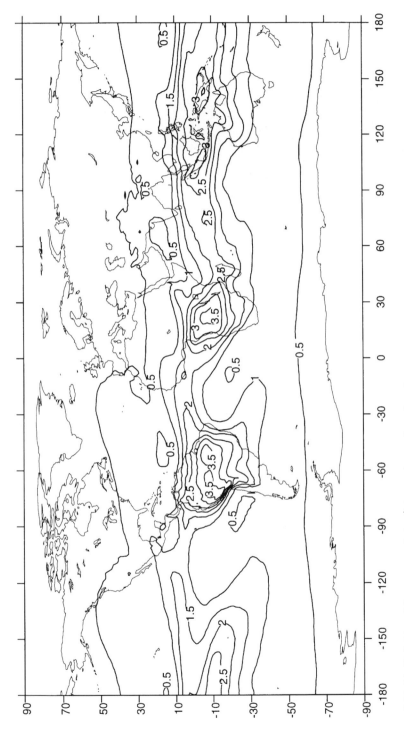

Figure 9.16 (a) Global humidity patterns (g kg^{-1}) in northern winter/southern summer.

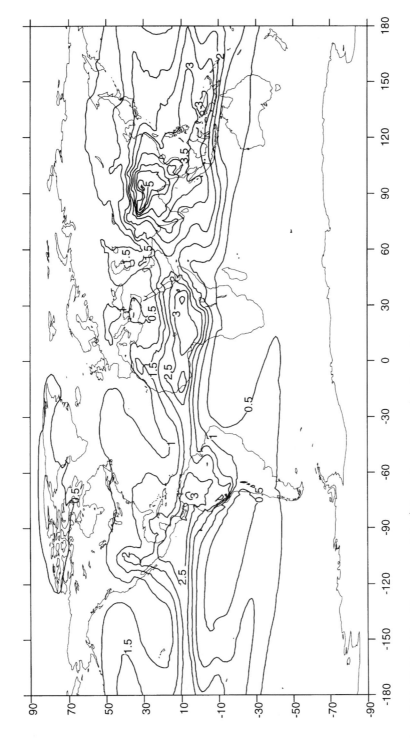

Figure 9.16 (b) Global humidity patterns (g kg^{-1}) in northern summer/southern winter.

Low values of dew point do however occur in the tropics, exemplified by the very dry conditions that characterize the Sahara. Dew points of around 0°C are recorded there (converting to about $4.0\,g\,kg^{-1}$). The winter extratropics illustrates the role of higher evaporation and poleward transport of water vapour over the windswept oceans where the frontal depressions track from south-west towards the north-east. The extensive southwesterlies are related to the significant north-eastward penetration of relatively high humidity into the northern Norwegian Sea and even as far as the White Sea, where the -6°C dew point equates to some $2.5\,g\,kg^{-1}$ of water vapour.

The driest surface air, on average, during the northern winter lies far to the east within the interior of eastern Siberia. Here, the water vapour levels are as low as $0.1\,g\,kg^{-1}$ and are generally below $1.0\,g\,kg^{-1}$ across all of Siberia. These extremes of dryness are partly responsible for the extended wintertime thermal energy loss to space and hence the prolonged cold conditions that are worst of all in north-east Siberia.

The North American continent is also dry under its extensive winter anticyclone, with relatively steep gradients along its coasts. Like the warmth of the eastern sides of the middle and high latitude continents, the moisture field indicates that these flanks are also more humid. In Britain the dew point is between 5 and 6°C while along the Labrador coast (lying at the same latitude) it is around -20°C! The associated mixing ratios are between 4.0 and $5.0\,g\,kg^{-1}$ for Britain and $<1.0\,g\,kg^{-1}$ for Labrador. The air therefore has a generally higher water content on the eastern side of the Atlantic. The North Pacific displays the same broad pattern from the British Columbian coast to the Russian coast of the Sea of Okhotsk.

The humidity pattern and gradients in the southern summer are far simpler than in the north because of the circumpolar Southern Ocean, the very broadly circumpolar Antarctic coast, and the fact that the continents narrow towards the poles. As mentioned above, maxima are observed from the western equatorial Pacific across to the 'Island Continent' of Malaysia and Indonesia and into northern Australia; across a zone that stretches from Madagascar to the Gulf of Guinea coast, and over Amazonia. The principal minimum lies deep within the interior of the Antarctic Continent, where dew points less than around -40°C equate to a humidity mixing ratio of $0.1\,g\,kg^{-1}$.

The dew point pattern around the Southern Ocean is essentially 'latitudinal' with no sign of marked poleward excursions of higher humidity like its northern counterpart oceans. This is due to the largely west to east (zonal) track of the frontal systems of the Roaring Forties. One notable feature however is the existence of lower dew point air along the western coasts of the subtropical continents, particularly off South America and Africa. These reflect the moisture transported by the equatorward wind currents in these areas (Figure 9.12). There are similar features within the Northern Hemisphere, e.g. along the Californian and north-west African coasts.

Northern summer/southern winter

In general, the dew point gradients in the northern summer/southern winter (Figure 9.16b) are weaker than those in the northern winter/southern summer (Figure 9.16a).

The maximum values have 'migrated' northwards in association with the similar seasonal movement of the ITCZ. Mean humidity mixing ratios of between 16.0 and $20.0\,g\,kg^{-1}$ stretch from South-East Asia through southern monsoon to India and across to the Red Sea. From there, high values run to West Africa and over the tropical North Atlantic to reach a 'peak' in the Caribbean.

The summertime extratropical continents are much moister than in the winter; evaporation is substantially increased from vegetation during the warmer season. The air above the oceans is also more moist during the summer season. The general increase in humidity is reflected by Britain's mean surface dew point rising to 12°C or about $9.0\,g\,kg^{-1}$ – roughly a doubling of the water vapour concentration compared with its wintertime value.

Although there are strong seasonal changes in humidity, the minima still reside in the coldest areas. Thus, during the northern summer, the lowest water vapour concentrations at the surface are not surprisingly found in the High Arctic and at higher elevations, for example, in Greenland and across the Himalayas.

As in the northern winter, the west-to-east change in dew point temperature across the Atlantic (and Pacific) around 50°N is notable. The eastward dew point increase of some 6°C (about 30°C in winter!) mirrors the transport of warm air towards the Poles by travelling frontal systems on the eastern side of the ocean, and the equatorward motion of cold air over eastern Canada. Even during summer, higher water vapour concentrations are measured as far north as the Norwegian Sea.

In the Southern Hemisphere winter there are even steeper gradients around the Antarctic flank than in the summer. The interior of the Antarctic continent is extremely cold, resulting in very low water vapour concentrations. The interiors of the subtropical continents are also relatively dry with dew point temperatures down to 0°C in the arid regions within Australia and western South Africa.

9.6.5 Cloud patterns

Cloud patterns are related to both the humidity and the wind patterns and indicate where precipitation levels may be high. Today, weather satellites monitor cloud patterns almost continuously, providing detailed information on the evolution and movement of weather fronts.

Northern winter/southern summer

Figure 9.17 shows a visible image from the European geosynchronous weather satellite (Meteosat) at 12:00 UTC (Universal Time Co-ordinated – the modern term for Greenwich Mean Time, GMT) on 5 January 1994 and is typical of a winter's day in the Northern Hemisphere. The image is an expression of the varying reflectivity (or albedo) of the surfaces in view, and is processed in such a way that features with the brightest albedo are white. Hence, cloud patterns stand out against other surfaces that are in general much poorer reflectors.

In the image, the ITCZ is revealed over the tropical Atlantic by the bright, globular west-east cloud mass, while over southern Africa the ITCZ is a much more complex feature partly because of the effect of surface heating from the warm continental land mass. The generally clear skies of the subtropical highs (Figure 9.17) are apparent over the subtropical North Atlantic and North Africa. The subtropical South Atlantic shows some scattered cloud, mainly low stratiform (layered) cloud trapped not far above the ocean surface.

There are also fine examples of elongated cloudbands that signify warm moist air streaming from the subtropics into mid-latitude regions along fronts. One runs south-west to north-east from the subtropical Atlantic towards and across western Europe while the

Figure 9.17 Meteosat image of the northern winter, 12:00 GMT, on 5 January 1994.

other stretches from central, eastern Brazil south-eastwards towards the central South Atlantic.

The extremely steep gradient in humidity between some of the moistest and driest air in the tropics lies across West Africa, and separates the dry easterly 'Harmattan' on the southern flank of the Sahara from the very damp south-westerlies on the coast. Figure 9.17 illustrates some of these mean features for an individual northern winter's day, for example the cloudy region from Madagascar into southern Africa. On this particular day it appears that the dry Saharan air had swept right across West Africa; the only hint of cloud in the vicinity is the very thin west–east line running across the Gulf of Guinea. Indeed, there was hardly any cloud across the whole of northern Africa.

Northern summer/southern winter

The Meteosat visible image for 12:00 UTC on 13 August 1983 (Figure 9.18) illustrates a typical northern summer picture. The Atlantic ITCZ cloud is less substantial than in the northern winter/southern summer and lies further north. The African ITCZ is located from Sierra Leone eastwards to northern Zaire then north-eastwards to Ethiopia and Eritrea. In complete contrast with Figure 9.17, southern Africa is almost completely cloud-free.

The broadly cloud-free zones poleward of this extensive ITCZ reflect the deep subsidence within the sinking branch of the Hadley cells that give way further poleward, as in the northern winter, to the trailing frontal cloudbands of the travelling mid-latitude frontal depressions.

The ITCZ is, by definition, a zone within which air converges from opposite directions near the Earth's surface. This mass convergence in the lowest kilometre or so of the

Figure 9.18 Meteosat image of the northern summer, 12:00 GMT, on 13 August 1983.

troposphere is associated with ascending air and is most obvious in those regions where the wind speed decreases along the line of the wind. This is perhaps clearest just to the west of West Africa where the strong Trades weaken really significantly towards a zone of very much lighter winds. This marked slowing down of the flow is related to a 'piling up' of mass, and therefore to widespread ascent, most often in the form of clusters of very deep cumulonimbus cloud.

Worked example 9.5   Moisture on the move.

The Inter-Tropical Convergence Zone is fed with huge amounts of water vapour by the Trade winds. Determine how much water is transported, on average, into this region by considering the rate of air mass flow through a vertical surface and the absolute concentration of water vapour in the same area.

Solution
The surface air flow over the Atlantic Ocean between West Africa and the Caribbean is typically $10 \, \text{m s}^{-1}$ in the northern winter. In the same region, the dew-point temperature is about 20°C, which translates into a concentration of some $16 \, \text{g m}^{-3}$ of water vapour.

Consider a surface 1000 km long and 100 m deep normal to the North-east Trades. The mean volume flow per second across this surface is $V \times A$ where V is the speed and A is the area normal to the air flow $= 10 \, \text{m s}^{-1} \times 10^6 \, \text{m} \times 100 \, \text{m} = 10^9 \, \text{m}^3 \text{s}^{-1}$.

So, only in the lowest 100 m, the mass flow rate of water vapour $= 16 \, \text{g m}^{-3} \times 10^9 \, \text{m}^3 \text{s}^{-1} = 16 \times 10^{-3} \, \text{kg m}^{-3} \times 10^9 \, \text{m}^3 \text{s}^{-1} = 1.6 \times 10^7 \, \text{kg s}^{-1} = 16000 \, \text{tonnes s}^{-1}$.

9.6.6   Precipitation

Clouds are a crude indication of where it may be raining; therefore rainfall is also meas-ured directly by precipitation radar. Precipitation radar uses centimetre wavelengths to measure the power of the backscattered radiation from raindrops and, unlike rain-gauges, provide spatially continuous and instantaneous rates averaged over $1 \, \text{km}^2$ boxes. Each radar has a 'viewing' radius of about 120 km.

Northern winter/southern summer

The highest precipitation occurs in the lowest latitudes (Figure 9.19a) with extensive seasonal mean daily rates of 5 mm (or a mean seasonal total of 460 mm). There are locally higher totals than this though within the tropics. Up to 1.4 m of rainfall has been recorded in Cameroon and eastern Zaire, north-western Madagascar and western Sumatra. These large falls are related to the huge flux of water vapour brought by the Trade winds into the ITCZ. Within this zone, vigorous ascent of extremely humid air leads to torren-tial downpours.

These regions of high precipitation are seen in South America, southern Africa and northern Australia. Immediately equatorward of these wet areas are extensive regions of dry weather. There is a strong, but not unique, relationship between the regions where <2 mm a day falls and the presence of extensive anticyclones (Figure 9.11). This is true over the subtropical oceans in both hemispheres and the cold wintertime continents of the Northern Hemisphere. It does not snow every day in the interior of Siberia; indeed in winter this is quite an arid region.

Other wet regions are those that see the daily passage of travelling weather fronts. Regions with >5 mm precipitation day^{-1} (460 mm for the season) stretch eastwards or north-eastwards across the northern Pacific and Atlantic Oceans and reach the adjacent higher latitude continents, such as western British Columbia and western Norway. Similar regions occur over the Southern Hemisphere oceans running south-eastwards into the central South Atlantic from southern Brazil and as the South Pacific Convergence Zone (SPCZ). Eastern Britain sees some 2 mm day^{-1} on average, or a seasonal total of about 180 mm.

Northern summer/southern winter

The most obvious pattern change from northern winter occurs within the tropics where the strong rainfall maximum shifts to a more northern location notably over the lower latitude tropical continents (Figure 9.19b). In fact, the location of the maximum precipitation over the oceans does not migrate very much seasonally, as the sea-surface temperature maximum is broadly in the same location round the year.

Within the heavy rain region there are localized areas of very much enhanced rates. These are found in the monsoon regions where marked low-level convergence stimulates very heavy downpours, especially where high ground is close by or at the coast. Mean daily rates of some 25 mm occur along coastal Sierra Leone and western India, for example. This means that the average total 3-month fall exceeds some 2.3 m!

Away from this maximum the intensity decreases markedly but the extensive, heated extratropical continents exhibit a significant increase in rainfall compared to their winter totals. The track of frontal depressions running north-east towards the Norwegian Sea remains but with generally smaller rainfall values than in the winter. The regional maxima over south-east USA and the China/Japan area are partly influenced by tropical disturbances, including hurricanes and typhoons respectively, that 'recurve' northwards or north-eastwards around the oceanic anticyclones.

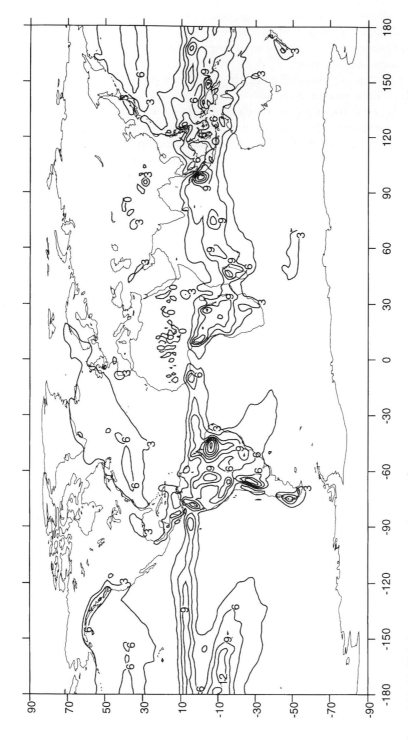

Figure 9.19 (a) Global precipitation patterns (mm day^{-1}) in northern winter/southern summer.

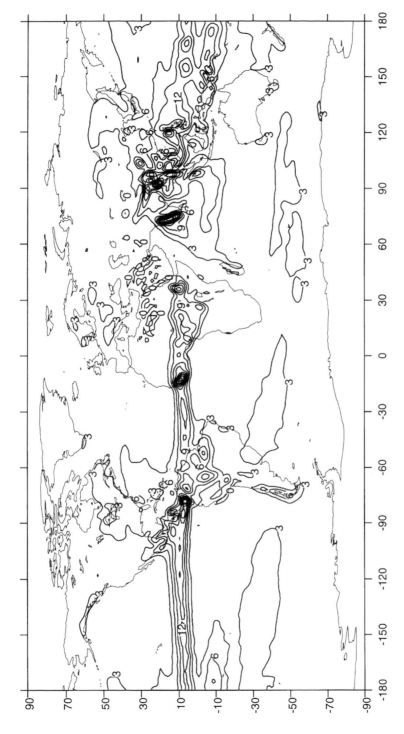

Figure 9.19 (b) Global precipitation patterns (mm day^{-1}) in northern summer/southern winter.

Case study 9.1　El Niño.

El Niño is a phenomenon responsible for producing unusual weather not only across the equatorial Pacific Ocean where it occurs, but also across many other parts of the Earth. It was first recorded a few centuries ago by fishermen working along the coasts of Equador and Peru. They noted that typically once every 3–9 years, around Christmas time, the usually plentiful supply of fish would suddenly vanish from their usual fishing grounds. This unusual event is related to the replacement of the very cool waters of the Peru Current, with much warmer water from the north. The inflow of such warm water suppresses the upwelling cold waters that are rich in the nutrients needed by the fish and leads to a rapid decline in fish stocks (particularly anchovies). Since it occurred near Christmas, the event was termed 'El Niño', which is Spanish for 'the baby boy' (i.e. the Christ child). Today, El Niño is used as a name for a much broader, equatorial Pacific-wide feature that occurs from time to time with varying intensity. The most recent intense El Niños occurred in 1982/83 and 1997/98.

The average equatorial Pacific sea surface temperature is quite low close to South America, but the waters are warm in the western Pacific around Indonesia (Figure 9.20). This pattern is related to the presence of North-east and South-east Trade winds that usually blow strongly towards and along the Equator. These winds 'build up' water in the western Pacific so that the equatorial sea surface actually slopes by approximately 1 m from east to west right across the vast Pacific Ocean basin. As the water moves slowly westwards, it is heated.

In the 1920s, Sir Gilbert Walker, the Director of the Indian Meteorological Service, wished to devise a scheme for predicting the total rainfall of the summer monsoon. He discovered that when the pressure in the South Pacific anticyclone (Figure 9.11) was unusually high, then the low over Indonesia was unusually deep. Conversely, when the Indonesian low was anomalously shallow, the South Pacific high was also less intense. This pressure see-saw is known as the 'Southern Oscillation'. The Southern Oscillation is routinely monitored using monthly mean pressures at two stations: one near the centre of the high (Tahiti) and one near the centre of the low (Darwin, North Australia).

The state of the Southern Oscillation is expressed as an *index*. It has a high index when the difference in pressure between Tahiti and Darwin is large, i.e. when the Trade winds are blowing strongly. It has a low index when the difference is small and the Trades are either very weak or even reversed. A low index is related to the evolution of an El Niño event. When the index is low, with much weaker, or even reversed Trades, the water that is normally built up in the western basin reverses direction and flows eastwards sending unusually warm water across to South America. The fact that the Southern Oscillation (SO) and the El Niño (EN) are linked has lead to the term '*El Niño Southern Oscillation*' (ENSO) also being used.

The large eastward flow of warm water stimulates the development of very deep convective cloud in regions where they are not usually experienced. These deep clouds often produce torrential downpours and flooding as they migrate along the equator. Indeed, such is their intensity that they transport significant amounts of thermal energy and moisture into the upper troposphere. This can enhance the

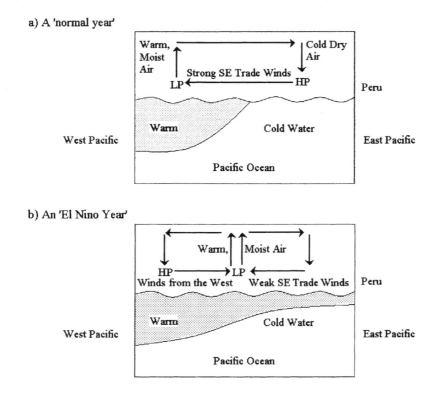

Figure 9.20 El Niño.

thermal gradient there which in turn will produce an increase in the winds of the jetstreams. Enhancement of the upper tropospheric jetstreams in both the Northern and Southern Hemispheres then leads to deeper depressions and unusually wet conditions over, for example, California and northern Argentina. Hence, El Niño events may affect the weather over large parts of the global surface far away from the equatorial regions.

The slow transport of very warm water across the equatorial Pacific is associated with a change in location and intensity of the *'Walker cells'*. These cells lie in the vicinity of the Equator and circulate in the vertical east–west plane, at right angles to the Hadley cells that circulate in the north–south plane. Walker cells are associated with the sinking air over the eastern equatorial Pacific and strong ascent above the western equatorial Pacific. During an ENSO event, however, the deep ascending branch migrates east with the warm water, generating more subsidence over northern South America/southern Caribbean. Hence, one effect of a well-developed ENSO is drought in these regions. The shift in the Walker circulation also leads to an increased drought risk across much of Indonesia and eastern Australia and increased rainfall over parts of eastern Africa. Indeed, during the 1998 El Niño, all of these effects were observed. It was largely to blame for the extreme pollution events related to forest fires in Borneo and serious flooding in Kenya.

9.7 Summary

In this and in Chapter 8 we have examined the physics and dynamics that underpin the global weather and climate patterns and discussed briefly how the weather can be predicted. However, weather forecasting is a very complicated scientific problem drawing on many facets of science which include aspects of chemistry, physical geography, mathematics and computing. In the next two chapters we will discuss the interaction between climate, weather and the biosphere.

References

Ahrens, C. D., *Meteorology Today: An Introduction to Weather, Climate and the Environment*, 6th edn. London: Brookes Cole, 1999.

Atkinson, B. W., *Dynamical Meteorology: An Introductory Selection*. London: Methuen Press, 1982.

Barry, R. G. and Chorley, R. J., *Atmosphere, Weather and Climate*, 7th edn. London: Routledge, 1998.

Bradley, R. S., *Palaeoclimatology: Reconstructing Climates of the Quaternary*, 2nd edn. San Diego: Harcourt/Academic, 1999.

Bryant, E., *Climate Process and Change*. Cambridge: Cambridge University Press, 1997.

Burroughs, W. J., *Does the Weather Really Matter?: The Social Implications of Climate Change*. Cambridge: Cambridge University Press, 1997.

Burroughs, W. J., *Watching the World's Weather*. Cambridge: Cambridge University Press, 1991.

Cotton, W. R. and Pielke, R. A., *Human Impacts on Weather and Climate*. Cambridge: Cambridge University Press, 1996.

Diaz, H. F. and Markgraf, V., eds, *El Niño: Historical and Paleoclimate Aspects of the Southern Oscillation*. Cambridge: Cambridge University Press, 1994.

Diaz, H. F. and Pulwarty, R. S., eds, *Hurricanes: Climate and Socioeconomic Impacts*. Berlin: Springer, 1997.

Dunlop, S. and Wilson, F., *Weather and Forecasting*, 2nd edn. London: Chancellor Press, 1998.

Glantz, M. H., *Currents of Change: El Niño's Impact on Climate and Society*. Cambridge: Cambridge University Press, 1997.

Hartmann, D. L., *Global Physical Climatology*. San Diego: Academic Press, 1994.

Hobbs, J. E., Lindesay, J. A. and Bridgman, H. A., *Climates of the Southern Continents: Present, Past and Future*. Chichester: Wiley, 1998.

Kondratyev, K. Y. and Cracknell, A. P., *Observing Global Climate Change*. London: Taylor and Francis, 1998.

McIlveen, R., *Fundamentals of Weather and Climate*, 2nd edn. London: Chapman and Hall, 1992.

McIntosh, D. H. and Thom, A. S., *Essentials of Meteorology*. London: Taylor and Francis, 1969.

National Research Council, *Decade to Century Scale Climate Variability and Change: A Science Strategy*. Washington, DC: National Academy Press, 1998.

Oke, T. R., *Boundary Layer Climates*, 2nd edn. London: Routledge, 1996.

Palmen, E. and Newton, C. W., *Atmospheric Circulation Systems*. New York: Academic Press, 1969.

Riehl, R., *Introduction to the Atmosphere*. New York: McGraw-Hill, 1978.

Robinson, P. J. and Henderson-Sellers, A., *Contemporary Climatology*, 2nd edn. Harlow: Longman, 1999.

Thompson, R. D. and Perry, A., eds, *Applied Climatology: Principles and Practice*. London: Routledge, 1997.

Web sites

Atmospheric circulation

www.earth.usc.edu/~stott/Catalina/circulation
www.envf.port.ac.uk/geog
www.tamu.edu/class/Metr151/tut/seabr/sea7

Hurricanes and tornadoes

http://members.tripod.com/~Michaelkruk/notes2.html
http://ww2010.atmos.uiuc.edu/(G1)/guides/mtr/hurr/stages/cane/home.rmxl
http://www.txdirect.net/~msattler/sweather.html
http://www.txdirect.net/~msatler/sform.html

Information and updates on the El Niño

http://www.crseo.ucsb.edu/geos/123.html
http://www.pmel.noas.gov.toga-tao/el-nino/impacts.html
http://www.pmel.noaa.gov/toga-tao/el-nino-report.html
http://www.vision.net.au/~daly/elnino.htm

Discussion questions

1   Consider a low-pressure system in the Southern Hemisphere. In which direction does air circulate around the low-pressure centre?
2   Near the Earth's surface, the observed wind often has a component directed from the high-pressure region toward the low-pressure region. Why?
3   Why do mid-latitude cyclones always produce precipitation?
4   When there is warm advection, i.e. when air blows from a warm region towards a cold one, the geostrophic wind rotates clockwise with height. Why?
5   Describe, in detail, the life cycle of a frontal depression.
6   Using the definition for the geostrophic wind, or otherwise, explain why hurricanes do not cross the Equator.
7   Discuss the basic reasons for the seasonal strengthening and weakening of the horizontal pressure gradient in mid-latitudes.
8   Why is the Equator not the hottest region of the Earth's surface?
9   Discuss the links between the mean surface wind flow (June–August) within the Tropics and the mean daily rainfall rate pattern in the same region.
10 (a) If the air were forced to move over the surface of the Earth by pressure difference alone in which direction would the wind blow compared to the isobars? (b) However, since the Earth rotates, any moving air mass is deflected by the Coriolis force. How does the Coriolis force change with (i) wind speed and (ii) latitude? (c) Hence or otherwise explain why the pressure difference and the rotation of the Earth make the wind blow parallel to the isobars at a height of 1 km above the surface of the Earth, (d) The figure below shows two isobars, and the resulting wind direction at a height of 1 km in the Northern Hemisphere. (i) Mark the wind direction on the ground and (ii) Explain why the wind direction at ground level is different from that at a height of 1 km?

1000 mb

at 1 km

1004 mb

(Oxford and Cambridge Examination Board: 'A' Level Physics: Physics in the Environment: June 1996)

11 Contrast the weather sequence during a depression with that of an anticyclone.
12 Discuss the possible climate consequences of global warming.
13 Describe the type of weather associated with the movement of a warm front of a mid-latitude depression.

Quantitative questions

1   Storm force northerly winds are blowing at $25\,\mathrm{ms}^{-1}$ over sea area Tyne east off North-east England at 55°N. Calculate the horizontal pressure gradient in Pa (Pa $10\,\mathrm{km}^{-1}$) associated with this speed, assuming geostrophic balance.

2   Assuming that the air in question 1 has travelled with a constant speed from the northern Norwegian Sea at 70°N, estimate its travel time hours to the location in sea area Tyne.

3   Calculate the geostrophic wind speed at two or three surface locations around the Southern Ocean at 50°N in the southern winter. Do your values confirm the title of 'Roaring Forties' for this region?

4   (a) Explain what is meant by (i) an isobar (ii) the Coriolis force
    (b) Discuss the Hadley, Ferrel and Polar cells as relevant to global convection.
    (c) Consider two isobaric surface pressures p_1 and p_2 at heights z_1 and z_2 (such that $z_2 > z_1$). Show that if T_{mean} is defined as the mean temperature of the layer that

$$z_2 - z_1 = [RT_{\mathrm{mean}}/Mg].\ln(p_1/p_2)$$

where M is the molecular mass.

Hence, or otherwise, describe why when we have advection, i.e. when air blows from a warm to a cold region, the geostrophic wind in the Northern Hemisphere rotates clockwise with height.

(University College London: BSc Physics: Environmental Physics: May 1995)

5   Define the term *geostrophic wind*.

Calculate the geostrophic wind speed for a pressure gradient of 0.03mb/km, assuming that the Coriolis parameter $f = 10^{-4}\,\mathrm{s}^{-1}$.

Why does the geostrophic wind approximation break down in (i) equatorial regions, (ii) near the Earth's surface, and (iii) when isobaric lines are curved and have small radii of curvature.

(University College London: BSc Physics: Environmental Physics: May 1998)

6   With the aid of diagrams discuss the major global wind systems.

Estimate the mass of air transported by a wind blowing at 8m/s over a depth of 1 km through 10 degrees latitude centred around 30°W, 15°N.

(University College London: BSc Physics: Environmental Physics: May 1999)

Chapter 10

Physics and soils

10.1 Introduction

The soil–water–atmosphere–vegetation continuum is of primary importance for the functioning of the biosphere and for agriculture. The fate of rain falling on the land surface is determined by the nature of the vegetation and soils present. The land surface, therefore, has a strong influence on the hydrological cycle, in terms of how much of the annual rainfall is diverted into surface waters by overland flow, how much is taken up by plants or otherwise lost through evaporation, and how much water percolates to depth into the soil.

An understanding of the fate of water arriving at the land surface via precipitation has a number of applications. Growing vegetation relies on water held within the soil, explored by roots in order for plant growth to be sustained. Water running off the soil surface not only creates the potential for soil erosion, but also transports sediments, nutrients and potential pollutants into surface waters. Deep percolation of water is of interest with respect to groundwater recharge, and the possible transport of potential pollutants from the land surface into groundwater. Evaporation from the land surface depends on weather variables as well as on the availability of water, and is an important mechanism for the dissipation of solar energy absorbed by the land surface. There are close links between the availability of water for evaporation and the temperature of the land surface, and the extent to which the land surface acts as a source of *sensible* heat and water vapour for the atmosphere.

This chapter will explore how physical ideas can be developed and applied to understand the retention and movement of water through soils, the movement of solutes through soils, and the effect of weather and the nature of the land surface on evaporation. Though much of the discussion relates to processes within the soil profile, consideration is also given to how the presence of vegetation influences the movement of water.

10.2 Soils

Most of the terrestrial land surface is covered by soil, though the thickness of the soil layer may vary from a few millimetres to several metres. Soils vary widely in their morphology (i.e. in their shape and structure) and characteristics both from region to region and over much smaller spatial scales (e.g. from one end of a field to another). The diversity of soils is remarkable, and often not immediately apparent unless one excavates pits to examine what lies below the soil surface. For example, deep sandy soils may appear to be little more than a pile of sand with little or no discernible structure (though perhaps with a very thin 'crust' a few mm thick evident at the surface). At the other extreme are soils with a high clay content with very well-developed structures, where the individual soil particles are aggregated into crumbs or very much larger structures separated from one another by relatively large voids or fissures. The composition of soils, and the sizes and arrangement of the particles and larger structural units are highly variable (Plate 13). Given that water is held within the spaces between the solid particles, this diversity in the nature and arrangement of the solid phase can have a dramatic influence on the way water is retained by, and moves through, the soil profile.

10.3   Water retention by soils

Water is stored within the spaces that exist between soil particles, known as *pores*. Soil solids usually consist predominantly of a mixture of mineral quartz (silica), clay minerals and organic matter, though other materials can be present such as calcium carbonate and amorphous oxides of iron and aluminium. The sizes of these particles and how they are arranged determines the distribution of the sizes of the voids or pores present, which is a major factor determining how tightly water is held in soils and how easily water moves through soils. The sizes of particles are classified rather arbitrarily into a number of fractions (Table 10.1) now a widely used system of classification).

The sand fraction is usually dominated by solid quartz crystals that have no internal pore space. The clay-sized fraction is usually dominated by clay minerals, which comprise stacks of aluminosilicate sheets a few atoms thick (*lamellae*) separated by lamellar spaces. In some soils, iron and aluminium oxides and hydroxides constitute an important part of the clay-sized fraction, either as discrete particles or as amorphous coatings on clay minerals. Carbonates (such as calcite) can be present in all size fractions.

The sizes of pores between adjacent particles can vary widely. The intra- and intercrystalline spaces in the clay fraction range from 1 nm to a fraction of a micrometre. By contrast, the voids between closely packed sand grains will be many orders of magnitude larger (typically 10–$500\,\mu m$, depending on the size and shape of the sand grains). Larger still are the pores that result from the actions of penetrating roots or burrowing animals such as earthworms. Large cracks, voids and channels can also be found between soil aggregates or they form as a result of processes such as cultivation, soil swelling and shrinking, freezing/thawing and activities of roots and earthworms. Such cracks can be very large indeed (several cm), for example, in swelling soils in the semi-arid tropics exposed to extremes of wetting and drying.

Soils are often classified according to their 'texture', which literally means how the soils 'feel' to the touch. When moist soil is rubbed between the fingers, the presence of sand will give rise to a 'gritty' feel, whereas silt and clay-sized fractions are evident as a 'silky' and a 'sticky' feel respectively. Soil texture is determined mostly by their distribution of particle size, and the triangular diagram (Figure 10.1) shows an example of the relation-

Table 10.1 US Department of Agriculture classification of soil particle size.

Fraction	Particle diameter (μm)
Gravel	>2000
Sand	60–2000
Silt	2–60
Clay	<2

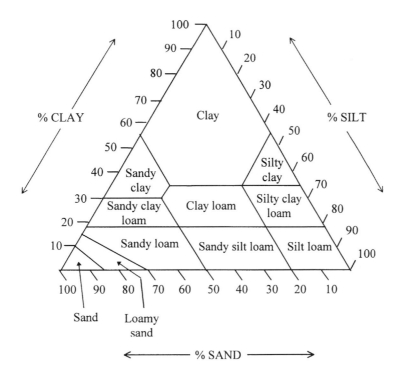

Figure 10.1 'Texture triangle'.

ship between the percentages of sand, silt and clay, and the soil textural class. The main textural classes are sand, silt, clay and loam (where loams are soils with a significant presence of sands, silts and clays), though there are subdivisions (for example, sandy loam).

The proportion of the soil volume occupied by pore space is highly variable (typically 35–70%), and depends on the degree of compaction of the soil. One measure of compaction is the *bulk density* (ρ_b) which is defined as the mass of soil solids per unit soil volume. A simple way of measuring bulk density is to push into the soil a steel cylinder of known volume, and then to dry the soil contained within the cylinder in an oven at 105°C to remove the water held in pores. The bulk density is the mass of oven-dry soil divided by the volume of the cylinder. Table 10.2 shows typical values of bulk density for a range of soils, though values for a given soil type will depend strongly on how the soil has been compressed (such as by trampling, traffic or consolidation by overlying soil) or loosened (for example, by cultivation).

The fraction of the soil volume occupied by soil solids is calculated by dividing the bulk density by the particle density of the soil solids, i.e.:

$$\text{Fraction of soil volume occupied by solids} = \frac{\text{mass of soil}}{\text{volume of soil}} \times \frac{\text{volume of solids}}{\text{mass of solids}}$$

$$= \text{bulk density} \times \frac{1}{\text{particle density}} \quad (10.1)$$

Table 10.2 Typical bulk densities and porosities of soils.

Soil texture	Bulk density ($Mg\,m^{-3}$ or $g\,cm^{-3}$)	Porosity
Sandstone	2.1	0.19
Sandy loam subsoil	1.65	0.36
Sandy loam plough layer	1.5	0.42
Clay loam subsoil	1.45	0.44
Recently ploughed clay loam	1.1	0.58

The particle densities of quartz and of most clay minerals are remarkably similar ($\sim 2.6\,Mg\,m^{-3}$ or $2600\,kg\,m^{-3}$), though the average soil particle density can be very different for soils with large components of organic matter, carbonates or iron/aluminium oxides. If one takes the particle density to be $2.6\,Mg\,m^{-3}$ ($= g\,cm^{-3}$) then the fraction of the soil volume occupied by pores (the *porosity*) is given by:

$$\text{Porosity} = 1 - (\text{bulk density}/2.6) \qquad (10.2)$$

where the bulk density divided by the particle density is the fraction of the soil volume occupied by solids. The third column in Table 10.2 uses equation (10.2) to calculate the typical porosities of the various soil types, and it is seen that the porosities of soil range from about 20 to 60%. The porosity provides a measure of the amount of water that a soil can hold when all of the pore space is saturated with water (i.e. when there are no air-filled pores present).

There are many factors that result in the effective size of the soil water reservoir being very much smaller than would be estimated from the total porosity. Figure 10.2 shows the

---

Worked example 10.1   How much water can soils hold?

Equation (10.2) suggests that a loamy top soil with a bulk density of, say, $1.3\,Mg\,m^{-3}$ would have a porosity of 50%. The roots of temperate annual crops growing in deep soils typically reach about 150 cm (1500 mm) depth, and so the volume of the pore space within the root zone for this soil would be sufficient to store $0.5 \times 1500 = 750$ mm equivalent depth of water. The 'equivalent depth' of water stored in a soil profile is the depth of the 'puddle' of water that would remain if the soil solid particles were removed, leaving the water behind. This is a convenient way of expressing water storage because it can be directly related to amounts of rainfall, which are also expressed in terms of depths of water. 750 mm of equivalent depth of water would be sufficient to meet the requirements of evaporation for over 200 days (about 7 months) assuming that a wet land surface would lose water at about $3.5\,mm\,day^{-1}$ through evaporation, which is a typical summer average for southern England. However, this is clearly a very misleading conclusion, because it is evident that the uptake of water by crops, and hence their rate of growth, becomes restricted by shortage of water during hot dry periods, usually within less than 2 weeks of the soil last being rewetted by heavy rainfall or irrigation.

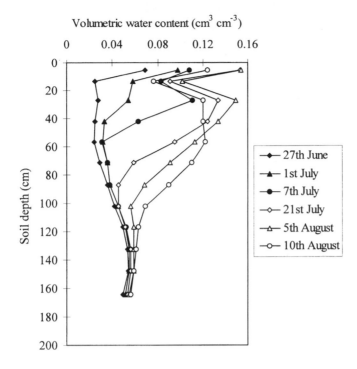

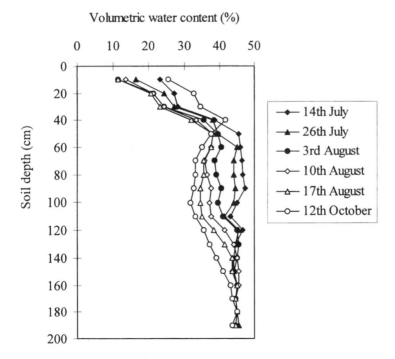

Figure 10.2 Profiles of soil water content at various stages during the growth of crops.

results of a series of measurements made in two fields of cereal crops (a very sandy soil in Niger, West Africa, and a clay loam soil in southern England) of the variation in soil water content with depth at various times during the growing season.

In each case, the measurements started shortly after a period when there was a large excess of rainfall over evaporation. Despite the heavy rainfall in the few weeks before the start of the measurements, the soil water contents throughout the profiles were substantially less than saturation. This is because water contained in pores and fissures $>60\,\mu m$ diameter can move sufficiently easily to be removed by gravity over a few days (see Case study 10.1). In the sandy soil, the loss of rapidly draining water had caused the water content of the soil to fall to $<16\%$ by volume, whereas the clay loam soil was much wetter shortly after heavy rain ($\sim40\%$ water content by volume). The difference between the soils in the amount of water lost via rapid drainage is explained because the sandy soil had a much greater quantity of large, rapidly draining pores than the clay loam soil.

Case study 10.1   Effect of pore radius on the rate of flow of water.

The effect of pore radius on the velocity of soil water movement in response to gravity can be investigated by representing soil pores as vertical capillary tubes with uniform radius (r), through which water is moving by gravity. The rate of flow of water through a capillary tube (assuming laminar rather than turbulent flow) is described by the *Hagen–Poiseuille equation*:

$$Q = \frac{\pi r^4}{8\eta} \frac{dP}{dx}$$

(10.3)

where Q is the rate of flow ($m^3\,s^{-1}$), dP/dx is the hydraulic head gradient, or pressure gradient ($Pa\,m^{-1}$), and η is the dynamic viscosity ($=10^{-3}\,Pa\,s$ for water at 20°C).

Note that Q is proportional to the fourth power of the radius. Hence, increasing the size of a pore by a factor of 10 will increase the volumetric flow rate by 10^4 (i.e. 10 000-fold). In soils, the sizes of pores can vary by many orders of magnitude (from $<0.1\,\mu m$ to $>1000\,\mu m$), so there will be a huge range of water flow rates through pores of different sizes in response to a given driving force (such as gravity).

The average velocity of flow (V) is equal to the volumetric flow rate ($Q\,m^3\,s^{-1}$) divided by the cross-sectional area of the tube. Dividing equation (10.3) by πr^2 gives:

$$V = \frac{r^2}{8\eta} \frac{dP}{dx}$$

(10.4)

Given that a 1 m-high column of water exerts a pressure of 10 kPa, the pressure gradient within a vertical capillary tube full of water due to the gravitational force acting on the water is $10^4\,Pa\,m^{-1}$. Substituting this, together with 0.001 Pa s for the dynamic viscosity of water, into equation (10.4) enables the average flow velocities to be calculated for vertical capillary tubes of different radius (Table 10.3).

Table 10.3 Water velocities through vertical capillary tubes due to gravity.

Tube radius (m)	(μm)	Flow velocity (m s⁻¹)	(cm d⁻¹)
10^{-3}	1000	1.25	1.08×10^7
10^{-4}	100	1.25×10^{-2}	108 000
10^{-5}	10	1.25×10^{-4}	1080
10^{-6}	1	1.25×10^{-6}	10.8
10^{-7}	0.1	1.25×10^{-8}	0.108

These calculations reveal that the water velocity through pores typical of the size between large sand grains (1000 μm) is 100 million times faster than through pores typical of the size within clay particles (0.1 μm). The objective of this analysis was to identify how small pores have to be before the gravitational flow of water can reasonably be regarded as negligible. However, the vertical capillary tube model used to obtain the velocities shown in Table 10.3 is a gross oversimplification of soil pores. Rather than being vertical tubes of constant radius, soil pores are tortuous and very irregular in size, having relatively large voids interlinked by narrow necks. Hence, in practice, the flow velocities through soil pores are orders of magnitude slower than predicted using the simple capillary tube model. Studies of water flow in the field suggest that the average velocity of the downward flow in response to gravity falls to negligible values (i.e. <0.1 mm day⁻¹) once the pores >30 μm radius have been drained of water.

Jean Poiseuille (1797–1869) was a French doctor who conducted research in physiology, and in particular of the circulation of blood in the arteries. He improved measurements of blood pressure and revealed that it increases and decreases on breathing out and in. His work on blood flow was extended to the mobility of water in capillary tubes and in 1840 his law was published. In 1839, G. Hagen had already come to the same conclusion and in 1925 it was renamed the Hagen–Poiseuille law.

In each of the two cases in Figure 10.2, the sequences of measurements of soil water content were obtained during a period when there was little rainfall, and so the soil profile was being virtually continuously dried via direct evaporation from the soil surface and water uptake by roots. In each case it is seen that the upper part of the profile tended to dry first. As time progressed, the depth from which water was being lost penetrated deeper into the profile, reflecting the progressive downward growth of roots. By the end of the measurement period there was evidence of roots extracting water from up to 140 cm depth (millet crop on the sandy soil) and 160 cm depth (wheat crop on the clay loam soil). Towards the end of the measurement period it appeared that little further water was being extracted from the upper part of the profile, even though there were still significant amounts of water present. This water remained unused because it was too tightly held in the soil to be extracted by roots. It is shown later that this 'residual' water is held in very fine pores. The clay loam soil has a much greater volume of very fine pores than the sandy soil, on account of having a much greater clay content. This explains why the minimum water content in the clay loam soil did not fall below about 20%, whereas the water content in the upper part of the sandy soil profile was reduced to about 4% by the end of the period of measurements.

One interpretation of the observations in Figure 10.2 is that there is, in effect, an upper limit to the amount of water that soil can hold in the long term. This effective upper limit is known as the *field capacity* (FC), and is the fraction of the soil volume occupied by pores small enough (and, therefore, have sufficient frictional resistance) for there to be negligible downward flow in response to gravity. Similarly, there appears to be an effective lower limit to the amount of water that can be extracted by plant roots, which reflects the proportion of the soil volume occupied by very fine pores that hold water too tightly for water to be extracted by plants. This lower limit is termed the *permanent wilting point* (PWP). The origin of this terminology was in the experiments conducted in the 1940s in the USA in which sunflower plants were grown in pots of soil left unwatered until the plants became permanently wilted. The water content of the soil at this time was termed the permanent wilting point. Experiments using soil with a range of texture revealed that in all cases the minimum size of pore from which plants could extract water was ~0.2 μm.

The values of the water contents corresponding to field capacity and the permanent wilting point vary widely between soils, depending on the distribution of pore sizes (Table 10.4). Hence, clay soils that have predominantly very small pores tend to have relatively high soil water contents at FC and PWP. The corresponding water contents are much lower for sandy soils, in which much of the porosity comprises relatively large voids between sand grains.

The difference between the volumetric water contents at field capacity and the permanent wilting point (last column of Table 10.4) provides a measure of the fraction of the soil volume occupied by pores small enough to hold water against gravity, but not so

Table 10.4 Typical water holding capacities of soils, expressed as volumetric water content.

Soil texture	Permanent wilting point (PWP)	Field capacity (FC)	Available water capacity (=FC − PWP)
Clay	0.28	0.44	0.16
Silty clay	0.28	0.44	0.16
Clay loam	0.23	0.44	0.21
Silty clay loam	0.20	0.42	0.22
Sandy clay loam	0.16	0.36	0.20
Loam	0.14	0.36	0.22
Silt loam	0.14	0.36	0.22
Sandy loam	0.08	0.22	0.14
Loamy sand	0.06	0.18	0.12
Sand	0.05	0.15	0.10

small that water is held too tightly for roots to extract it. This difference is termed the *available water capacity*, and provides an indication of the effective size of the soil water reservoir. Table 10.4 shows that although soils differ widely in their water contents at FC and the PWP, there is much less variation between soils in the values of the available water capacity. Loamy soils with a preponderance of mid-sized pores tend to have larger available water capacities than very sandy or very clayey soils. This is one of the major reasons why some of the most productive agricultural soils in Europe are the silty loam soils in East Anglia, Lincolnshire and the Dutch polders that were reclaimed from the sea, giving rise to their high silt content.

The available water capacities of soils typically range from about 10–20% by volume. Hence, a soil profile with average available water capacity and with roots present in reasonable quantities to 1000 mm depth would be expected to store about 150 mm of plant-available water within the root zone following heavy rainfall, once the readily drainable water had been lost during the few days following rewetting. This is consistent with the experimental evidence shown in Table 10.5 that a range of crops in a variety of locations managed to extract no more than 120 mm water from the soil profile during prolonged periods when the crops were reliant on water stored in the soil.

Table 10.5 shows examples from various studies of agricultural fields around the world of the amounts of rainfall, and the amounts of water used by the crops during the period of rapid crop growth. In each case, there was negligible loss of water through surface runoff, so all rainfall infiltrated the soil. Each of the examples shows that during the period of rapid crop growth (e.g. during late spring and summer in the UK) the amount of rainfall was very much less than the amount of water that was lost through the combination of direct evaporation from the soil surface and water uptake by plants. The difference between evaporation and rainfall was accounted for by depletion of water from the soil profile. It is evident that the actual amounts of water that were evaporated were substantially less than the *potential evaporation*. This is the evaporation that would be expected if the land surface was continually wet (i.e. is freely evaporating), and depends on the prevailing meteorological conditions, as discussed later. An interesting feature of the data in Table 10.5 is that the amount of evaporation achieved in excess of rainfall was generally in the region of 60–120 mm. This represents the amount of water that was withdrawn from the reservoir of water stored in the soil profile at the start of the period. The implication is that in each of these situations the amount of water taken up by the crops during the main period of growth was limited by the amount of water that could be stored within the soil profile.

Table 10.5 Seasonal evaporation from cropped fields.

Crop	Location	Season	Actual evaporation (E, mm)	Rainfall (P, mm)	Potential evaporation (mm)	E − P (mm)
Wheat	UK	May–July	188	68	270	120
Wheat	UK	June–August	230	134	280	96
Barley	UK	May–July	220	125	270	95
Barley	Syria	March–May	154	93	360	61
Millet	India	November–February	87	30	480	57
Barley	Syria	March–May	103	39	360	64
Peanut	India	December–February	102	0	400	102

10.4   Soil water suction

Section 10.3 introduced the idea that as a soil dries, so the larger pores empty first because they are most easily emptied, and the water remaining in the soil is held increasingly tightly. This section considers the physical basis for this phenomenon and discusses some of the implications.

A start is made by considering a simple experiment in which soil is placed on the ceramic plate in a filter funnel flooded with water and connected to a flexible manometer (Figure 10.3). At the start of the experiment, the water level in the manometer is set to be level with the soil surface (Figure 10.3a). In this state, the pores within the soil are completely saturated with water, and the water level in the manometer tells us that the water in the soil is at atmospheric pressure.

Consider what happens if the manometer is now lowered such that the water level in the manometer is some distance below the soil surface. This causes a suction to be applied to the soil water, because the soil water has a downward force applied to it due to the weight of water in the funnel side of the manometer which is not counterbalanced by water in the open side. Some water, therefore, drains from the soil, allowing air to enter to replace the water that is lost. The water draining from the soil moves into the manometer, causing the water level in the open side to rise. Eventually an equilibrium state is reached when there is no further water loss from the soil (Figure 10.3b). The soil is partially desaturated, and the water remaining in the soil is subject to a suction (S) equivalent to the hydraulic head of the column of water that is not being counterbalanced by water in the open side of the manometer. S, therefore, represents the soil water suction, and is the amount by which the pressure of the water in the soil is reduced below atmospheric pressure.

If the manometer were to be lowered further, then more water would move out of the soil until a new equilibrium is established, with a somewhat higher soil water suction. Pro-

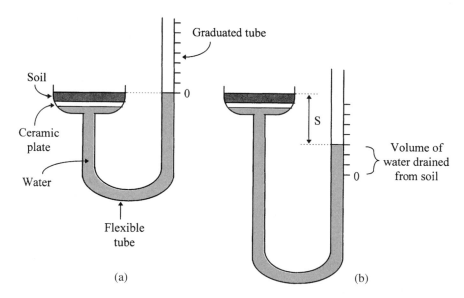

Figure 10.3 Experimental procedure for determining the water release curve for a soil.
(a) Soil water at zero suction (soil saturated).
(b) Soil water at suction 'S' (soil partly desaturated).

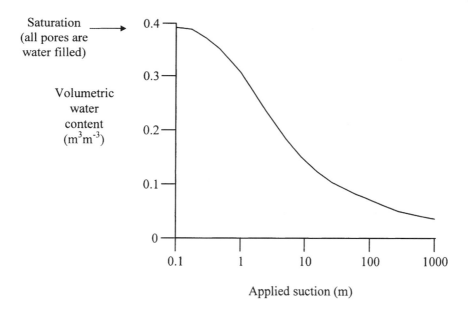

Figure 10.4 Water release curve of a soil.

gressive lowering of the manometer, allowing sufficient time for equilibration at each stage, would allow a graph to be drawn showing how the volumetric water content of the soil decreased as the suction increased (Figure 10.4). This is known as the *water release curve* for the soil.

Soil water suction is a rather abstract concept that can be difficult to visualize, but which has important implications for how soils behave. One implication is that as soil dries, so increasing suction needs to be applied to extract any further water, which is one of the reasons why plants experience increasing water stress as soil dries. This can be visualized by considering another porous medium with which you are familiar – a bath sponge. Water can easily be removed from a very wet sponge, but very much more work needs to done to extract further water from the sponge once the easily removable water has gone.

Another manifestation of soil water suction is appreciated at a very early age, when building sand castles on the beach. When sand is moist, but not too wet, then a strong sand castle can be built. The force that holds the sand grains together arises because the water between the sand grains is at lower pressure than the surrounding air (i.e. the water is under suction). The excess of the air pressure over the water pressure forces the sand grains together such that they interlock and form a strong structure. However, you may remember experiencing dismay when the tide starts to come in, because the moment that free water (at atmospheric pressure) makes contact with the sand castle, the water sandwiched between the sand grains is returned to atmospheric pressure. The binding force due to the soil water suction is lost, and the sand castle collapses. Understanding of the principles of soil physics begins at an early age!

Microscopic examination of water in soil pores provides a clue as to what is going on.

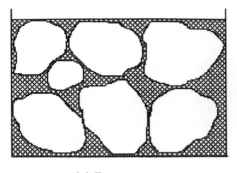
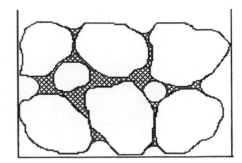

(a) Zero suction   (b) Positive suction

Figure 10.5 Distribution of water in soil pores.

Figure 10.5 shows how water might be distributed between soil particles when the soil is saturated and at zero suction (Figure 10.5a) and when the soil is partially desaturated (Figure 10.5b) and the soil water is at less than atmospheric pressure (i.e. there is a positive soil water suction).

Close inspection of the saturated case reveals that all the pore space is full of water, and that the shape of the *meniscus* at the air–water interface is flat. In the unsaturated case, water has been lost from the largest voids, remaining only in the finer pores, and the menisci at the air–water interfaces are now curved. The meniscus at the air–water interface behaves in effect like an elastic skin. Evidence of this skin-like behaviour, with its surface tension properties, is its ability to support objects denser than water (e.g. insects that skim on the surface of ponds, or floating sewing needles on the surface of water to make a home-made compass). The curvature of the menisci in the unsaturated soil is evidence that the pressure of the water is less than atmospheric pressure. If the meniscus acts, in effect, as a taut skin, then increasing the soil water suction will cause the menisci to become more tightly curved, because the greater will be the excess of the air pressure on the one side of the meniscus relative to the pressure in the soil water on the other side.

The curvature of the meniscus provides the clue about why it is the largest pores that are emptied by relatively small suctions whereas very fine pores require large suctions to be applied before they are emptied of water. Consider a simple, cylindrical soil pore full of water (Figure 10.6). If the water in the pore is at atmospheric pressure, that is at zero suction, the meniscus will be flat (A). Applying suction to the water causes the mensicus to curve, with the degree of curvature increasing as the suction is increased (B).

Once the radius of curvature is equal to the radius of the pore (i.e. when the meniscus forms a complete hemisphere – C), then any further increase in suction will cause the pore to be drained of water, as it cannot support a meniscus with a radius of curvature less than the radius of the pore itself. Consideration of the simple physics of the forces acting (see Case study 10.2) show that the suction (S – expressed in cm of hydraulic head) required to drain a pore of radius r (mm) is given by:

$$S = 1.5/r \qquad (10.5)$$

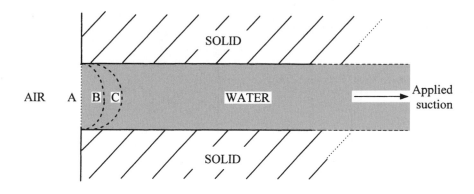

A = Zero suction
B = Moderate suction
C = Maximum suction at which pore can retain water

Figure 10.6 Effect of applied suction on the radius of curvature of the air–water interface.

Case study 10.2   Calculating the suction required to drain a pore of given size.

To extract water from a soil pore, sufficient suction has to be applied to overcome the forces of attraction that exist between water and the pore walls. Let us look again at the very simplistic soil pore given in Figure 10.6 in the case where the applied suction has been increased to the point where water is just about to be emptied from the pore. In this case, the meniscus has curved such that the angle of contact between the edge of the meniscus and the pore wall is zero (Figure 10.7).

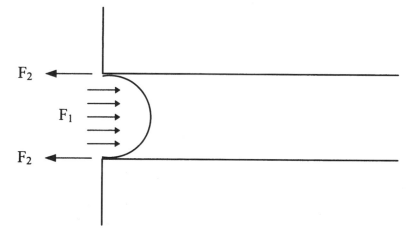

Figure 10.7 Forces acting on water in a soil pore.

Consider now the forces acting on the water in the pore. The water within the pore is at lower pressure than the air, and this pressure difference is, by definition, equal to the soil water suction. Hence, the excess of air pressure above the water pressure gives rise to the force F_1 being applied to the water in the pore. As force = pressure × area, F_1 is given by:

$$F_1 = S\pi r^2 \qquad (10.6)$$

where S is the applied suction and r is the radius of the pore. Water remains in the pore because the force due to the applied suction is exactly balanced by an opposing force (F_2) attributable to the force of attraction between the water at the rim of the meniscus and the pore wall. The magnitude of this force depends on the surface tension of water, which is a measure of the magnitude of the force per unit length of contact between the rim of the water meniscus and the pore wall. Surface tension is, in effect, a measure of the strength of the adhesive forces that exist between the water molecules and the rim of the pore. The surface tension of water (γ) is ~0.075 N m^{-1} at 20°C. Hence, the magnitude of F_2 depends on the surface tension, and the length of contact at the rim of the pore, i.e.:

$$F_2 = 2\pi r \gamma \qquad (10.7)$$

At the critical suction when a pore is just about to empty, F_1 is balanced by F_2, so that one can equate the above two expressions:

$$S\pi r^2 = 2\pi r \gamma \qquad (10.8)$$

which simplifies to the following, assuming that $\gamma = 0.075$ N m^{-1}, and that S and r are in SI units (Pa and m respectively):

$$S = \frac{0.15}{r} \qquad (10.9)$$

Rather than working in the SI units of pressure or suction (i.e. Pascals), it is often convenient in soil physics to express pressures and suctions in units of 'hydraulic head', where a unit of hydraulic head is the pressure exerted by the weight of a unit height of water column (1 cm hydraulic head exerts a pressure of 100 Pa). Converting the suction term in equation (10.9) into units of cm hydraulic head, and expressing the pore radius in mm leads to equation (10.5).

Table 10.6 uses this equation to calculate the suction required to empty the water from pores of different sizes. Note that the suctions are expressed in two units, either cm hydraulic head, or in more conventional units of pressure (in this case, kPa). Remember that suction is defined as the amount by which the water pressure is reduced below atmospheric pressure. Remember that 1 kPa pressure is equivalent to the pressure exerted by a

Table 10.6 Relationship between pore radius, and the suction required to drain water from the pore.

Pore radius (μm)	Minimum suction required to empty water from the pore (kPa)	cm hydraulic head	Comments
1500	0.1	1	3 mm diameter earthworm channel
100	1.5	15	radius of a fine grass root
15	10	100	maximum radius of a pore retaining water at field capacity
1	150	1500	radius of a large bacterium
0.1	1500	15 000	maximum radius of a pore retaining water at the permanent wilting point

column of water 10 cm high, and that atmospheric pressure is equivalent approximately to 1000 cm hydraulic head, or 100 kPa.

Large fissures, or large pores resulting from the activity of agencies such as roots or earthworms, are emptied of water at very low suctions. It was shown above that pores >15–30 μm radius offer such low frictional resistance to water movement that water is rapidly drained by gravity following heavy rainfall. Only pores smaller than this can retain water for long periods. Hence, in a soil at field capacity, Table 10.6 suggests that the soil water suction would be expected to be in the range 50–100 cm (5–10 kPa). This is confirmed by field measurements of the soil water suction a few days after heavy rainfall.

In soil at the permanent wilting point, where the only water remaining is held in pores <0.1 μm, the soil water suction is very large indeed (15 000 cm or 1500 kPa). This implies that a suction equivalent to 15 atm is required to extract any further water. Hence, soil at the PWP will feel very dry to the touch. This suction represents the maximum suction that plant roots typically can exert on the soil water.

It is notable that roots grow along relatively large soil pores that cannot retain water, even when the soil is at field capacity. This is of benefit to plants because the roots of most plants require a good oxygen supply. However, except in very dry soils, the roots are in good contact with the finer pores in the surrounding soil matrix responsible for storing water. This contact is maintained by a variety of mechanisms, including the production of very fine hairs on the surface of roots just visible to the naked eye (many species have them – try pulling up some weeds and see if you can see any). Also, many roots excrete a gel-like mucilage (which consists of long-range polysaccharide material) which maintains contact with the soil water, which is why soil particles often adhere to roots when roots are pulled up from the soil.

Figure 10.4 showed that for a given soil there will be a characteristic relationship between the water content of the soil and the soil water suction. The shape of this curve reflects the distribution of pore size, which has such a major effect on the hydraulic behaviour of the soil. Hence, the shape of the so-called *water-release curve* (otherwise known as the *moisture characteristic* of the soil) is an important soil property to measure when trying to predict the movement and retention of water in soils. Figure 10.8a shows examples of the water-release curves for three soils which have very different distributions of pore size (Figure 10.8b).

In each case, as the suction is increased from zero up to some threshold value (known

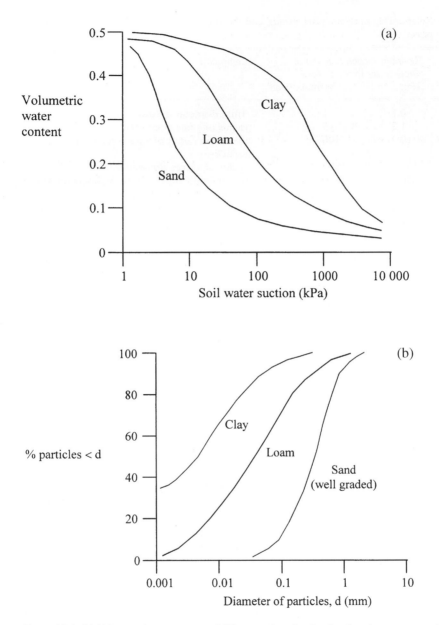

Figure 10.8 (a) Water release curve and (b) pore size distribution for three contrasting soils.

as the *air-entry suction*) there is little loss of water because insufficient suction has been applied to empty even the largest pores present. A small increase in suction above the air-entry suction results in a large loss of water from the sandy soil, but very much less drying of the clay. This is because the clay tends to have much finer pores than the sand. Substantial amounts of water are only released from the clay when much higher suctions are applied. The suctions corresponding approximately to the field capacity and permanent wilting points are 10 and 1500 kPa respectively. The water contents for the three soils at these suctions, and the water-holding capacities for the two soils, are consistent with the results presented in Table 10.4, and they suggest that the loam can hold much more water between field capacity and the permanent wilting point than either the clay or the sandy soil.

10.5   Movement of water through soils

The earlier sections have touched on factors controlling the movement of water through soils, and have introduced the idea that the velocity of water flow through soil pores depends very strongly on the sizes of pores through which water is flowing. In this section the factors controlling the rate of water flow through soils will be discussed in more detail.

Soil water, like any other physical body, moves in response to the application of a resultant force. In the case of water flowing through soils (whether it be flow towards the root surface, upward flow to replace water evaporating from the soil surface or downward flow driven by gravity), there are usually two forces acting on the soil water. The first is gravity, which provides a constant downward force of 9.81 Newtons acting on each kg of water (equivalent to $9810\,\mathrm{N\,m^{-3}}$ water, assuming water has a density of $1000\,\mathrm{kg\,m^{-3}}$). The second force that usually applies to water within the soil is that which arises as a consequence of a gradient in hydrostatic pressure (i.e. a gradient in suction), whereby water tends to move from regions of wet soil (low suction) towards drier soil. Bearing in mind that pressure is force per unit area, one can see that the units of a gradient in suction can be expressed, like the gravitational force, in terms of a force acting per unit volume of soil water:

$$\text{Suction gradient} = \frac{\text{change in suction}}{\text{distance}} \equiv \frac{\mathrm{Pa}}{\mathrm{m}} \equiv \frac{\mathrm{N\,m^{-2}}}{\mathrm{m}} \equiv \mathrm{N\,m^{-3}} \qquad (10.10)$$

Very often, these forces act in opposition. For example, imagine the case of a wet soil profile that has been subject to progressive drying close to the surface (e.g. through water uptake by roots or by direct evaporation from the soil surface). Figure 10.2 provides an example of two such cases. In such a situation, the soil at the top of the profile will be much drier than at depth, and so will have a higher suction. In the absence of gravity, water would be expected to move upwards from the relatively wet soil to the drier soil, being driven by the gradient in suction. In other words, water would be expected to move in the direction of decreasing pressure. However, gravity would be acting in opposition, applying a downward force. Whether water flows upwards or downwards depends on which of these forces is the stronger. Case study 10.3 shows how the direction of water flow can be analysed in a situation where there is a 'tug of war' between the suction gradient and gravity.

Case study 10.3   In what direction does water move in soil?

(a)

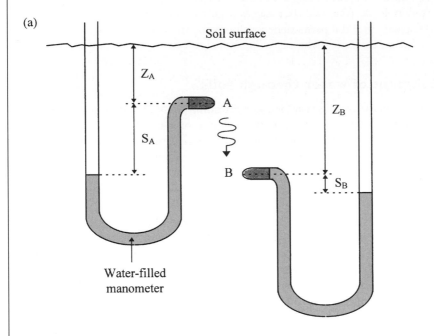

(b)

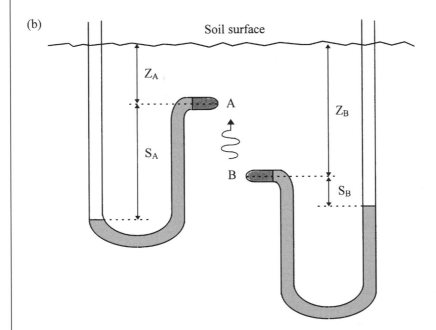

Figure 10.9 Use of tensiometers to determine the direction of water movement in soil.

Consider the locations A and B in Figure 10.9. In each location, an instrument called a *tensiometer* has been inserted to measure the soil water suction. The tensiometer is in effect a variant of the filter funnel device (Figure 10.3) that was used to apply a suction to a soil sample. It consists of a porous ceramic cup containing water placed in contact with the soil. The principle is that water exchanges with the soil via the pores in the ceramic until the suction of the water within the cup is at equilibrium with the soil water suction. For the design of the tensiometer shown in Figure 10.9, the water pressure within the ceramic cup is measured using a very simple water manometer. Hence, the soil water suction at A is given by the hydraulic head (S_A) indicated by the manometer.

The tensiometers indicate that the soil at A is drier than that at B (i.e. has higher soil suction), so in the absence of gravity water would be expected to move upward from B to A. However, gravity is applying a downward force trying to move water from A to B. In which direction does the water move? The answer is obtained by comparing the relative elevations of the water levels in the open ends of the two manometers. As in Figure 10.9a, water would be expected to move from A to B in order for the water levels in the two manometers to move towards equilibration. By contrast, if the soil at A was much drier (Figure 10.9b), then the flow of water would be upward (from B to A).

In effect, what happens is that water moves in the direction of decreasing potential energy. The potential energy of the soil water can be thought of as the capacity of the water to do work, which depends on both elevation and suction. Water near the top of a soil profile will be more energetic than water at greater depth on the basis of elevation. However, this effect might be offset if the water close to the top of the soil profile is held very tightly (i.e. is at high suction), which would considerably reduce the capacity of the water to do work.

A convenient measure of the potential energy of the soil water (usually abbreviated to the *soil water potential*), taking account of both the elevation and the suction, is indicated by the water level in the open side of the manometer relative to the soil surface (Figure 10.9), which is used as an arbitrary reference. Measured in this way, the soil water potential is calculated as:

Soil water potential $= -(\text{depth} + \text{suction})$ (10.11)

Note the minus sign, which indicates that the soil water potential decreases (i.e. becomes more negative) as either the depth or the suction increases.

Worked example 10.2   Calculating the soil water potential.

Table 10.7 shows the case of a relatively dry soil (i.e. it has a relatively large suction) close to the surface, but is much wetter at depth. The third column uses equation (10.11) to calculate the soil water potential at each of the depths that the suction was measured.

What can be concluded about the direction that water is moving through this soil profile?

Table 10.7 Calculating soil water potential.

Depth (cm)	Suction (cm)	Water potential (cm)	Water lost from each layer during the preceding week (mm)
10	300	−310	18
20	120	−140	10
30	95	−125	7
40	88	−128	4
50	83	−133	3
60	83	−143	2

Solution

Assuming that water moves in the direction of decreasing potential energy, then water at any point in the soil profile would be expected to move in the direction of increasingly negative soil water potential. Hence, water held between 10 and 20 cm depth, and between 20 and 30 cm depth, would be expected to be moving upwards, because the downward-acting gravitational force is being overwhelmed by the gradient in soil water suction that is forcing water upwards. By contrast, the gradient in soil water suction below 30 cm depth is relatively small, and insufficient to overcome gravity. Hence, below 30 cm depth water would be moving downwards.

Worked example 10.3   Estimating evaporation and deep drainage from soil water records.

One application of the ideas in Worked example 10.2 is in field studies to investigate how much water is being lost from soil profiles by evaporation and water uptake by plants, and how much is being lost by deep drainage (which will ultimately contribute either to groundwater or to surface water via a spring at a lower position in the landscape). To illustrate this, imagine that the soil profile measurements in Table 10.7 were made after 1 week where there had been no rainfall. Hence, the soil profile had been continuously drying during the preceding week both as a result of downward movement via gravity and the upward movement via water uptake by roots and direct evaporation from the soil surface. The fourth column in Table 10.7 is based on measurements of the change in soil water content in each layer over the preceding week and it shows the amount of water (mm) that had been lost from each layer. From the evidence of the measurements of soil water potential, one can conclude that somewhere between 30 and 40 cm depth there is, in effect, a water-shed that separates the upward- and downward-flowing water. Hence, one might reasonably assume that the water lost from the upper three layers ($18 + 10 + 7 = 35$ mm) was lost by upward flow (i.e. by evaporation from plant leaves or the soil surface), whereas the water lost from the bottom three layers ($4 + 3 + 2 = 9$ mm) was lost as drainage. Note that in this example, the evaporation of 35 mm during the week is equivalent to 5 mm day^{-1}, which is typical of the potential evaporation rate for a hot summer's day in Northern Europe. By regular monitoring of the water contents and water potentials down a soil profile and carrying out these calculations, it is possible to deduce the amounts of water lost from the land surface by deep drainage and evaporation over weeks, months or years. This provides a valuable tool for studying issues such as how changes in land use influence hydrological processes such as groundwater recharge. Examples include the effects of clearing natural savannah vegetation to grow agricultural crops in Sub-Saharan Africa, or the effects of returning agricultural land to woodland.

So far, the concept of soil water potential has been used to investigate the *direction* of water movement through soil, making use of the idea that water moves along gradients of water potential in the direction of decreasing energy. A simple extension of this idea is that the *rate* of water movement is proportional to the gradient in water potential, because the magnitude of the gradient in water potential provides a measure of the driving force for water flow (Worked example 10.4). This leads to an expression for describing the rate of water flow, that resembles an equation developed in 1856 by the French hydraulics engineer Henry Darcy, who was consulted to improve the water supply of the city of Dijon. The expression is now known as *Darcy's law*:

$$Q/A = F = -K.d\psi/dz \tag{10.12}$$

where Q is the volume flow-rate (m^3 s^{-1}), A is the area, ψ is the soil water potential and z is depth, and so $d\psi/dz$ is the gradient in soil water potential. The rate of flow of water (F) is usually expressed in terms of the volume of water flowing per unit area per unit time, which has the dimensions of velocity (e.g. m s^{-1} or mm h^{-1}). The minus sign indicates that the flow of water is in the direction of decreasing (i.e. increasingly negative) soil water potential. The coefficient of proportionality in Darcy's law (K) is known as the *hydraulic conductivity*, and provides a measure of the ability of the soil to conduct water. It shall be shown below that hydraulic conductivity depends strongly on the sizes of the water-filled pores responsible for conducting the water, because as already shown (Case study 10.1), the frictional resistance to water flow through small pores can be orders of magnitude greater than in larger pores. Notice how equation (10.12) compares with other transport phenomena encountered already, such as Fourier's law in Chapter 2.

Worked example 10.4   Calculation of the gradient in water potential.

The gradient in water potential between two points is the difference in water potential between the points divided by the distance between them, and can be illustrated using the measurements in Table 10.7, which shows the water potentials measured at a number of depths in a soil profile. Determine the gradient in soil water potential between (a) 10 and 20 cm and (b) 50 and 60 cm depth.

Solution

(a) The gradient in soil water potential in the soil between 10 and 20 cm depth is given by:

$$\frac{[\text{water potential at 20 cm depth}] - [\text{Water potential at 10 cm depth}]}{[\text{depth at 20 cm depth}] - [\text{Depth at 10 cm depth}]}$$

$$= \frac{[-140] - [-310]}{[20] - [10]} = \frac{170}{10} = 17 \, \text{cm cm}^{-1} \tag{10.13}$$

Note that the sign of the potential gradient is positive, so this will lead to a negative F (i.e. upward flow) when the water flux is calculated using Darcy's law.

(b) Consider now the gradient in water potential that exists between 50 and 60 cm depth in the case of the soil profile in Table 10.7. In this case, the soil water suction at 50 and 60 cm depth is the same (83 cm), and so it is only the difference in elevation that causes the soil water potentials at 50 and 60 cm depth to be different. Following the example above, the gradient of water potential between 50 and 60 cm depth is calculated as:

$$\frac{[-143] - [133]}{[60] - [50]} = \frac{-10}{10} = -1 \, \text{cm cm}^{-1} \tag{10.14}$$

In this case the sign of the potential gradient is negative, indicating that the flow rate will be positive (i.e. downward, in the direction of increasing depth). Note that in this particular case, where gravity is the only driving force, the water potential gradient is $-1 \, \text{cm cm}^{-1}$.

In the special case where there are no suction gradients contributing to the gradient in soil water potential (i.e. gravity is the only driving force for water flow), then $d\psi/dz$ is $-1 \, \text{cm cm}^{-1}$ (see Worked example 10.4). Substituting $d\psi/dz = -1$ into Darcy's law (equation (10.12)) shows that when gravity is the only driving force for flow, then the rate of flow of water is numerically equal to the hydraulic conductivity of the soil. Hence, one interpretation of the hydraulic conductivity is that it represents the rate of water flow that would occur due to gravity alone.

One of the features of soil that has a huge impact on the way that soils behave is that the hydraulic conductivity depends very dramatically on how wet the soil is. When the soil is saturated with water, then all of the soil pores will be water-filled and contributing to the flow of water. If the soil dries a little, then water will be lost from the largest pores, which will no longer be contributing to the flow of water. It was shown earlier by analogy with the flow of water along capillary tubes (see Case study 10.1) that the velocity of water flow along a tube, in response to a given driving force, depends on the square of the radius of the tube. Table 10.6 indicated that the sizes of pores in soils can range over four orders of magnitude, and hence the velocity of water flow through the different sized pores is likely to range over eight orders of magnitude in response to a given driving force (i.e. be 100 000 000 times faster through an earthworm channel than through the pores within clay crystals). It is, therefore, not surprising that as the water content of a soil is decreased, so there is a dramatic decrease in their ability of soil to conduct water.

Figure 10.10 shows how the hydraulic conductivity of two soils is related to soil water content (note the logarithmic scale for hydraulic conductivity). In both cases the hydraulic conductivity is greatest when the soils are saturated, as all of the pores (including the largest pores) are water-filled. Note that the hydraulic conductivity at saturation is much greater for the sandy soil (almost 1000 mm h^{-1}) than for the clay soil (5 mm h^{-1}). This is because the sandy soil has a significant proportion of its pore space contributed to by the relatively large voids between sand grains. By contrast, the clay soil in question has little by way of relatively large pores which offer low resistance to water flow.

These values for the hydraulic conductivity at saturation imply that the sandy soil can conduct water downwards at almost 1000 mm h^{-1} in response to gravity alone, compared

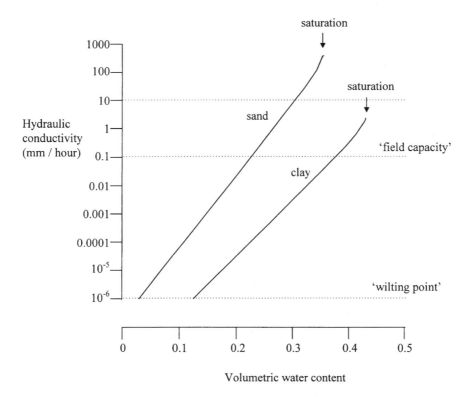

Figure 10.10 Effect of soil water content on hydraulic conductivity of two contrasting soils.

with $5\,mm\,h^{-1}$ in the case of the clay. Hence, the sandy soil could easily accommodate the water falling during a heavy shower of rain with a peak intensity of $10\,mm\,h^{-1}$ (see the upper dashed line in Figure 10.10). Indeed, the sandy soil could cope with such a rain event without having to get wetter than about 30% water content. Hence, a puddle would be most unlikely to form on the sand. By contrast, even with all of its pores water-filled, the clay soil would be unable to conduct water by gravitational flow at $>5\,mm\,h^{-1}$, so it would be expected that water would accumulate on the soil surface leading potentially to surface run-off and possible erosion.

For both soils, the hydraulic conductivity decreases by many orders of magnitude as the soil dries. One application of this dramatic decrease in hydraulic conductivity as the soil dries is in the interpretation of the *field capacity* concept used earlier. It was shown above that K represents the gravitational water flux. If one assumes, rather arbitrarily, that soils are effectively no longer losing water through drainage (and are hence at field capacity) when K falls to about $0.1\,mm\,h^{-1}$, then one can deduce from Figure 10.10 that the field capacity for the sandy and the clay soils correspond to water contents of about 25 and 40% respectively. This broadly agrees with the field capacity values presented in Table 10.4.

When the soil is very dry, the hydraulic conductivity decreases to extremely low values because the only water-filled pores remaining are very small and, therefore, have extremely high frictional resistance to water flow. It was noted in Table 10.6 that roots grow in pores that are much larger than those responsible for holding the plant-available water, implying that water has to move through a matrix of water-filled pores to reach the root surface. If the soil hydraulic conductivity falls below about $10^{-5}\,mm\,h^{-1}$, then the movement of water through the soil immediately around roots becomes a major factor limitation to the water uptake by plant roots. When $K < 10^{-7}\,mm\,h^{-1}$, then the rate of water uptake by root systems is likely to be only a tiny fraction of the water required to meet the evaporative demand. It has been shown that for a wide range of soils that K this small occurs when the soil water suction is of a similar order of magnitude to the permanent wilting point ($15\,000\,cm$ suction). This provides a physically based explanation why it is that plants permanently wilt once the soil water suction is in this range. Figure 10.10 shows that the water contents corresponding to $K = 10^{-6}\,mm\,h^{-1}$ for the sandy and clay soils (3 and 13% respectively) are similar to the 'typical' permanent wilting point soil water contents listed in Table 10.4.

10.6   Soil–water balance

This section considers the fate of water once it arrives at the soil surface through precipitation or is applied to the soil surface artificially as irrigation. The possible fates of water are illustrated in Figure 10.11, which shows that precipitation or irrigation arriving at the soil surface either runs off or infiltrates the soil. Water that infiltrates the soil obviously increases, at least temporarily, the amount of water stored within the soil profile. Water held within the soil profile will then at some stage be lost from the profile either by deep drainage in response to the gravitational pull, uptake by roots, or evaporation from the soil surface.

The difference between the various gains and losses must equal the change in the amount of water stored in the soil profile. Hence, the water balance for a soil profile can be expressed as:

$$P = E_s + T + D + R + \Delta S \qquad (10.15)$$

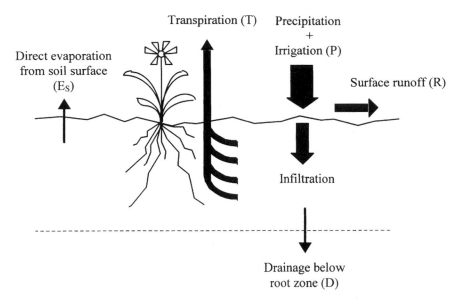

Figure 10.11 Soil–water balance components.

where P is precipitation (including any irrigation), E_s is direct evaporation from the soil surface, T is transpiration (the loss of water through evaporation from leaves which is replaced by uptake of water from the soil by the roots, see Section 11.4), D is deep drainage of water out of the base of the soil profile, R is surface runoff (negative values indicate run on) and ΔS is the change in water content of the soil profile.

Soil water balances can be calculated over any period. For example, an annual balance would be useful to know when studying the effect of changes in land use on the amount of deep drainage that contributes potentially to annual groundwater recharge. Daily balances are of interest to farmers who need to know when it is necessary to irrigate crops to maximize yield. Diurnal patterns of evaporation might be of interest to meteorologists interested in hour-by-hour changes in the fluxes of water vapour and sensible heat into the atmosphere from the land surface.

When it rains, it is possible that not all of the water will infiltrate into the soil. During prolonged and intense rain events puddles can form on the land surface, which can lead to overland flow of water (known as *surface runoff*). Surface runoff might result in increased water deposition in downhill areas where overland flow is collected, or else might flow directly into surface water bodies (e.g. drainage ditches, streams or lakes). Surface runoff is also an important agency responsible for erosion of soil. The amount of runoff that occurs during a rain event clearly depends on the intensity and the amount of rainfall, given that a light shower of rain might cause no runoff, whereas an intense storm could lead to a large proportion of the rainfall running off and causing severe loss of soil through water erosion. It has already been noted that the generation of runoff depends also on the soil hydraulic conductivity (see Section 10.5).

Once water infiltrates into the soil profile, it is stored within the pores that exist between the soil particles and in water films covering the surfaces of particles. This store of water is depleted by direct evaporation of water from the soil surface, or by evaporative loss of water from the leaves of vegetation, which is replaced by water being extracted from the soil by roots. These evaporation processes are continuous, though the actual rate of evaporation depends in part on the prevailing weather (as discussed below) and in part on the availability of water in the soil. Evaporation from the soil surface and the uptake of water by roots tends to decrease as the soil dries and water becomes less available.

The amount of water taken up by roots and then evaporated from the leaves of vegetation is linked closely to the rate of growth of vegetation. For example, in the UK most gardeners find that the growth of lawns is most vigorous during late spring/early summer when the soil profile is still wet following the wetting up of the soil profile during the winter. However, during a dry summer the soil becomes much drier, and the grass grows more slowly (and may even turn brown during periods of extreme water stress). This reduction in growth is linked to the reduced availability of water, and hence the reduced rate of evaporation from leaves.

The pore space within the soil profile obviously has a finite volume, so there is a limit to how much water a soil profile can hold. The soil profile gets progressively wetter during periods where the amount of infiltration exceeds the rate at which water is evaporated. When the soil becomes very wet, water is lost from the soil profile as drainage resulting from the downward movement of water caused by gravity. Deep percolation of water is the primary source of water for recharging aquifers (see Chapter 5), and so is of considerable interest with respect to the maintenance of groundwater supplies. Deep percolation of water can also be responsible for transporting solutes from the soil surface into groundwater. This is the subject of the next section.

10.7 Leaching of solutes through soil profiles

The leaching of solutes is currently the subject of much attention because of the possible contamination of groundwater that is pumped out of boreholes for domestic water supplies (Chapter 4). The soil solution contains a wide variety of solutes both as a result of natural processes (e.g. the decomposition of nitrogen-rich organic matter resulting in the production of nitrates) and of applications of materials to the land surface by people, including fertilizers, pesticides, slurry and waste. If water is percolating through the soil, then this creates the potential for the leaching of any solutes present in the soil solution.

This can be illustrated by considering a simple experiment in which a pulse of a solute (such as bromide) is applied to the surface of a column of soil which is then continuously irrigated using a gentle sprinkler to generate a slow downward flow of water through the column. The dotted line in Figure 10.12 shows the variation in concentration of the solute through the profile immediately after the pulse was applied.

How will this pulse have moved after 50 mm of water has passed downwards through the column? If the 'new' water had acted as a simple piston pushing the 'old' water within the soil downwards, then the profile of solute concentration would simply have been displaced downwards as shown by the dashed line. The amount of the displacement would be the depth of soil required to store the 50 mm of water that had been applied since the addition of the solute to the soil surface. With the gentle irrigation regime used, the volumetric water content of the soil is likely to be close to the field capacity value for the soil. Let us assume that this value is 25% by volume, which Table 10.4 suggests is typical of a sandy loam soil. If so, then $50/0.25 = 200$ mm is the depth of soil required to accommodate

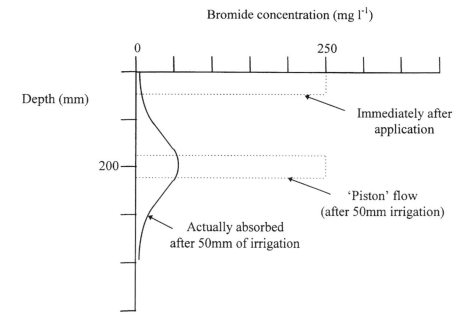

Figure 10.12 Downward movement through the soil profile of a pulse of solute applied to the soil surface.

50 mm of water, and so one would expect the pulse of solute to appear at 200 mm depth in the soil column.

However, in practice the water that infiltrates into the soil does not act as a simple piston. Instead, some of the water moves relatively quickly downwards through large pores, whereas the water entering small pores moves at much slower velocity (see Case study 10.1). The result is that most of the solute moves at close to the average pore water velocity (which is equivalent to the 'piston flow' velocity predicted by the dashed line), but some of the solute moves much more quickly whereas some hardly moves at all. The resulting profile of solute concentration is given by the solid line, which shows that the solute pulse tends to disperse as it is moved down the profile by the leaching process. This is known as *hydrodynamic dispersion* because it is caused by water (and therefore any solute it contains) moving at much higher velocity through large diameter pores than through small pores. The greater the range of pore size, the more dispersion that occurs.

Figure 10.13 shows the profile of nitrate concentration measured through the upper 30 m of chalk on the South Downs in Hampshire in 1976. One of the features is that there are localized, slightly dispersed peaks of nitrate evident at various depths in the profile. These are associated with pulses of nitrate, which were introduced into the thin mantle of soil overlaying the chalk at various times during the preceding decades. Of particular note is the large peak in nitrate concentration that is evident at around 30 m depth. This has been attributed to the release of a large pulse of nitrate into the soil that occurred when grassland was ploughed up in the autumn of 1947 in order to increase the land area used for the production of arable crops during the period of food shortage following the Second World War. Incorporation of the grass into the soil stimulated rapid microbial breakdown

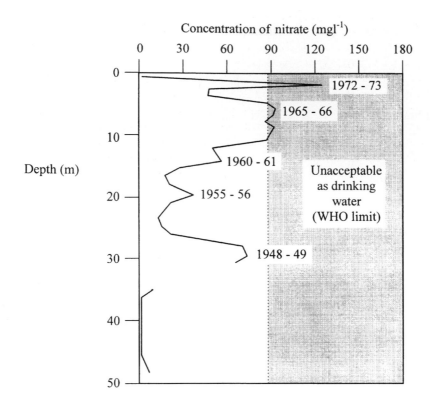

Figure 10.13 Profile of nitrate concentration measured through a core of chalk sampled in 1976 from the South Downs of England. (WHO is the World Health Organisation)

of the grass residue, which released large quantities of nitrate that were leached downwards into the chalk during the following winter. After 29 years this pulse had reached 33 m depth, and so has been travelling $33/29 = 1.2\,\mathrm{m\,year^{-1}}$ on average.

In this location, the average annual rainfall is about 500 mm, of which about 400 mm is likely to have been lost through evapotranspiration, leaving 100 mm being lost from the soil profile as deep drainage. Hence, 100 mm of deep drainage each year appears to have moved the nitrate pulse downwards by an average of $1.2\,\mathrm{m\,year^{-1}}$, implying that the 'effective' volumetric water content of the chalk is about 8%.

Another feature of the chalk example in Figure 10.13 is that there appears to be remarkably little dispersion of the nitrate pulse, even after 29 years. This is because the chalk matrix contains pores of remarkably uniform size. A very different situation occurs in soils that have large continuous fissures through a matrix of fine pores. In this situation, water can move through the fissures at velocities many orders of magnitude faster than occurs in the fine pore matrix. This can enable a small proportion of the solute present in the soil solution to be transported very rapidly to a great depth. Such fissure flow is likely to be generated during intense rain events during which the hydraulic conductivity of the fine pore matrix is insufficient to accommodate the intense input of water.

Rapid flow through fissures can be an important mechanism for the leaching of potential groundwater contaminants. Many potential contaminants are, to some degree, adsorbed onto soil particles (principally clay minerals or soil organic matter) which has the effect of retarding their downward flow relative to the flow of water. Also, organic compounds are generally broken down as a result of microbial degradation. Retardation of movement as a result of adsorption processes means that organic compounds are normally resident in the microbially active upper part of the soil profile for a sufficient time for the compounds to be degraded naturally. However, rapid flow through fissures can result in small quantities of potential groundwater contaminants 'bypassing' the adsorbing and biologically active soil horizons. The application of the physics of water and solute flow through soil pores is making an important contribution to understanding how to assess and manage the risk of contamination of groundwater and surface waters.

10.8 Evaporation from the land surface

It is evident that the evaporation of water from the land surface (whether from the leaves of vegetation or direct from the soil surface) is an important component of the soil water budget. This section explores the factors controlling the amount of water lost through evaporation from the land surface.

There are requirements, as encountered in Chapters 2 and 3, that need to be fulfilled if water is to continue evaporating from a surface:

- Supply of energy
- Mechanism for the transfer of water vapour away from the surface
- Supply of water for evaporation.

10.8.1 Energy requirement for evaporation

Energy is required to break the intermolecular bonds that exist in liquid water in order to produce vapour. The amount of energy required to convert a unit mass of liquid water into vapour (the latent heat of vaporization, L) is unusually large for water by comparison with other liquids, being around $2.5\,\mathrm{MJ\,kg^{-1}}$ at room temperature. In the case of evaporation from the land surface, this energy is provided, directly or indirectly, from solar radiation. On a clear summer's day in the UK, solar radiation supplies about $20\,\mathrm{MJ\,m^{-2}}$ of land surface during the course of the day. If *all* of this energy was used to vaporize water, then the amount evaporated would be $20/2.5 = 8\,\mathrm{kg}$ water $\mathrm{m^{-2}}$ of land surface. Eight kilograms of water spread over $1\,\mathrm{m^2}$ is equivalent to $8\,\mathrm{mm}$ depth of water evaporating from a pool of water in a day.

In practice, not all of the incident solar radiation is used in evaporation. Some is reflected. Some is lost via longwave (i.e. thermal) radiation that depends on the temperature of the surface. The difference between the incoming and outgoing radiation is termed the *net radiation* (R_n) (see Section 7.7.1), and represents the radiant energy available for evaporating water and for heating the soil and the overlying air. The radiant energy absorbed by the land surface must be dissipated as either sensible or latent heat, so we can write:

$$R_n = H + LE + G \tag{10.16}$$

where G (*ground heat flux*) is the rate at which heat is conducted downwards into the ground and H (*sensible heat flux*) is the rate at which sensible heat is transferred into the overlying air by conduction or convection. LE (*latent heat flux*) is the rate at which energy is dissipated by the vaporization of water, and is the product of the rate of evaporation (E, in SI units of $kg\,m^{-2}\,s^{-1}$) and amount of energy required to evaporate each kg of water. Each term in equation (10.16) represents an energy flux per unit land area ($J\,m^{-2}\,s^{-1} \equiv W\,m^{-2}$).

Equation (10.16) suggests that not all of the radiant energy incident on the land surface is used to evaporate water. The fraction used to evaporate water is rarely more than about 60% in the UK, even for a very wet surface such as a lake. Hence, although the amount of solar energy incident on the land surface during a clear July day in the UK might be sufficient to evaporate 8 mm depth of water if *all* of it was used to evaporate water, the amount of water that would actually evaporate from a water body is very much less, and rarely exceeds 5 mm day^{-1}.

It is evident that there is a weather-determined maximum rate at which evaporation can occur, even for a wet surface with an abundant supply of water, and that this is determined to a large extent by the amount of radiant energy absorbed by the surface in question. However, other weather factors also influence evaporation, particularly the wind speed and the temperature and humidity of the air which arise mainly through their effects on the rate at which water vapour is transported away from the evaporating surface.

10.8.2   *Energy balance of wet and dry land surfaces*

Figure 10.14 illustrates how the components of the surface energy budget might change during the course of a clear spring day in southern England for either a wet surface such as a sodden field (Figure 10.14a) or a very dry surface, such as very dry, bare soil (Figure 10.14b).

Net radiation is slightly negative at night, because there is negligible incoming solar radiation although long-wave radiation continues to be emitted from the soil surface. At night, the energy lost from the surface is supplied by heat being conducted upwards from relatively warm soil at depth. Vertical conduction of heat through soils can be described by:

$$G = -K_{heat}.dT/dz \tag{10.17}$$

where T is soil temperature and z is depth, so that dT/dz is the vertical gradient in temperature that drives thermal energy conduction to or from the surface. K_{heat} is the thermal conductivity of soil, which is defined as the rate of flow of energy per unit land area ($J\,m^{-2}\,s^{-1} = W\,m^{-2}$) that is generated by a temperature gradient of $-1°C\,m^{-1}$.

After dawn, R_n becomes positive as a result of incoming solar radiation. The temperature of the soil surface rises (middle graphs in Figure 10.14) thereby reversing the direction of G, indicating that heat is now moving downwards into the soil profile. Energy is also dissipated by evaporation (*LE*) and the convection of sensible heat into the overlying air (*H*). The fraction of R_n accounted for by G tends to be less in the afternoon than in the morning. This is because the deeper soil layers warm up in the afternoon, thereby reducing the steepness of the temperature gradient (dT/dz) close to the soil surface.

There are interesting contrasts between the energy balances for the wet and dry soil cases illustrated in Figure 10.14. The relatively slow dissipation of energy from the dry soil

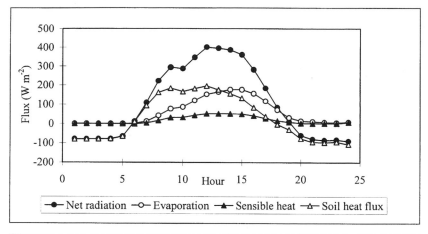

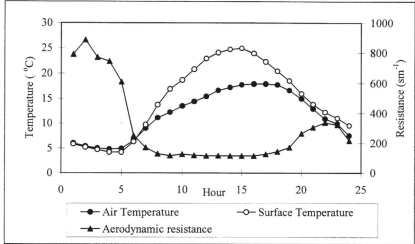

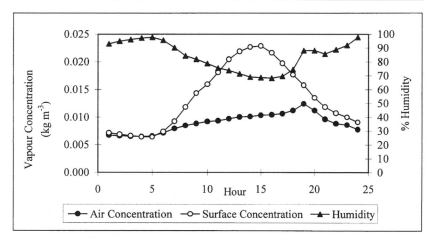

Figure 10.14 (a) Wet soil surface energy balance.

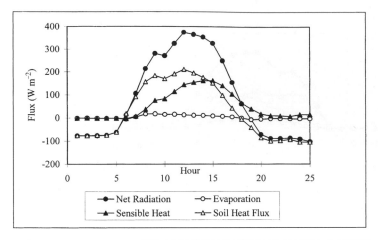

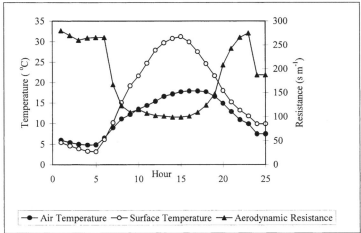

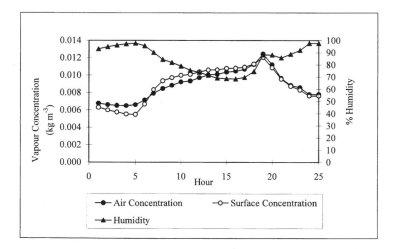

Figure 10.14 (b) Dry soil surface energy balance.

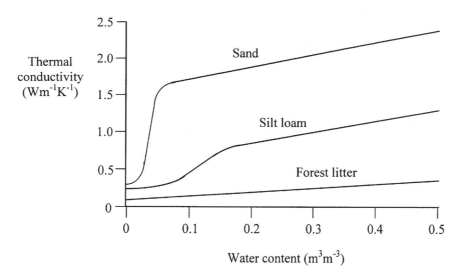

Figure 10.15 Effect of soil water content on the thermal conductivity of soil.

via evaporation was largely compensated for by an increase in the sensible heat flux (H). This was brought about by the soil surface temperature becoming hotter in the dry soil (32°C maximum) than the wet soil (25°C maximum), thereby enhancing the conduction and convection of sensible heat into the overlying air.

Given that the surface temperature tended to be greater in the dry soil, it is perhaps surprising that this did not result in faster conduction of energy downwards into the soil profile (i.e. a higher G). This is because dry soil is a poor conductor of heat compared with wet soil. When a soil becomes wetter, some of the air (a very poor conductor of heat) in the voids between soil particles is replaced by water, which is a good heat conductor and acts as a thermal 'bridge' for heat conduction between adjoining soil particles. The thermal conductivity of a very wet soil might be several times larger than for dry soil (Figure 10.15).

There are interesting practical applications of the effect of soil wetness on soil thermal properties. One example is that grape vines are irrigated during cold, dry periods when there is risk of ground frost. Wetting the soil increases both the thermal conductivity and the heat storage capacity (given that the specific heat of unit volume of water is very much greater than that of air). The combined effect is that a wet soil profile will take in more heat during the daytime, and release more heat into the overlying air at night, thereby reducing the frost risk.

10.8.3 Mechanisms for the transfer of latent and sensible heat away from the evaporating surface

The supply of energy for evaporation is only part of the story of how atmospheric conditions influence the evaporation rate. Temperature, humidity and wind speed also influence evaporation through their influence on the transfer of water vapour away from the evaporating surface. This is an important part of the explanation of why evaporation is so rapid in hot, arid environments.

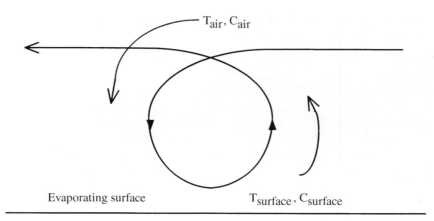

T_{air}, C_{air}

Evaporating surface $T_{surface}, C_{surface}$

Figure 10.16 Turbulent transfer.

The principal mechanism for transporting water vapour and sensible heat between the land surface and the overlying air is *turbulent transfer* (Chapter 5). Eddies caused by moving air (and also by thermal convection) sweep away humid air (at the surface temperature) immediately adjacent to the evaporating surface and replace it with drier air drawn down from above (Figure 10.16).

The larger the difference between the humidity of the air immediately adjacent to the free water ($C_{surface}$) and the overlying air (C_{air}), then the faster will be the rate of transfer of water vapour (E). Similarly, the sensible heat flux (H) will be proportional to the difference in temperature between the evaporating surface and the overlying air. These aerodynamic transfers of water vapour and sensible heat can be described by:

$$\lambda E = \lambda K_1 (C_{surface} - C_{air}) \text{ and } H = K_2 (T_{surface} - T_{air}) \qquad (10.18)$$

where K_1 and K_2 are measures of the effectiveness of the air eddies in transporting water vapour and heat, and so depend on the wind speed.

We can get clues about the effect of air temperature and humidity on the evaporation process by considering the factors that control the magnitude of the difference in humidity between the air next to the surface and the overlying air. The air immediately adjacent to the wet surface will be saturated with water vapour. Hence, if the surface and air temperatures are similar, then the difference in concentration represented in equation 10.18 ($C_{surface} - C_{air}$) is approximately the *saturation deficit* of the air, as defined in Section 3.6.2. The saturation deficit is a measure of the amount of water that needs to be added to the air in order to bring it to saturation. Hence, for a given relative humidity, the saturation deficit (and hence the evaporation rate) will be very much faster at higher temperatures, because of the greater capacity of warm air to hold water vapour.

> ## Worked example 10.5   Effect of temperature on the saturation deficit of the air.
>
> Assess the effect of rising temperature on the saturation deficit of air.
>
> ### Solution
> Consider air at 20°C with a vapour concentration of $10.2\,\mathrm{g\,m^{-3}}$ (equivalent to 60% RH). The saturated vapour concentration at 20°C is $17\,\mathrm{g\,m^{-3}}$, so the saturation deficit is $6.8\,\mathrm{g\,m^{-3}}$. If the temperature of the air were to be raised to 35°C, then the saturated vapour concentration would rise to $36\,\mathrm{g\,m^{-3}}$, thereby raising the saturation deficit to $25.8\,\mathrm{g\,m^{-3}}$, which is over three times larger than at 20°C. This is the main reason why evaporation rates tend to be faster on days with high air temperature.

10.8.4   Potential evaporation and the Penman equation

The above sections have considered the energy requirement for evaporation, and the factors affecting the exchanges of water vapour and sensible heat between the land surface and the overlying air via turbulent transfer. The conclusions drawn include:

- Net radiation provides the energy for evaporation, and the energy absorbed by the land surface as net radiation is dissipated by a combination of the ground energy flux and the exchanges of sensible and latent heat between the land surface and the overlying air (equation (10.16)).
- Movement of water vapour and sensible heat away from the evaporating surface into the overlying air depends on the wind speed, surface temperature, and temperature and humidity of the overlying air.

These ideas can be combined to predict how measurements of radiation, wind speed, and air temperature and humidity affect the rate of evaporation that would be expected from a wet surface that is freely supplied with water (such as a lake). This is known as the *potential evaporation rate*. This was done by Howard Penman in a famous paper written in 1948, which proposed the so-called *Penman equation*, which is used widely to estimate the potential evaporation rate using data available from meteorological stations. It is beyond the scope of this chapter to derive the Penman equation and to explain in detail how it can be applied. However, this is covered in an excellent publication by the Food and Agriculture Organisation, much of which is available from their Web pages, including a software package called *CROPWAT*, which is an irrigation scheduling programme based on the use of the Penman equation to calculate the amount of water used by crops.

One of the many versions of the Penman equation is:

$$E = AR_n + B(a + bW)D \tag{10.19}$$

where E is the evaporation rate, R_n is net radiation, D is the saturation deficit of the air and W is the wind speed. A and B are coefficients that incorporate a number of physical properties of air and water vapour (such as air density, specific heat capacity, latent heat of

Howard Latimer Penman (1909–84) was a major forerunner in the development of environmental physics. He was head of physics at the Rothamsted Experimental Station, which continues to specialize in agricultural research. His work revolutionized our understanding of water transport and evaporation from land surfaces. John Monteith joined him in 1954, and one of the results of their studies was the famous Penman–Monteith equation, which is used in a wide variety of applications, from irrigation scheduling on sugar cane plantations to global atmospheric circulation models used to predict climate change. Monteith was subsequently appointed the first Professor of Environmental Physics in Britain at Nottingham University in 1967. This became a major centre for the subject and it was here that he wrote the original edition of the seminal *Principles of Environmental Physics*.

vaporization and the slope of the saturated vapour pressure/temperature curve), some of which vary with temperature. Hence, A and B are slightly, but predictably, temperature-dependent. The coefficients in the wind speed term (a and b) are empirically determined.

An interesting feature of the Penman equation is that the evaporation rate appears to be the sum of two terms. The first (known as the radiation term) is proportional to the radiant energy available for evaporation. The form of the Penman equation presented here is that used for average daily evaporation, and so ignores the ground heat flux. G is usually negligible over a day because the heating of the soil profile during the day is usually offset by cooling at night.

The second term in the Penman equation (the aerodynamic term) depends on the temperature and humidity of the air (both of which influence the saturation deficit, D) and on the wind speed. The aerodynamic term reflects the influence on evaporation of those factors which control the movement of water vapour and sensible heat away from the evaporating surface. Hot, dry air (i.e. large D) will result in faster evaporation than when the overlying air is cold and humid. The effect of wind speed is much greater when the air is hot and dry, because of the greater contribution of the aerodynamic term to potential evaporation.

Figure 10.17 shows examples for contrasting locations of how the mean daily potential evaporation rate varies from month to month. It also shows how the overall potential evaporation rate is partitioned between the radiation and aerodynamic components. The last graph in Figure 10.17 shows the monthly rainfall in each location.

In southern England (e.g. Reading) the potential evaporation rate during mid-winter is virtually zero, and it increases dramatically during the summer months because of more intense radiation and also because the air has a larger D (mainly because the air is hotter). In a humid tropical environment (e.g. Alor Star Airport, Singapore), the potential evaporation varies little during the year, although it tends to be less during the latter part of the year when the sky is more cloudy (reducing the amount of solar radiation reaching the land surface) and the air is more humid. At Alor Star, it is the radiation term that dominates because the air is very humid (i.e. low D). The site at Harare (in Zimbabwe) is very

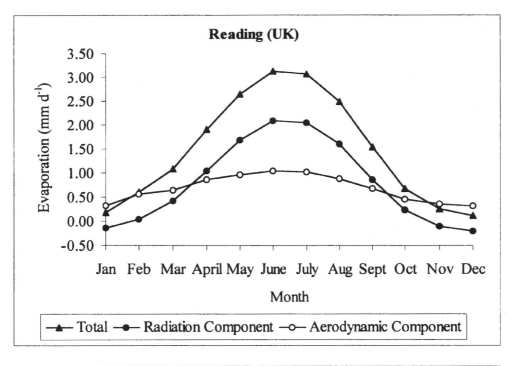

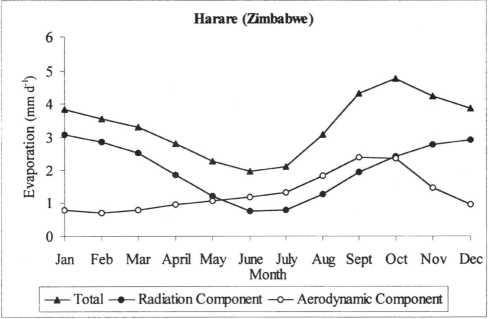

Figure 10.17 Monthly evaporation rates for various locations.

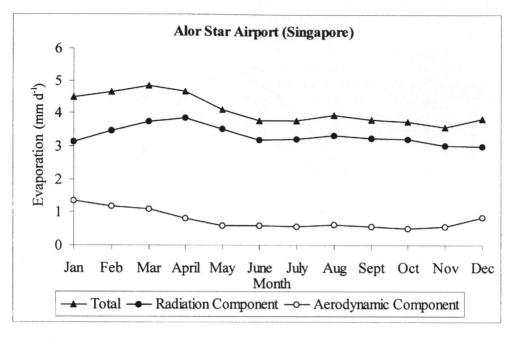

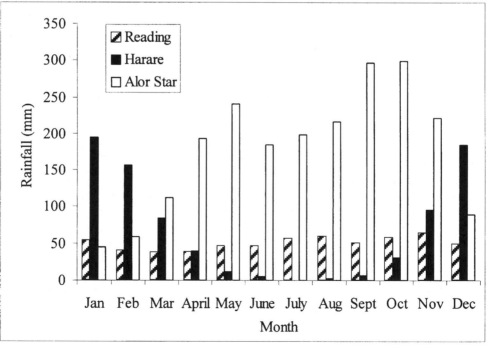

Figure 10.17 (Contd).

different to the humid tropical location. Harare is in the semi-arid tropics where virtually all the rain falls between October and April, with May–September being increasingly hot and dry. During the rainy season, the air is relatively cool and humid, and so potential evaporation is dominated by the radiation term. The radiation input decreases considerably during the winter months (remember that Zimbabwe is in the Southern Hemisphere, so that July is mid-winter), causing the potential evaporation to decrease. However, in the spring (July–October) there is a sharp increase in potential evaporation. This is contributed to largely by the aerodynamic term, because of the air being hot and dry as a result of there being little rain during this period.

10.8.5   Evaporation from the land surface

The rate of evaporation from the land surface is often very different to the rate of evaporation from a free water surface, which is what is estimated using the Penman equation. Evaporation from the land surface takes place direct from the soil surface and also from the leaves of vegetation. Evaporation from a unit area of leaf is usually less than from free water, because the humid air within the leaf has to pass through the stomata, which, in effect, are perforations in the 'skin' of tightly packed cells that form the surface of the leaf. Stomata enable CO_2 to diffuse into the leaf to sustain photosynthesis (see Chapter 11). However, the leaves covering $1\,m^2$ of land surface may have a total area $>1\,m^2$ (bearing in mind that evaporation also occurs from both sides of leaves) so the total area of evaporating surface might be greater than for a surface such as a lake. Hence, evaporation from the leaves of vegetation above unit land area might be less, or even slightly greater, than the rate predicted using the Penman equation, depending in part on the area of leaf that is present.

For land surfaces where there is a sparse cover of leaves (i.e. there is a lot of bare soil visible from above) the evaporation rate is likely to be considerably less than would be the case from a lake, even though the soil profile might contain a large amount of water. This is because the top few mm of soil can dry out very quickly following rewetting by rain. This has the effect of reducing the soil hydraulic conductivity (Figure 10.10) of the soil close to the surface which then acts as a significant barrier to the upward flow of water to replace that lost through evaporation.

One approach that is widely used to estimate evaporation from the land surface from the Penman equation is to multiply the potential evaporation, as measured by the Penman equation, by a coefficient (called the *crop factor*) that takes account of the amount of leaves present. Figure 10.18 shows a typical example of how the crop factor might change during the year for a field used to grow sugar beet. Just after the crop is sown, the crop coefficient is at its lowest because of the large area of exposed bare soil. By the time the land surface is completely covered by leaves the crop factor rises to close to 1, indicating that evaporation is similar to that expected from a free water surface. During the early stages, the crop factor depends strongly on how frequently the soil surface is rewetted.

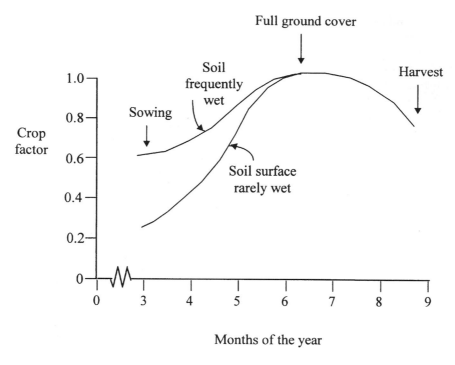

Figure 10.18 Change in crop factor during a growing season.

Case study 10.4   Effect of land use on the soil water balance in Sub-Saharan Africa.

To complete this chapter, we consider a case study of the annual water balance of deep sandy soils at a site just south of Niamey (the capital of Niger, West Africa). The average annual rainfall is about 550 mm and is compressed into a few months of rainy season (May–October). It is an area of the world where there is concern that the climate is changing so as to reduce the annual rainfall, because the rainfall in most years of the past decade or so has been below the previous long-term year average. It is also an area with a dramatically increasing population, resulting in increased abstraction of water from boreholes and wells (groundwater is the main source of water for villages). The demands of increased population pressure have also led to an increase in the clearance of natural savannah vegetation (bushes with a herbaceous understorey) in order to grow arable crops with minimal inputs of fertilizers. Given the changes in the past decade in rainfall, water abstraction and land use, it is not surprising that there has been a change in groundwater levels, though it is not clear the extent to which each cause has contributed to the change.

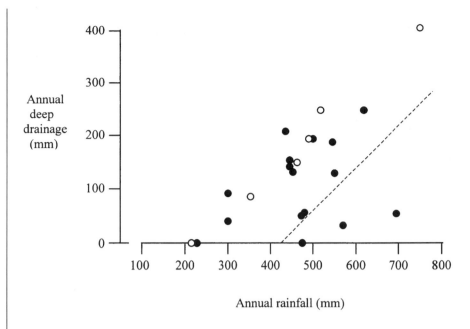

Figure 10.19 Effect of land use and annual rainfall on annual deep drainage from deep sandy soils in Niger.

Various studies have applied techniques similar to those described in Worked example 10.3 to estimate, under different types of land use, how much of the annual rainfall was lost from the soil profile as deep drainage or as evaporation (where evaporation combines direct evaporation from the soil surface as well as water taken up by plants and evaporated from leaves). Figure 10.19 shows the results obtained in different locations and in different years by plotting the annual drainage against annual rainfall, as measured at a number of sites on deep sandy soils in West Africa. Distinction is made between fallow savannah land, cereal crops grown with little or no inputs by subsistence farmers, and cereal crops grown intensively at agricultural research stations using the best crop varieties and inputs of fertilizers and pesticides.

The open symbols in Figure 10.19 refer to measurements made on farmers' fields, while the closed symbols are measurements on trials plots at agricultural research stations, where high levels on inputs are used giving rise to better crop growth. Hence, more water is used in transpiration (and, therefore, less in deep drainage). The dashed line shows the relationship for sites with natural savannah vegetation.

For each land use type it is seen, not surprisingly, that there is more drainage (hence, more potential groundwater recharge) in wet years than in dry years. In each case it seems that there is a 'threshold' annual rainfall needed to generate any drainage at all. Below the threshold, all of the rainfall is lost through a combination of direct evaporation from the soil surface and plant water uptake. This threshold is much higher (around 400 mm) when there is dense vegetation (i.e. intensive cropping or natural savannah) rather than the sparsely vegetated fields managed by subsistence farmers, where significant drainage is observed when rainfall >250 mm.

The limited water uptake by poor cereal crops in subsistence farmers' fields results in much more drainage than when vegetation is more dense. For example, in a year with 500 mm rainfall, the drainage from a subsistence farmer's field is likely to be around 200 mm, almost three times that from a neighbouring savannah area. This implies that the clearance of natural vegetation to grow crops has actually enhanced groundwater recharge, though the low rainfall and greater water abstraction from wells in recent years has caused water tables to drop. The corollary is that annual evaporation from subsistence farmers' fields is less than from natural savannah, with the result that the land surface will be hotter and provide less water vapour into the atmosphere. It has been argued that this might be one of the reasons for the apparent reduction in rainfall during the past decade or so, though this is still being hotly debated.

As a footnote, one might have concerns about what might happen to groundwater recharge if the improved crop management practices adopted by agricultural research stations were implemented more widely. Presumably this would reduce groundwater recharge, leading to more rapid falling of water tables. This provides an interesting example of the conflicts that exist when managing the use of land.

10.9 Summary

In introducing aspects of the physics of the biosphere, this chapter has been concerned with some of the principal features of soil physics and, in particular, the nature of the soil–water interface. It has focused on water retention and transport through soils, and the means by which the weather and the land surface can influence evaporation. In the next chapter we will see how the conditions already described can facilitate the growth of plants, and thus provide the food that humanity needs to survive.

References

Freeze, R. A. and Cherry, J. A., *Groundwater*. Englewood Cliffs: Prentice-Hall, 1979.

Hillel, D., *Introduction to Soil Physics*. New York: Academic Press, 1982.

Jones, J. A. A., *Global Hydrology: Processes, Resources and Environmental Management*. Harlow: Addison Wesley Longman, 1997.

Marshall, T. J., Holmes, J. W. and Rose, C. W., *Soil Physics*, 3rd edn. Cambridge: Cambridge University Press, 1999.

McLaren, R. G. and Cameron, K. C., *Soil Science: Sustainable Production and Environmental Protection*. Auckland: Oxford University Press, 1997.

Paton, T. R., Humphreys, G. S. and Mitchell, P. B., *Soils: A New Global View*. London: UCL Press, 1995.

Rowell, D. L., *Soil Science: Methods and Applications*. Harlow: Addison Wesley Longman, 1997.

Shaw, E. M., *Hydrology in Practice*, 3rd edn. London: Chapman and Hall, 1998.

Tivy, J., *Biogeography: A Study of Plants in the Ecosphere*, 3rd edn. Harlow: Longman, 1998.

Wheater, H. and Kirby, C., eds, *Hydrology in a Changing Environment: Proceedings of the British Hydrological Society International Conference, Exeter, 1998*, 3 vols. Chichester: Wiley, 1998.

White, R. E., *Principles and Practice of Soil Science*, 3rd edn. Oxford: Blackwell, 1997.

Wild, A., *Soils and the Environment: An Introduction*. Cambridge: Cambridge University Press, 1995.

Discussion questions

1   Explain the principles you might use to apportion measured losses of water from a soil profile between drainage and evaporation.
2   Discuss how characteristics of the soil pore space influence the movement of water and solutes through soils.
3   The water content has a large influence on many of the physical properties of soils. Prepare a list of soil physical properties likely to change by >10% as soil is wetted from an air-dry state to saturation.
4   In terms of soil erosion and degradation, describe how desertification can result.
5   Describe the soils, vegetation and types of agriculture found in (a) temperate regions and (b) semi-arid areas.
6   Discuss how variations in water supplies to agricultural land can influence changes in the patterns of land use.
7   Is there any substance in the contention that in the twenty-first-century water may be a possible source of conflict?

Quantitative questions

1   Describe Darcy's law for the movement of water through soils.
     The following table shows measurements of water content made at a number of depths on consecutive days during a rain-free period in an unvegetated soil profile. Also shown is the average soil water suction measured during the 24 hours from noon on day 1 to noon on day 2.

Depth (cm)	Volumetric water content, day 1	Volumetric water content, day 2	Average soil water suction during period (cm)	Layer dimensions (cm)
5	0.363	0.351	171	0–10
15	0.376	0.365	140	10–20
25	0.382	0.372	125	20–30
35	0.387	0.377	113	30–40
45	0.391	0.382	101	40–50
55	0.395	0.385	94	50–60

(a)   Plot graphs showing how the matric, gravitational and hydraulic potentials vary with depth.
(b)   Determine the depth of the zero flux plane.
(c)   Calculate the amounts (mm) of water lost from each layer during the 24 hours (the layer dimensions are shown in the last column).
(d)   Estimate the average rates of evaporation and of drainage below 60 cm depth during 24 hours.

2   A sandy loam soil profile has an average bulk density of $1.5 \, g \, cm^{-3}$ and an average gravimetric water content of 18% at field capacity. In each of the following, state your reasoning and any assumptions made.
(a)   Calculate the amount of water (mm) corresponding to one water-filled pore volume for a 1 m-deep profile during periods when water is draining (i.e. is very close to field capacity).
(b)   Imagine that a pulse of a non-reactive solute (e.g. bromide) was applied to the surface of a 1 m-deep core containing the above soil, and that the concentration of bromide in the water draining at 1 m depth was monitored.
     Sketch the likely shape of the graph relating the bromide concentration in the leachate to the amount (mm) of water leached.

3   The sizes of pores within soil range over many orders of magnitude, often from fractions of a micron to 1 mm diameter or more. Calculate by how much the following would differ between pores that had a 100-fold difference in diameter:

(a) The suction required to drain the pore of water.
(b) The rate of flow of water $(m^3 s^{-1})$ under conditions of a given pressure gradient.
(c) The average velocity of a solute ion (assuming mass flow) under conditions of a given pressure gradient.

In each case, discuss the theoretical basis of your answer, and any assumptions you are making.

4   The table below shows the average volumetric water content, soil water suction and hydraulic conductivity of three layers of a soil profile.

Layer (cm)	Volumetric water content	Soil water suction (cm)	Hydraulic conductivity (ms^{-1})
45–60	0.479	174	1.26×10^{-8}
60–75	0.484	157	1.58×10^{-8}
75–90	0.488	145	2.51×10^{-8}

Stating any assumptions you make:
(a) Calculate the hydraulic potential at the mid point of each layer.
(b) Calculate the hydraulic potential gradient between the first and second layers and between the second and third layers, based on your answers to (a).
(c) Calculate the water flux, using Darcy's Law, between the first and second layers and between the second and third layers.
(d) Is the second layer getting wetter or drier?

(University of Reading: BSc Soils and the Environment)

5   The original Penman combination equation was derived to estimate E_0, the evaporation rate from an open water surface. Explain the significance of the term "combination" in the derivation of this equation and indicate what modifications to the original equation are commonly applied to estimate E_T, the potential evaporation from a short green crop.

The evaporation rate from vegetation, E, can be represented by the Penman-Monteith combination equation:

$$\lambda E = \{\Delta R_n + (\rho c_p (e_s - e)/r_A)\}./\{\Delta + \gamma(1 + r_s/r_A)\}$$

where λ is the latent heat of vaporization of water, ρ the density and c_p the specific heat of air, Δ the slope of the saturated vapour pressure-temperature curve and γ the psychrometer constant, R_n is the net radiation input rate, e_s and e the saturated and absolute humidity respectively at a reference height and r_s, r_A the surface resistance and aerodynamic resistance to the reference height, respectively.

If under given atmospheric conditions the values of r_A and r_S (water non-limiting) are as follows:

	r_A (sm^{-1})	r_S (sm^{-1})
Short grass	30	45
Conifer forest	10	150

Calculate the ratio of the rate of evaporation of intercepted water to the rate of evaporation of freely transpired water for each crop. Hence, discuss the influence of climate on the effect of afforestation on catchment water yield. (Assume typical values of Δ and γ of 0.83 and 0.65 mbar/°C respectively).

(Imperial College, London University: MSc Engineering Hydrology: April 1991)

(Hint: 'Intercepted water' refers to water sitting on the surface of a leaf following rainfall. In this case, the surface resistance, r_s, will be zero).

Chapter 11

Vegetation growth and the carbon balance

11.1   Introduction

The central theme of this chapter is to explore how the nature and properties of the land surface influence the growth of vegetation and the cycling of carbon. This understanding can then be applied to examine how land use influences the global carbon budget (7.3.2), and any implications there might be for processes such as atmospheric circulation and climate change, food production, and the quality and quantity of water resources.

A very important factor influencing the exchanges of energy, water and carbon at the land surface is the amount of vegetation present. Much of the land surface of the Earth has large seasonal variation in vegetation cover. An extreme case is the semi-arid tropics where there are prolonged dry seasons, during which there will be very little coverage of the land by a vegetation canopy, interspersed by rainy seasons, during which vegetation will be actively growing. Much of the Earth's land surface is used to produce arable crops, which involves repeated cycles of clearance of vegetation from the land before planting the next crop.

It was shown in Chapter 10 that the presence of vegetation can have a large effect on the amount of water evaporating from the land surface. This has consequences for the temperature of the surface and the amounts of sensible heat and water vapour returned to the atmosphere, thereby affecting atmospheric circulation and weather. The large effect of the presence of vegetation on evaporation from the land surface has consequences for the other components of the annual water budget, such as the amount of water that percolates into groundwater. Hence, an understanding of the factors controlling the growth of vegetation is a very important aspect of how the land surface affects the energy and hydrological cycles (see Chapter 8).

The growth of vegetation is also, of course, a major element of the global carbon cycle, because vegetation growth involves the removal of CO_2 from the atmosphere for the assimilation of carbohydrates. This is of major interest for two reasons. First, there is the role of vegetation as a factor influencing the CO_2 content of the atmosphere, which has implications for global warming (see Chapter 7). Second, the process of photosynthesis uses energy from solar radiation to produce energy-rich carbohydrate from CO_2 and water is the foundation of the food chain. Much research has been directed towards understanding the factors affecting the growth of plants in the context of finding ways of making our agricultural systems more productive. This continues to be an important area of research as the population of the world continues to grow, thereby increasing the demand for food.

The purpose of this chapter is to investigate how the physical environment affects plant growth. Such understanding has potential applications, including:

- Predicting the effects of weather and climate change on the growth of agricultural crops and natural vegetation
- Developing mathematical models to estimate at the global scale the contribution of photosynthesis to the global atmospheric CO_2 budget, and how this might be affected by large scale changes in land use
- Predicting how changes in land use (e.g. clearing natural vegetation to grow cereal crops) affect the hydrological cycle.

The growth of plants involves very complex processes, and is influenced not only just by the physical environment (such as the input of solar radiation, air temperature and humidity, and the availability of soil water for uptake by roots), but also by a wide range of chemical factors (e.g. the supply of nutrients and the effects of soil pH) and biotic factors (e.g. pests, diseases and the role of soil micro-organisms in nutrient cycling processes). It is clearly beyond the scope of this chapter to embrace all these issues. Instead, attention will be restricted to how the weather affects the growth of plants.

Weather is clearly a major factor affecting plant growth. Farmers are very aware of year-to-year variation in yield that is predominantly attributable to variation in weather. Figure 11.1 shows an interesting example of interannual variation in the response of the yield of spring barley to the amount of nitrogen fertilizer applied per hectare (ha^{-1}). When large amounts of fertilizer were applied, the yields approached a plateau, implying that the supply of nutrients was not a limiting factor. However, the grain yields of well-fertilized crops ranged from as low as 2 tonnes hectare^{-1} in 1976 (a drought year) to as much as

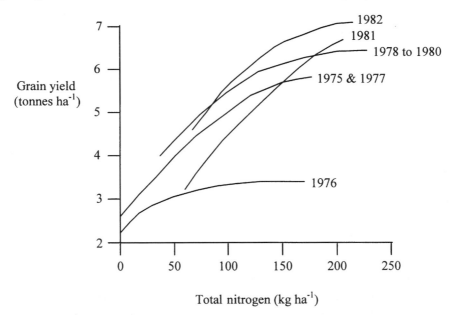

Figure 11.1 Differences between years in the response of spring barley yield to the amount of nitrogen fertilizer applied.

7 tonnes ha^{-1} in 1982. There are also interesting weather-nutrient interactions evident. For example, well-fertilized crops yielded more in 1981 than 1977, whereas the 1977 crop was the more productive when little fertilizer was applied.

Weather has a complex influence on plant growth. Temperature generally influences the rates of the biochemical and physiological processes involved in the growth and development of plants. The amount of solar radiation incident on green foliage will influence the rate of photosynthesis, which uses radiation in the blue and red parts of the spectrum as the source of energy (Section 11.3). The amount and distribution of rainfall, and the meteorological factors affecting evaporation (solar radiation, wind speed, and the temperature and humidity of the air) combine to affect the availability of water in the soil. The extent to which these weather factors induce plant water stress (and thereby affect plant growth) depend strongly on the properties of the soil and the vegetation.

11.2   Plant development

The starting point for analysis is to recognize that it is important to distinguish between weather factors that influence the *development* of plants, and those which influence the *growth* of plants. Plants go through a number of stages of development in their lives. For example, if we take the case of a cereal plant such as wheat, key developmental events can be identified, including:

- Germination of the seed
- Emergence of the seedling above the soil surface
- Appearance of the first leaf, second leaf, third leaf, and so on
- First appearance of the stem
- Formation of a flower (i.e. the ear)
- Production of pollen by the male parts of the flower (the anthers) to fertilize the female organs that ultimately form seeds (the time for this to occur is known as *anthesis*)
- Seed maturity (when farmers would ideally harvest the crop).

Each stage being defined by a period known as *developmental phase* (in units of days).

11.2.1   Weather

Weather has an important influence in the timing of these events. For example, development tends to take place faster at higher temperatures, provided temperatures are not excessive. Plant *growth* refers to the increase in size of the plant, and may be largely independent of plant development. For example, all of the barley crops shown in Figure 11.1 reached maturity, but the sizes of the plants at maturity varied three-fold. The production of carbohydrates through photosynthesis is one of a number of weather-dependent processes that influence plant growth, as distinct from the plant development.

11.2.2   Rate of plant development

The rate of plant development is a measure of the speed with which the plant moves through its life cycle. Consider, for example, some results from a controlled environment glasshouse experiment (Table 11.1) that investigated the effect of air temperature on the time taken for pearl millet (a tropical cereal crop) to reach various stages of development.

Table 11.1 Effect of temperature on the time from floral initiation to heading for pearl millet plants.

Mean air temperature (°C)	Days from floral initiation to heading	Rate of development (day⁻¹)
19	34	0.0294
22	28	0.0357
25	23	0.0435
28	20	0.0500
31	16	0.0625

As the temperature increased, so duration of the phase of development became shorter. The rate of development, shown in the third column, can be thought of as the amount of development that occurs per unit time (just as rate of movement is expressed as distance moved per unit time). It is calculated as the inverse of the duration of the developmental phase in question, and has units of day^{-1}.

Figure 11.2 shows examples of how the rate of development varies with temperature during two phases of development of a particular variety of pearl millet. The first is the rate of germination (expressed as the inverse of the time taken for seeds to germinate),

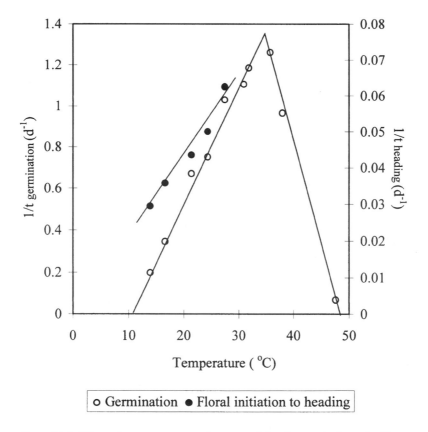

o Germination ● Floral initiation to heading

Figure 11.2 Effect of temperature on the rates of development of pearl millet.

and the second is the rate of development during the period from floral initiation to heading (as shown in column 3 of Table 11.1). In both cases there seems to be a remarkably simple relationship between temperature and the rate of development. There is a minimum temperature below which development stops (the so-called base temperature, T_{base}). Above the base temperature there is a linear phase where the rate of development is proportional to the increase in temperature until the optimum temperature (T_{opt}) is reached at which the rate of development is at its maximum. The results of the germination experiment show that increasing temperature above T_{opt} causes the rate of development to slow down in a linear manner, and ultimately to cease at some critical maximum temperature (T_{max}). In the case of this particular variety of pearl millet, T_{base} (~10°C), T_{opt} (~32°C), and T_{max} (~44°C), were similar for all phases of development. The 'wig-wam' shape of the temperature–response curve has been found in other areas of biology – an interesting example being the effect of temperature on the frequency of 'chirruping' of crickets, which is caused by vibration of their legs.

The apparently linear relationship between the rate of development and temperature (provided the temperature falls between T_{base} and T_{opt}) implies that the time it takes for a given developmental phase to occur can be predicted by:

$$\frac{1}{\text{days to complete phase}} = \frac{1}{\theta_{dd}}[T - T_{base}] \tag{11.1}$$

where T is the air temperature and $1/\theta_{dd}$ is the slope of the line in Figure 11.2. The physical meaning of θ_{dd} can be seen by rearranging equation (11.1):

$$\theta_{dd} = (\text{days to complete phase}) \times (T - T_{base}) \tag{11.2}$$

It can be seen from equation (11.2) that θ_{dd} is a measure of the amount of heat that needs to be accumulated in order for a given phase of development to be completed, and is calculated as the product of the number of days and the number of °C that the temperature is above the base temperature. This quantity is known as the *thermal time requirement* or the *day-degree requirement* and is expressed in units of day °C.

Worked example 11.1   Predicting the development of maize at a location in southern England.

The use of the thermal time concept can be illustrated in a case study to predict the development of a maize crop growing in southern England. Figure 11.3 shows some of the easily identifiable stages in the development of the maize crop.

For each stage, typical values are given for the amount of thermal time that needs to be accumulated since the time of sowing the seed for each stage of development to be completed. Table 11.2 shows the mean daily temperature over each 10-day period during the year when it was warm enough for maize to develop and grow, based on the average daily temperatures recorded between 1968 and 1991 at the University of Reading meteorological station in southern England. The base temperature *base* for the development of the maize variety in question is 9°C, so the winter period with average daily temperatures <9°C has been omitted.

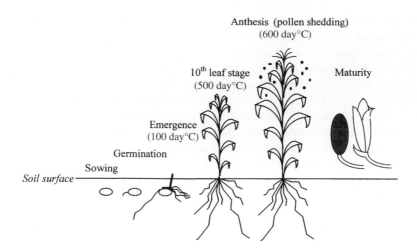

Figure 11.3 Developmental stages of maize.

Table 11.2 Calculation of thermal time from daily temperature records.

Date (day/month)	Mean temperature (T) (°C)	T−T$_{base}$ (°C)	Accumulated thermal time since start of year (day °C)
19/4	8.48	0	0
29/4	8.975	0	0
9/5	10.48	1.48	14.8
19/5	11.67	2.67	41.5
29/5	12.565	3.565	77.15
8/6	13.67	4.67	123.85
18/6	14.535	5.535	179.2
28/6	15.245	6.245	241.65
8/7	16.775	7.775	319.4
18/7	16.99	7.99	399.3
28/7	17.09	8.09	480.2
7/8	17.095	8.095	561.15
17/8	16.99	7.99	641.05
27/8	16.42	7.42	715.25
6/9	15.445	6.445	779.7
16/9	14.43	5.43	834
26/9	13.585	4.585	879.85
6/10	12.465	3.465	914.5
16/10	11.245	2.245	936.95
26/10	10.38	1.38	950.75
5/11	8.905	0	950.75
15/11	7.3	0	950.75

The 10-day mean temperature first rose >9°C between 30 April and 9 May, during which time the average temperature was 10.48°C, which is 1.48°C above the base temperature. Hence, during this 10-day period, there were $10 \times 1.48 = 14.8$ day°C of thermal time accumulated. The next 10-day period was an average 2.67°C above T_{base}, and so a further 26.7 day °C of thermal time was accumulated between 10 and 19 May. Hence, by 19 May, 41.5 day °C of thermal time was accumulated in total. Continuing these calculations through the remainder of the year showed that by the time (in November) when mean temperatures fell below T_{base}, a seasonal total of 950 day °C of thermal time had been accumulated.

The data presented in Table 11.2 and Figure 11.4 suggest that 1200 day °C of thermal time needs to be accumulated in order for short-duration maize varieties such as the one in question to reach maturity. Hence, it is concluded that in an average year in the vicinity of Reading, it will not be possible for farmers to grow a mature maize crop. In practice, some farmers grow maize in southern England for use as animal fodder where it is not necessary for the cobs to mature. Maize crops are not generally sown until at least mid-May, which is consistent with the prediction in Table 11.2 that no significant amounts of thermal time are accumulated before this date. Though farmers will leave maize as late as possible in the autumn before harvesting, the decision of when to harvest is usually based on the time of the first frosts, or avoiding harvesting when the soil is too wet.

11.2.3   Impact of global warming on crop distribution

An interesting application of the thermal time concept is to investigate possible impacts of global warming on the geographical distribution of crops. For example, if temperatures in northern Europe were to increase by 2°C, would it now be possible for maize to reach maturity in southern England? Curve A in Figure 11.4 shows the seasonal pattern of thermal time accumulation assuming a base temperature of 9°C. The data used to produce curve A were those presented in Table 11.2. Curve B shows the corresponding time-

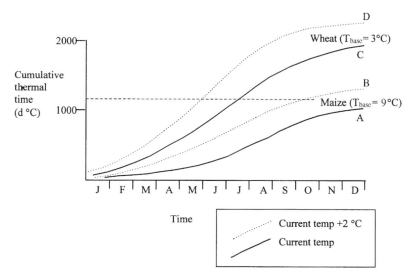

Figure 11.4 Accumulation of thermal time at a location in southern England, showing effects of crop type and elevation of temperature.

course of thermal time accumulation assuming that the mean daily temperatures shown in the second column of Table 11.2 are each increased by 2°C. It is seen that the effect of the increase in temperature is that it would enable maize to be sown three weeks earlier, and that the 1200 day°C of thermal time required to reach maturity would be accumulated by the end of September.

Tropical crops, such as maize, typically have base temperatures ~10°C, whereas the T_{base} for temperate crops (such as wheat, barley and potatoes) is generally ~3°C. Curve C in Figure 11.4 shows the seasonal pattern of thermal time accumulation for a wheat crop assuming current average temperatures and a base temperature of 3°C. In this case, thermal time starts to be accumulated rapidly from March onwards. Hence, mid-March would be an appropriate time for spring sowing, and crop varieties which require 1200 day°C of thermal time to mature would be ready for harvesting around the end of July, which is the time when combine harvesters start to be seen in action in southern England. A 2°C temperature elevation would result in very much earlier maturity of wheat curve D. One of the consequences of such predicted shortening of the growth period would be a reduction in yields as a result of leaves having less time to photosynthesize.

It is evident from the above discussion that such simple thermal time analysis is a very powerful tool for helping farmers to identify suitable crops for growing in given locations, and to help predict the impact of future climate change on the geographical distribution of crops. Though the discussion has been centred on annual crops, the principles of thermal time are widely applicable to many types of vegetation. However, in many plant species, the timing of development can be influenced by environmental factors other than this simple thermal time concept. One example is that some plants require specific environmental conditions to be experienced before developmental processes leading to seed germination or flowering are triggered. Examples include:

- The requirement to experience periods of long nights before flowering to be triggered. This prevents plants that normally flower in spring beginning their flowering process too early in the event of very mild winters.
- The requirement of some plants for seeds to experience prolonged periods of very low temperature before seeds start to germinate. This reduces the risk of seeds produced the previous summer germinating prematurely in the event of unusually warm conditions in late autumn or winter, rather than germinating in spring.

However, in many cases these additional environmental controls on plant development are well known, and can be incorporated into models used to predict the influence of weather on plant development.

11.3   Plant growth

In Section 11.2, a clear distinction was made between the effects of environmental conditions on plant development and on plant growth. In this context plant growth was defined as an increase in size. The primary interest of this chapter is in understanding how environmental factors influence:

- the size of vegetation canopy (particularly in the context of the effect of vegetation on evaporation); and

- the amount of organic material that is produced by growing vegetation (in the context of the carbon cycle and food production).

For these purposes, useful indicators of vegetation growth are the area of leaf per unit land area (the *leaf area index, L*) or the mass of oven-dry plant matter per unit land area (W). The reason for choosing oven-dry rather than fresh weight is that >80% of the fresh weight of herbaceous plant material is water, and the water content can fluctuate hour by hour depending on the degree of tissue hydration. Oven-dry weight, therefore, provides a more stable indicator of the amount of structural plant material.

J Robert Mayer (1814–1878). Mayer is one of the shadowy geniuses of modern science, and in him we come full circle, through his bridging of physiology and physics.

In Chapter 2 we examined the paramount role of metabolism in survival, and Mayer made major contributions to our understanding of the nature of heat, thermal energy transfers and the work that a system can do.

Born in Heilbronn, SW Germany, he trained initially in an evangelical seminary and then as a medical doctor. It was while as a ship's doctor en route to Java in the Dutch East Indies, that he observed that the venal blood of the crewmen was much brigher in the hotter climates than in the colder. He came to the conclusion that their body's metabolism required less oxygen in the hotter areas and that the metabolic rate was lower. From reflections on the food eaten, the heat produced and the work done he inferred that heat and work are interchangeable. By developing ideas for the mechanical equivalent of heat and the principle of the conservation of energy he helped to lay the foundations of the First Law of Thermodynamics. This was set against an acrimonious dispute between Mayer and Joule, and though Mayer was the first to formulate these ideas it was Joule, through the classic investigations conducted in the cellar of his Salford brewery, who won recognition.

Mayer peceived that his ideas were fundamental not only to mechanical but also to biological systems, and in 1845 he suggested that in the photosynthetic process plants use sunlight in the creation of chemical energy.

At the age of 36, the combination of the lack of encouragement of his pioneering work by the scientific establishment and catastrophe in his personal life (the death of two of his children) caused him to jump from the second floor of his apartment. Fortunately, he survived, and though temporarily consigned to a lunatic asylum, he subsequently gained the recognition that he rightly deserved.

11.3.1 Photosynthesis by individual leaves

One of the most marvellous physical processes is the transformation of sunshine into stored energy for life processes – *photosynthesis*. It is a theme that is often neglected in standard physics textbooks. In Chapter 2 it was pointed out that the laws of physics underlie the biochemical processes in our bodies and the thermal radiation emitted. Photosynthesis, too, is a physical process and it is paramount for the survival of living organisms. Both food and fossil-fuels are the products of photosynthesis, and it shall be shown that the quantum theory that explained black body radiation (see Chapter 2) and the photoelectric effect (see Chapter 5) has application in photosynthesis.

In 1772, the British clergyman and chemist Joseph Priestley performed the following experiment. He placed a mouse in a closed glass chamber containing a living mint plant, and another mouse and mint plant separately in two other glass chambers. The mouse and plant sharing the same chamber survived, while the mouse and plant each in their own chamber died. Clearly, the plant produced an essential ingredient to sustain the life of the mouse, and vice versa, such that when one was absent both would die. We now know that these 'ingredients' are CO_2 and oxygen. The mouse breathes in oxygen and expels CO_2, whereas the plant absorbs CO_2 and releases oxygen. The former process is known as respiration, and that in the plant is called *photosynthesis*.

Photosynthesis is the process where CO_2 and water combine under the influence of sunlight to produce oxygen and the chemical energy that is stored in carbohydrates. It is, in effect, the ultimate energy source.

Within the leaves of plants are cells containing *chloroplasts*, within which the absorbed CO_2 and water react to form a carbohydrate, with the simultaneous release of oxygen. The energy necessary for this process comes from the sunlight absorbed by the molecule chlorophyll within the chloroplasts. Thus, without sunlight photosynthesis cannot occur and the plant cannot produce the carbohydrates used as both building blocks and as an energy source. Let us look at how this happens in more detail.

Light and photosynthesis

The wave–particle duality of light suggests that light can have both wave-like and particulate characteristics. Case study 2.1 showed that Planck's formulation of quantum theory incorporated the idea that discrete packets of energy could be quantized. We also saw in Chapter 5 that Einstein in his theory of special relativity suggested that each quantum of energy called a photon had an energy, $E = h\nu = hc/\lambda$, where h is Planck's constant, ν is the frequency of the radiation, λ is the wavelength of the radiation and c is the velocity of light. This equation implies that the energy is inversely proportional to the wavelength and that the radiation with shorter wavelengths is more energetic.

Worked example 11.2   Energy associated with blue light.

Calculate the energy associated with blue light ($\lambda = 500$ nm) needed photosynthetically to activate one mole of a compound.

Solution

The frequency of blue light, $\nu = c/\lambda = 3 \times 10^8/500 \times 10^{-9}$ Hz.

Then the energy associated with light at this frequency for 1 mole of the substance (i.e. 1 Einstein) $= N_A.h\nu = 6.02 \times 10^{23} \times 6.63 \times 10^{-34} \times 3 \times 10^8/500 \times 10^{-9}$ $= 2.4 \times 10^5$ J.

This means that the energy needed to activate a given amount of substance can be determined from the energy associated with a particular wavelength.

In photosynthesis the green pigment (chlorophyll) absorbs the photon's energy at specific wavelengths and there is electron transfer from the chlorophyll in the process. The photon's quantum of energy has to be sufficient to excite an electron.

For a photosynthetic reaction to be initiated $nh\nu$ joules of energy are required, where n

is the number of moles. For one mole of the substance, n becomes N_A, the Avogadro number ($=6.02 \times 10^{23}$). The total energy needed to initiate the reaction for 1 mole $= N_A.h\nu$, and this is called one Einstein, i.e. 1 Einstein $= N_A$. quanta.

Chloroplasts are the pigmented organelles mainly responsible for leaf colour. The pigments are predominately chlorophyll a and b (though there are exceptions), which respectively absorb particularly strongly in the blue and red parts of the visible light spectrum (Figure 11.5). Chlorophyll absorbs all wavelengths of light except green. It is this predominance of green in the reflected light that gives leaves their characteristic colour. Wavelengths in the visible region of the electromagnetic spectrum are appropriate for exciting electrons to higher energy levels. At longer wavelengths, such as in the infrared, vibrations and rotations of the molecules may increase. Conversely, if the wavelengths are too short (such as in the UV region) the system would be too energetic and possibly rupture the bonds. How effectively the pigments can absorb various wavelengths can be seen in an absorption spectrum of chlorophyll a and b (Figure 11.5).

In contrast, an action spectrum (Figure 11.6) is a measure of the efficiency with which a molecule can absorb light of various wavelengths so that photosynthesis can occur. It is a measure of the photosynthetic rate and the ability to produce electrons.

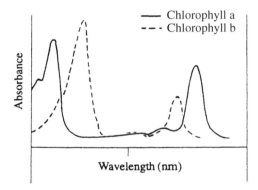

Figure 11.5 Absorption spectrum of chlorophyll compounds [a, b].

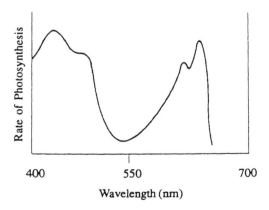

Figure 11.6 Action spectrum.

It can be observed that not all the wavelengths of the solar spectrum are utilized in photosynthesis; in fact, the graph indicates that photosynthetic efficiency is a function of the wavelength of the incident radiation. Those wavelengths that give leaves their green colour are reflected. Other wavelengths are absorbed and raise the temperature within the leaf. Red light (650 nm) is the most effective in photosynthesis, although blue, orange and yellow wavelengths also function. Perhaps, you have seen greenhouses where red light is used to stimulate growth in younger plants?

It is the absorption of light energy at the blue and red wavelengths that provides the energy required in the photosynthetic process to carry out the endothermic (i.e. energy requiring) reaction of converting CO_2 and water into carbohydrates. In essence, the biochemical endothermic reaction involved in photosynthesis is the reverse of that of the exothermic combustion reaction seen in Chapter 2. The photosynthetic process can be described by:

$$6CO_2 + 6H_2O + h\nu \rightarrow C_6H_{12}O_2 + 6O_2 \qquad (11.3)$$

Some of the consequences of this most remarkable reaction are that:

- The carbon in atmospheric carbon dioxide is transformed into the solid carbohydrate matter of plants
- The oxygen in water (O) is transformed into atmospheric oxygen gas (O_2)
- The photon energy of sunshine, $h\nu$, is stored in the internal energy of reactive oxygen molecules
- Therefore, held within the reactive oxygen molecules, the stored energy moves around the Earth in the atmosphere to become available in metabolism, respiration and combustion.

Paired with photosynthesis, therefore, is the general equation of metabolism and combustion:

$$[CH_2O] + O_2 = CO_2 + H_2O + \text{heat}$$

where $[CH_2O]$ is indicative of a wide range of organic material.

Chlorophyll is necessary for photosynthesis and consists of molecules that provide locations for the absorption of solar energy in the light reactions (see below). The chloroplasts, which contain the chlorophyll, engage in energy metabolism, and photosynthesis involves two stages:

- *Light reactions*, which involve the photolysis of water by sunlight and the evolution of oxygen. In this process light energy is absorbed by molecules of chlorophyll A and electrons become excited and are then transported. In the ensuing reactions the energy is transformed, through the mobility of electrons, into ATP (see Chapter 2) and the reduction by electrons of NADP$^+$ produces NADPH. The energies of the electrons can be utilized to 'pump' protons (H$^+$). The proton gradient plays an important role in the synthesis of ATP.
- *Dark reactions*, so-called because they can be light-independent, involve the fixation of atmospheric CO_2 and the inclusion of hydrogen to form the C—C covalent bonds

that lie at the heart of the carbohydrates $(CH_2O)_n$. Carbon fixation entails the inclusion and reduction of CO_2 into more complex compounds.

We saw in Chapter 2 how carbohydrates can be used to provide energy through glycolysis in metabolism. What marvellous physico-biochemical parallels lie between human beings and plants! The energy generated in the light reactions, through the formation of ATP and NADPH (which is an electron transporting molecule), is now used to produce glucose from CO_2 and H_2O. This is referred to as the *Calvin cycle*.

We will now examine one of many mathematical models that has been developed to predict the effect of weather on the growth of plants. Most of this research work has been done in the context of predicting the growth of arable crops, though the principles apply more widely. Crop growth models vary from the extremely simple to the highly complex. The one described below is intermediate in complexity, and incorporates an understanding of the basic processes involved in crop growth.

In the photosynthetic process approximately 476 kJ of Gibbs free energy is stored for each mole of CO_2 (i.e. each 12 g carbon) that is incorporated into carbohydrate. A typical rate of growth of a well-watered, well-fertilized crop with a dense canopy on a warm, sunny day is 25 g dry-matter m^{-2} land area day^{-1}. This is equivalent to about 14.5 g carbon $m^{-2} day^{-1}$ (14.5/12 = 12.1 moles carbon $m^{-2} day^{-1}$), assuming tissue dry-matter is 58% carbon. Such a growth rate would store $476 \times 14.5/12 = 575$ kJ energy m^{-2} land area day^{-1}. This represents about 2% of the solar energy incident on the land surface (typically 25 MJ $m^{-2} day^{-1}$) on a clear summer's day in Northern Europe, and so is usually ignored when considering the energy balance of the land surface (Section 10.8).

The rate of photosynthesis by unit area of leaf depends strongly on the intensity of radiation incident on the leaf. In photosynthetic studies it is usually the irradiance in the photosynthetically active part of the spectrum that is measured. Figure 11.7 shows the typical

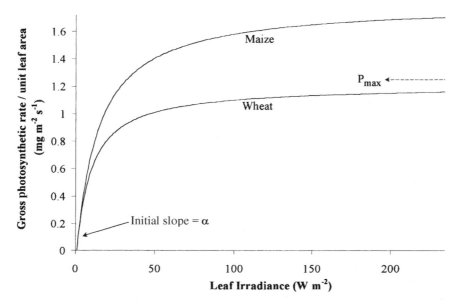

Figure 11.7 Light response curves for the gross photosynthesis of maize and wheat.

shapes of the light response curves for the gross photosynthesis of maize and wheat leaves. Such data are usually obtained from measurements of the rate of uptake of CO_2 from a transparent chamber enclosing a known area of leaf, with an adjustment made to take account of any liberation of CO_2 through respiration during the period of measurement.

At low irradiances, the slope of the response curve is very steep, but at high irradiances the rate of photosynthesis approaches a maximum, light-saturated rate (P_{max}). The following function has been found to describe adequately the response of gross photosynthetic rate per unit leaf area (P_g) to leaf irradiance (I_{leaf}) for a wide range of species, under a variety of conditions:

$$P_g = \frac{\alpha I_{leaf} P_{max}}{\alpha I_{leaf} + P_{leaf}} \tag{11.4}$$

Note that when I_{leaf} is very large, equation (11.4) approximates to $P_g = P_{max}$. When I_{leaf} is very small, equation (11.4) approximates to $P_g = \alpha I_{leaf}$. Hence, α is the initial slope of the light response curve (Figure 11.7).

Species differ in their photosynthetic light response curves. Of particular importance in this respect is that most plant species have one of two alternative biochemical photosynthetic pathways. In most plants, the immediate product of photosynthesis is pyruvate (a 3-carbon compound). These are known as the 'C3' plants. An alternative pathway (the C4 pathway) has a 4-carbon compound (malate) as the immediate product of photosynthesis. Tropical grasses (including cereals such as maize, sorghum and millet) dominate the C4 group. C4 plants are generally capable of much faster photosynthetic rates at high light intensities, and the rate of photosynthesis does not become light-saturated until much higher irradiances than C3 plants. As is discussed in Section 11.4, C4 species are also more efficient in their water use, as a result of having a more efficient photosynthetic system able to maintain a lower CO_2 concentration in the air within the leaf. A few species (e.g. cacti and pineapples) have other photosynthetic mechanisms, but it is not possible to deal with these relatively rare cases here. Typical α and P_{max} for C3 and C4 species are given in Table 11.3.

Photosynthesis also depends on temperature. The main effect of temperature is to modify P_{max}. There is usually a fairly broad band of temperature over which photosynthesis is hardly sensitive to temperature, but when temperatures are excessively high or low the rate of photosynthesis decreases sharply (Figure 11.8). Typical optimum temperatures are about 20°C for temperate species and 30°C for tropical species.

11.3.2   Photosynthesis by a vegetation canopy

Equation (11.4) provides the basis for modelling the uptake of CO_2 by a unit area of leaf as a

Table 11.3 Light response curve parameters for C3 and C4 plants.

Crop	α (mg J^{-1})	P_{max} (mg m^{-2} s^{-1})	Optimum temperature for photosynthesis (°C)
Maize (C4)	0.013	1.8	30
Wheat (C3)	0.013	1.2	20

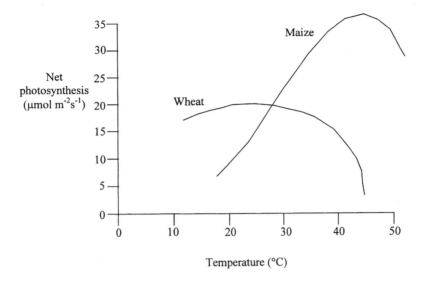

Figure 11.8 Effect of temperature on the maximum photosynthetic rate.

function of the light intensity falling on the leaf in question. How can one then predict the photosynthesis of a vegetation canopy as a whole, which is made up of individual leaves that have very different leaf irradiances depending on their orientation and the degree to which they are shaded by other leaves in the canopy? One approach is to use the physics of the transfer of light through a plant canopy to predict how the average leaf irradiance decreases as you move from fully insolated leaves at the top to heavily shaded leaves at the bottom.

Consider a thin horizontal slice taken through a vegetation canopy that contains a small area of leaf (ΔL) m^{-2} of land area. In other words, the leaf area index within the slice is ΔL (Figure 11.9). In the simplest case shown (Figure 11.9a) the leaf segment is horizontal and is completely opaque. In this case, the radiation that is intercepted by the leaves is simply $I.\Delta L$, where I is the irradiance at the upper surface, which represents the change in the average irradiance (ΔI) between the upper and lower boundaries of the slice. In reality, leaves are usually inclined to some degree (Figure 11.9b). In this case the radiation intercepted by the leaves is reduced by a fraction, f, which is between 0 (vertical leaves) and 1 (horizontal leaves). Hence, the change in irradiance between the upper and lower boundaries of the slice is now $fI.\Delta L$.

A further complication is that leaves are not perfectly opaque, but transmit a small fraction of the irradiance they intercept (Figure 11.9c). This fraction is the transmission coefficient, m, which typically is 0.1 (i.e. 10% of the radiation incident on a leaf is transmitted through it). In this case, the amount of light absorbed by the leaves within the slice (ignoring any reflection), and hence the change in irradiance between the upper and lower boundaries of the slice is:

$$\Delta I = \frac{f}{1-m} I \Delta L \qquad (11.5)$$

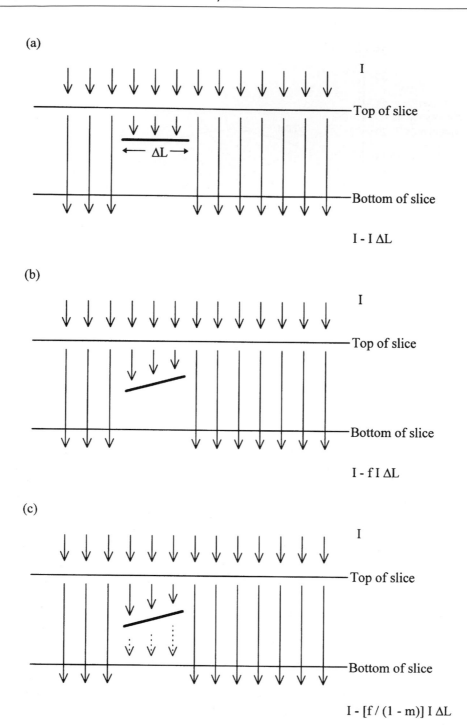

Figure 11.9 Transmission of light through a slice of vegetation canopy.

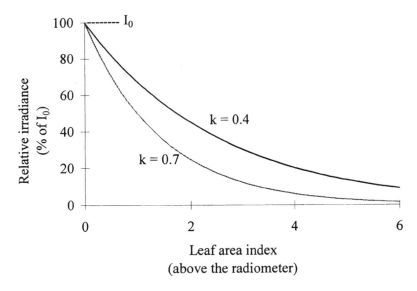

Figure 11.10 Effect of leaf area index on the transmission of light through a vegetation canopy.

which can be rearranged to give the change in irradiance expressed per unit of leaf area:

$$\frac{\Delta I}{\Delta L} = \frac{f}{1-m} I = KI \tag{11.6}$$

where $K = f/(1-m)$ is known as the extinction coefficient which depends on a combination of the orientation and the transmission properties of the leaves. Vegetation canopies with predominantly erect leaves (such as cereals and grasses) have lower extinction coefficients (typically $K = 0.4$) than canopies dominated by leaves that are close to horizontal (typically $K = 0.7$), which results in less attenuation of light per unit leaf area index.

Figure 11.10 shows the graph that would be expected if measurements were made using a horizontally oriented radiometer placed at various heights in a dense vegetation canopy that has an extinction coefficient of 0.4. At the top of the canopy, the radiometer would record the radiation incident on the top of the canopy (I_o). When the radiometer is moved down through the canopy to the point where there is $1 \, m^2$ leaf area m^{-2} land area above the radiometer (i.e. below a leaf area index of 1), the horizontal irradiance recorded would be 67% of I_o. Below $L = 2$, the irradiance would be reduced to 67% of that recorded beneath $L = 1$, and so on. This gives rise to an exponential relationship between irradiance measured on a horizontal plane and leaf area index that is described by the *Beer–Lambert law* (see Section 7.5.1). Equation (11.6) can be solved using the technique of integration to show that the irradiance measured on a horizontal plane (I) beneath a leaf area index of L is given by the exponential relationship:

$$I = I_o.e^{-KL} \tag{11.7}$$

If $K = 0.4$ is substituted into equation (11.7), you should satisfy yourself that the irradiance

I should decrease 33% for each successive increase of unit leaf area index as you move a radiometer down through a vegetation canopy (Figure 11.10).

What we are particularly interested in from the viewpoint of modelling canopy photosynthesis is the average irradiance of the leaf surface within each layer of the canopy, rather than the irradiance measured on a horizontal plane. If leaves are inclined with respect to the direction of incoming radiation, then the leaf irradiance (i.e. the average light intensity per unit leaf area) will be less than horizontal irradiance. If we refer to the discussion of Figure 11.9b, we are reminded that the amount of radiation incident on the surface of a leaf within a thin slice of canopy is $fI.\Delta L$. Given that the radiation incident on the segment of leaf is distributed over ΔL of leaf surface per unit land area, then the average irradiance on the inclined leaf surface is:

$$I_{leaf} = \frac{fI\Delta L}{\Delta L} = fI = \frac{K}{1-m}I \qquad (11.8)$$

Combining equations (11.7) and (11.8) shows that the average leaf irradiance beneath a leaf area index of L is given by:

$$I_{leaf} = \frac{K}{1-m}I_o e^{-KL} \qquad (11.9)$$

Now that the average leaf irradiance in each layer of the canopy is known, one can combine equations (11.4) and (11.9) to calculate, for any position in the canopy, the rate of gross photosynthesis:

$$P_g = \frac{\alpha\dfrac{K}{1-m}I_o e^{-KL}P_{max}}{\alpha\dfrac{K}{1-m}I_o e^{-KL} + P_{max}} \qquad (11.10)$$

The gross photosynthetic rate of the canopy as a whole (P_{canopy}) could be calculated by dividing the canopy up into a large number of thin layers and following the procedure in Table 11.5 above to obtain the contribution from each layer. P_{canopy} is given by the sum of the photosynthetic rates of the individual layers. Table 11.4 shows P_{canopy} for the following four plant types:

Table 11.4 Overall gross rates of photosynthesis ($mg\,m^{-2}$ land area s^{-1}) for contrasting canopies exposed to high ($800\,W\,m^{-2}$) or low ($100\,W\,m^{-2}$) incident irradiance.

	Irradiance ($100\,W\,m^{-2}$)		Irradiance ($800\,W\,m^{-2}$)	
	L = 2	L = 5	L = 2	L = 5
C3 plants				
Erect	0.65	1.04	1.89	3.62
Planophile	0.80	1.06	2.01	3.31
C4 plants				
Erect	0.71	1.11	2.49	4.56
Planophile	0.90	1.17	2.70	4.22

Worked example 11.3 Estimating the gross photosynthesis of a vegetation canopy.

To illustrate the implications of equation (11.10), consider a canopy of a wheat crop that has a leaf area index of 5, and which is exposed to an incident irradiance (I_o) of $600\,W\,m^{-2}$. Assume that α, K and P_{max} for wheat are $0.013\,mgJ^{-1}$, 0.4 and $1.5\,mg\,m^{-2}\,s^{-1}$ respectively. In Table 11.5, the canopy has been divided into five layers, each with leaf area index of 1. Equations (11.7), (11.9) and (11.10) are used to calculate the horizontal irradiance, leaf irradiance and gross photosynthetic rate at each of the nodes corresponding to the layer boundaries (i.e. at leaf area indices of 0, 1, 2, 3, 4 and 5 below the top of the canopy).

As one moves down the canopy, notice that the horizontal irradiance decreases exponentially and that the leaf irradiance for wheat is only 44% ($=K/1-m$) of the horizontal irradiance, because of the inclination of the leaves. The gross photosynthetic rate decreases from 1.047 to $0.358\,mg\,m^{-2}\,s^{-1}$ from the top to the bottom of the canopy.

Also shown are the results of similar calculations for a canopy with more horizontally inclined leaves ($K=0.7$). Note that the horizontal irradiance decreases more sharply as one moves down the canopy because of the greater effectiveness of planophile leaves in intercepting radiation. In the upper two layers, the leaves in the more horizontally inclined canopy have greater leaf irradiance and consequently greater photosynthetic rate than the erectophile canopy. However, the greater degree of shading deeper in the canopy causes the leaf irradiance to be less than in the more erectophile canopy, and the photosynthetic rates are consequently lower.

Table 11.5 Calculating the variation of light intensity and gross photosynthesis through a vegetation canopy.

Leaf area index above node L	Erectophile canopy K = 0.4			Planophile canopy K = 0.7		
	Horizontal irradiance $(W\,m^{-2})$ I, equation (11.7)	Leaf irradiance $(W\,m^{-2})$ I_{leaf}, equation (11.9)	Gross photosynthesis $(mg\,m^{-2}\,s^{-1})$ P_g, equation (11.10)	Horizontal irradiance $(W\,m^{-2})$ I, equation (11.7)	Leaf irradiance $(W\,m^{-2})$ I_{leaf}, equation (11.9)	Gross photosynthesis $(mg\,m^{-2}\,s^{-1})$ P_g, equation (11.10)
0	600	267	1.047	600	467	1.203
1	402	178	0.912	298	232	1.001
2	270	120	0.764	148	115	0.749
3	181	80	0.616	73	57	0.497
4	121	54	0.477	36	28	0.296
5	81	36	0.358	18	14	0.168
Total			4.173			3.909

- Erectophile C3 grass $(P_{max} = 1.2 \, \text{mg m}^{-2} \text{s}^{-1}; K = 0.4; \alpha = 0.013 \, \text{mg J}^{-1})$
- Planophile C3 plant $(P_{max} = 1.2 \, \text{mg m}^{-2} \text{s}^{-1}; K = 0.7; \alpha = 0.013 \, \text{mg J}^{-1})$
- Erectophile C4 grass $(P_{max} = 1.8 \, \text{mg m}^{-2} \text{s}^{-1}; K = 0.4; \alpha = 0.013 \, \text{mg J}^{-1})$
- Planophile C4 plant $(P_{max} = 1.8 \, \text{mg m}^{-2} \text{s}^{-1}; K = 0.7; \alpha = 0.013 \, \text{mg J}^{-1})$.

In each case, P_{canopy} has been calculated for vegetation with a small canopy ($L = 2$) and a dense canopy ($L = 5$). Calculations are shown assuming low ($I_o = 100 \, \text{W m}^{-2}$) and high ($I_o = 800 \, \text{W m}^{-2}$) incident irradiance.

Note the following:

- When the canopy is small ($L = 2$), planophile canopies have faster gross photosynthesis than erect canopies, because they intercept more light.
- In the case of large canopies exposed to high irradiance, most of the incident radiation is intercepted by foliage, irrespective of leaf inclination. In such circumstances, erectophile canopies photosynthesize faster because light penetrates deeper into the canopy.
- C4 plants photosynthesize faster than C3 plants, though the effect is small at low light intensities.
- Canopy photosynthesis is faster at the higher incident irradiance.

The observations help explain why plants adapted to low light environments (e.g. forest floors) tend to have horizontally oriented leaves to maximize the interception of radiation. On the other hand, more erectophile canopies (such as grasses) would be expected to have a competitive edge in open areas, particularly in conditions of high light intensity. The conclusion that C4 plants photosynthesize faster than C3 plants explains why C4 plants (maize, sorghum, millet, rice) are the main providers of staple food in tropical environments with temperatures high enough for C4 plants to develop and grow.

11.3.3   Respiration

As well as assimilating CO_2 into carbohydrates via photosynthesis, plants also consume carbohydrates in respiration to provide the energy required for the maintenance of existing tissues and the work involved in the construction of new tissues. The amount of

---

Worked example 11.4   Daily carbon balance of growing vegetation.

Imagine a crop of wheat that comprises 600 g of dry-matter m^{-2} land area, and which produces 20 g of carbohydrate m^{-2} land area via gross photosynthesis in a day.

- Maintenance respiration would use $1.25\% \times 600 = 7.5$ g carbohydrate m^{-2}, leaving 12.5 g of the daily assimilate available for producing new tissue.
- Of that, $25\% \times 12.5 = 3.1$ g m^{-2} would be lost as growth respiration.
- Hence, of the original 20 g m^{-2} of carbohydrate produced during the day by gross photosynthesis, only $20 - 7.5 - 3.1 = 9.4$ g m^{-2} would end up as new plant tissue.

carbohydrate respired in order to maintain existing plant tissue is typically about 1.25% of the weight of the plant per day, and depends strongly on temperature. In the case of the construction of new tissues, about 25% of the carbohydrate invested in the production of new tissue is consumed in respiration. Hence, vegetation acts as both a source and a sink for CO_2. These ideas provide a simple framework for incorporating respiratory losses into a model of crop growth that considers the daily carbon balance of growing vegetation.

11.3.4   Allocation of new growth between the various plant parts

The difference between gross photosynthesis and respiration represents the net accumulation of dry-matter in the plant. The allocation of new growth between the various plant organs (such as roots, leaves, stems, and seeds) varies widely between species, and depends strongly on the stage of development. Furthermore, the distribution of dry-matter between the various plant organs can be influenced by the environment. For example, water stress tends to encourage the production of roots rather than leaves. In the early stages of development of a grass plant about 70% of new dry-matter is typically allocated to leaves and 25% to roots, whereas during the latter part of the seed-filling phase it is likely that the great majority of new dry-matter would be allocated to seeds with little or no growth of roots and leaves. Vegetation growth models usually contain rules about the fraction of new growth that is allocated to leaves, roots and stems, from which it is possible to calculate, each day, the amount of new leaf area that is produced. Continuing the case discussed in Worked example 11.1, if 75% of growth is allocated to leaves, then the amount of new leaf produced would have been $75\% \times 9.4 = 7.05$ g new leaf produced m^{-2} land area. Given that the leaves of wheat typically have a dry-matter content per unit area of 20 g m^{-2}, then this would represent an increase in the leaf area index of $7.05/20 = 0.35$.

One conclusion that could be drawn from the above is that very dense stands of vegetation (such as an old forest) will reach an equilibrium state in which the production of new dry-matter through photosynthesis is exactly matched by the loss of carbohydrate through maintenance respiration and the loss of either organs (such as senescing leaves or dead branches) or whole plants from the community of plants present. This has important implications for the overall carbon budget of stands of vegetation, which are of interest in understanding the role of vegetation in the global carbon cycle and as a factor controlling the amount of CO_2 in the atmosphere. Such issues are discussed in Section 11.5.

11.4   Water stress and vegetation growth

It is well known that the rate of growth of vegetation slows down during periods of drought; hence, the need for farmers to irrigate crops to maintain rapid growth during periods with low rainfall. There is a very close link between the growth of plants and the amount of water that is taken up by the plant roots and ultimately lost through evaporation from leaves. If the uptake of water by roots becomes limited by shortage of water in the soil, then the rate of growth will be reduced.

The reason for the close link between growth and water use by plants can be understood by considering the way in which CO_2 enters the plant to provide the substrate for photosynthesis. CO_2 is taken up from the air within the intercellular spaces of the leaf by chloroplasts at a rate that depends on the leaf irradiance and temperature, as discussed earlier. This results in the concentration of CO_2 in the air within the leaf falling below the concentration in the atmosphere. The surfaces of leaves contain apertures (*stomata*, plural; *stoma*, singular) which allow the passage of gases. CO_2, therefore, diffuses into the leaf from the atmosphere along the gradient in CO_2 concentration that is generated by the uptake of CO_2 in photosynthesis. However, the air in the intercellular spaces of the leaf is saturated with water vapour, and therefore usually has a higher vapour concentration than the air surrounding the leaf. Hence, it is inevitable that water vapour will tend to diffuse out of the leaf, causing the leaf partially to dehydrate. The water lost through evaporation from the leaf will be replaced by water taken up from the soil via the root system. This process is known as *transpiration*.

One of the mechanisms that plants have developed to minimize the 'wastage' of water through transpiration is the ability to vary the size of the stomatal apertures such that the stomata are only open so far as is necessary to maintain the supply of CO_2 required to meet the demand of photosynthesis. One of the consequences of this strategy is that the concentration of CO_2 within the intercellular spaces of the leaf is maintained approximately constant at values that are typically about 70% of ambient in the case of C3 plants, and 30% of ambient in the case of C4 plants.

Figure 11.11 shows a schematic diagram of a stoma. CO_2 diffuses into the leaf at a rate that is proportional to the difference in CO_2 concentration between the external air (C_{air}) and the air in the substomatal cavity within the leaf (C_{leaf}). Similarly, water vapour

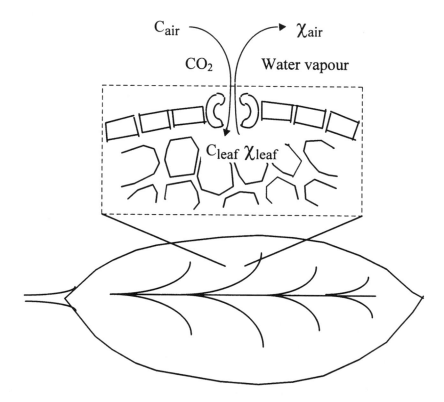

Figure 11.11 Stomata showing the relationship between the uptake of CO_2 and the loss of water vapour.

diffuses from the leaf proportionally to the corresponding difference in vapour concentration ($\chi_{leaf} - \chi_{air}$). The influx of CO_2 (F_c) and the efflux of water vapour (F_w) can, therefore, be described by the following equations (note the similarity with equation 10.18):

$$F_c = K_1(C_{air} - C_{leaf}) \tag{11.11}$$

$$F_w = K_2(\chi_{leaf} - \chi_{air}) \tag{11.12}$$

where K_1 and K_2 are related constants that depend on the degree of opening of the stomata. It follows that the ratio of $F_c:F_w$, which is a measure of the amount of CO_2 taken up per unit of water transpired, is proportional to the ratio of $C_{air} - C_{leaf}$ to $\chi_{leaf} - \chi_{air}$:

$$\frac{F_c}{F_w} = \frac{K_1}{K_2}\left(\frac{C_{air} - C_{leaf}}{\chi_{leaf} - \chi_{air}}\right) \tag{11.13}$$

This conclusion has important implications for the relationship between vegetation growth and water use. It was stated above that stomata vary their aperture to in effect control $C_{air} - C_{leaf}$ to a fairly constant value that is typically $0.3 \times C_{air}$ for C3 plants and $0.6 \times C_{air}$ for C4 plants. Hence, the numerator of the right hand side of equation (11.13) is effectively fixed, depending on the photosynthetic pathway. However, the difference in water vapour concentration between the substomatal cavity and the ambient air is highly variable and depends on the weather. χ_{air} is simply the ambient humidity. Plants growing in very dry air are, therefore, expected to use much more water per unit uptake of CO_2 than plants growing in humid air. χ_{leaf} is also weather-dependent. The air within the leaf is very close to being saturated with water vapour, and so χ_{leaf} depends uniquely on the temperature of the leaf, as described by the relationship shown in Figure 3.14. However, it should be noted that Figure 3.14 expresses the humidity in terms of a vapour pressure, rather than concentration, though the two are directly linked via the gas laws. If it is assumed that the leaf temperature is close to the temperature of the surrounding air, then $\chi_{leaf} - \chi_{air}$ approximates to the saturation deficit of the air, D (see Section 3.6.2).

Given that the growth of vegetation is closely linked to the uptake of CO_2, then one might predict that:

- the growth of vegetation is proportional to the loss of water in transpiration and
- the amount of vegetation growth produced per unit of water transpired will be approximately inversely proportional to the saturation deficit of the air, and will be approximately twice as great for C4 than for C3 species.

Table 11.6 is based on results obtained from a variety of glasshouse and field experiments in the UK, India and Africa, and shows the amount of dry-matter produced by crops of pearl millet (a C4 plant) and groundnuts (a C3 plant) together with the amounts of water transpired by the crops during the growing season. The fourth column shows the amount of dry-matter produced per unit of water transpired (the dry-matter:water ratio, q), and the final column shows the mean daytime saturation deficit of the air during the growing season.

There were large differences between the crops in the amount of dry-matter produced that were due largely to differences between the locations in the amount of rainfall and irrigation. For example, experiments 3 and 4 were unirrigated millet crops grown in the

The close link between D and the amount of growth produced per unit water transpired has important implications for the effects of weather on the productivity of agricultural crops. For example, much of the crop-growing region in the semi-arid areas of Sub-Saharan Africa has an annual rainfall of 500–800 mm, which is similar to the range of annual rainfall in southern England, yet crop yields are substantially lower, even when crops are well-fertilized.

11.5   Carbon balance of the land surface

Perhaps the greatest global environmental issue debated over the past decade or so has been the progressive elevation of atmospheric CO_2 concentration, and the effect this has on the so-called greenhouse effect and the consequences for global warming. The impact of green-house gases on radiative transfer processes in the atmosphere and global warming were discussed in Chapter 7. One aspect of this debate has been the role that land surface vegetation has on atmospheric CO_2 concentrations, and the possible impacts that might result from large-scale changes of land use, such as clearance of extensive areas of rain forest to develop land for agricultural use. It is well known that vegetation removes CO_2 from the atmosphere via the process of photosynthesis, which gives rise to the widely held belief that changes in land use involving the destruction of vegetation will exacerbate the accumulation of CO_2 in the atmosphere. However, the uptake of CO_2 by vegetation via photosynthesis is only one part of the land surface carbon balance story. Plant material produced as a result of photosynthesis will also be acting as a source of CO_2 either directly as a result of respiration, or indirectly when dead plant material becomes incorporated into the soil and decays as a result of microbial activity. This section considers how vegetation and soil processes interact to influence the extent to which the land surface acts as a net source or sink for atmospheric CO_2, and the implications for how land use interacts with atmospheric CO_2.

11.5.1   Terrestrial carbon store

It is useful at this stage to consider the relative amounts of carbon stored within growing vegetation and within the soil as soil organic matter. Indeed, one of the tasks set at the Kyoto Earth Summit (1997) was that countries would make an assessment of their national carbon stores, which are usually dominated by the carbon locked up in standing vegetation and in soil organic matter.

The amount of carbon stored per unit land area within growing vegetation is highly variable (Table 11.7) and depends, not surprisingly, on the nature of the vegetation, and the extent to which vegetation is periodically removed (e.g. through harvesting agricultural products or by grazing). Values range from virtually zero to a maximum of about 16–20 kg carbon m^{-2} land area in the case of tropical forest. Global estimates of the amount of carbon stored in vegetation (last column of Table 11.7) are obtained by multiplying the amount per unit land area by the area of coverage of each vegetation type, which is usually obtained from analysis of satellite images (such as Landsat, see Chapter 6). This reveals that over half of the global carbon store within vegetation is associated with tropical forest, and that cultivated land contributes <1% to the total!

The amount of organic matter stored in soil is highly variable and depends on the relative rates of organic matter input and the rate at which soil organic matter is lost through microbial degradation (Table 11.8). Arable fields in temperate regions typically have 1–3%

Table 11.7 Estimates of total amount of carbon stored in vegetation around the land surface of the world.

Vegetation type	Land area $(\times 10^{12}\,km)$	Typical carbon stored per unit land area $(kg\,carbon\,m^{-2})$	Global carbon storage in vegetation $(Gt\,(\times 10^{12}\,kg))$
Tropical forest	24.5	16–20	460
Temperate forest	12.0	13–16	175
Boreal forest	12.0	9	108
Woodland/shrub	8.0	2.7	22
Tropical savannah	15.0	1.8	27
Temperate grassland	9.0	0.7	6.3
Tundra	8.0	0.3	2.4
Desert scrub	18.0	0.3	5.4
Swamp/marsh	2.0	6.8	13.6
Cultivated land	14.0	0.05–0.8	7.0
Total	149	55.5	827

Table 11.8 Examples of amounts of carbon stored in soil profiles.

Soil	Total carbon in whole profile $(kg\,m^{-2}\ land\ area)$
Deep peat	>500
Grassland prairie soil (Nebraska)	23
Australian 'black earth' soils	20
'Wilderness' brown earth (woodland since 1883)	7 (top 23 cm only)
'Hoosfield' brown earth (with annual manuring for 130 years)*	8
'Hoosfield' brown earth (no manuring in last 130 years)*	2.7
Arable sandy soils in semi-arid tropics	<1

*The Hoosfield experiment at Rothamsted, in which barley has been grown continuously since 1852. The nearby Wilderness experiment was an arable field that was allowed to revert to woodland since 1883.

carbon by mass in the upper 30 cm of the profile, with the amount of carbon decreasing dramatically with depth. Forest and grassland soils usually have somewhat higher organic matter contents, particularly if they are poorly drained. Peat soils store extremely large quantities of carbon, the amount depending mainly on the depth of peat present.

A number of points arise from Tables 11.7 and 11.8:

- On a 'per unit land area' basis, the amount of carbon stored as organic matter in soil is usually around an order of magnitude (or larger) than the rather more obvious store of carbon in standing vegetation.

- Upland peat soils contain very large amounts of carbon, and form the dominant contribution to the terrestrial carbon store in northern Europe. Hence, the management and dynamics of peat uplands is likely to be a major factor influencing the extent to which the land surface in high latitudes acts as a source or sink for atmospheric carbon.
- The data from the Rothamsted experiments illustrate that agricultural fields have less carbon stored in vegetation and lower levels of soil organic matter than areas such as woodland where there is less removal of vegetation.
- Low input agricultural fields in tropical environments can have very low soil organic matter. This can be an important factor limiting the agricultural productivity in the tropics, because of the role of organic matter in the retention and release of nutrients for plants. Also, organic matter acts as a 'binding agent' helping to maintain the structural stability of soil aggregates, which is important for processes such as maintaining rapid infiltration and preventing the collapse of the soil surface structure to form a relatively impermeable 'crust' which encourages surface run-off and erosion.

11.5.2   Degradation of soil organic matter

We can understand better the factors underlying the variability between soils in organic matter content by considering the fate of organic matter that is added to soil as a result of processes such as the fall of leaf litter, the death of roots and the ploughing in of crop residues. One experiment to investigate this involved incorporating into soil a quantity of cuttings from ryegrass that had been grown in an atmosphere containing CO_2 that had been labelled with the 14 molar mass radioactive isotope of carbon (^{14}C). Some of the CO_2 was taken up by photosynthesis and incorporated into the plant tissues, thereby providing a source of radioactive labelled plant material. By monitoring the levels of radioactivity in the soil, it was possible to monitor the amount of the introduced ryegrass carbon that remained in the soil. The amount remaining decreased with time (Figure 11.14) because of oxidation of the organic matter to CO_2 and water vapour as a result of microbial respiration using this organic matter as an energy source. Identical experiments, using the same soil, were carried out in the UK and in Nigeria (where the mean temperature was about 20°C warmer). In both cases the initial loss was rapid, but the loss rate decreased with time. The rates of loss were about four times faster in Nigeria than in the UK, presumably because of higher temperature.

The progressive decrease in the rate of loss of soil organic matter is expected if the rate of degradation is limited by the amount of substrate. Indeed, if one assumes that the rate of degradation of organic matter is proportional to the amount of organic matter present (i.e. it is assumed that degradation obeys the law of first order kinetics), then:

$$\frac{dC}{dt} = -kC \qquad (11.15)$$

where dC/dt is the rate of change of concentration, C is the carbon content of the soil, and k is the coefficient of proportionality known as the *turnover rate*. If the time units used in the rate of change term are years, then the units of k will be 'per year', and can be thought of as 'the number of turnovers per year'.

Equation (11.15) predicts that the concentration of radioactive ryegrass remaining in

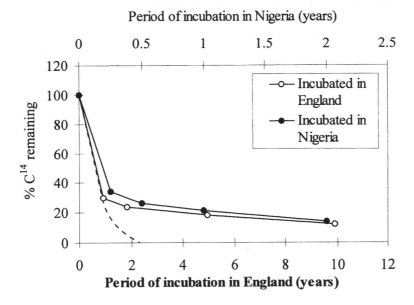

Figure 11.14 Degradation of labelled ryegrass cuttings incorporated into soil.

the soil in Figure 11.14 will decrease exponentially. An exponential decline means that after a certain time (i.e. one 'half-life') the amount remaining would be reduced to one half. After two half-lives, one-quarter of the original amount would remain. After three half-lives, one-eighth would remain, and so on. Simple calculus based on equation (11.15) shows that the half-life is related to k by:

$$\text{Half-life} = \frac{0.693}{k} \qquad\qquad (11.16)$$

Hence, a turnover rate of 2 per year implies a half-life of 0.346 years.

But does the decay of soil organic matter obey first-order kinetics? The dotted line on Figure 11.14 shows the result of fitting equation (11.15) to the first part of the decay curve determined experimentally. The dotted line (equation (11.15), assuming $k = 2.5$ year^{-1}) fits the experimental data well initially, but predicts that the soil organic matter has virtually disappeared after 2 years. However, the experimental data suggested that there was an abrupt slowing down of the rate of disappearance of organic matter after about 1 year. This is because ryegrass (like any plant material) is not made up of a single compound, but consists of a range of organic materials which decay at different rates. For example, soluble sugars present in the cell sap will be consumed very rapidly, whereas compounds such as cellulose in cell walls will have a much longer turnover time. Plant material consists of a large number of compounds, each with their own turnover rate, so you might imagine that this would defy any attempt to create a simple mathematical model to describe the decomposition process. However, a very good approximation to the experimental data in Figure 11.14 can be made by assuming that the ryegrass cuttings consisted of two fractions: a rapidly decaying fraction comprising 70% of the original material and a more slowly decaying fraction comprised of the other 30%. The appropriate turnover rates for the fast and slow fractions in England are 2.6 and 0.5 per year respectively. If the time courses of the amounts of rapidly-decaying and slowly-decaying fractions are calculated separately, using equation 15, and then added together the resulting totals match the 'dog-leg' time course observed in Figure 11.14.

Had the experiment reported in Figure 11.14 been carried out for a much longer period (many decades), it would have been found that there was evidence of a small, very slowly decaying fraction that would have remained largely undecomposed for many decades. This consists of either very indigestible compounds, or else organic material that is protected from decomposition as a result of being chemically combined with inorganic compounds present in the soil or physically protected as a result of being coated with material such as clay minerals or iron and aluminium oxides. k for this very slowly decaying fraction can be <0.001 per year^{-1}, implying that some of the organic matter in soils might be centuries old.

11.5.3 Modelling soil organic matter dynamics

The rate at which the soil organic matter changes with time can be expressed as the difference between the rate of addition of organic matter to the soil (A) and the rate of decomposition ($=kC$; equation (11.15)), in which case one can write:

$$\frac{dC}{dt} = A - kC \tag{11.17}$$

However, equation (11.17) is an oversimplification because it has been shown that degradation of soil organic matter obeys simple first-order kinetics only if we consider the organic matter added to soil to be made up of a number of different fractions, each with their own turnover rate. In this case, equation (11.17) should be expanded to consider each of the fractions individually:

$$\frac{dC}{dt} = [A_1 - k_1 C_1] + [A_2 - k_2 C_2] + [A_3 - k_3 C_3] + \ldots \tag{11.18}$$

where the subscripts 1, 2 and 3 refer to the different organic matter fractions. It can be seen from equation (11.18) that the amount of soil organic matter will move towards an equilibrium state where the rate of degradation of each fraction balances the rate at which the organic matter fraction in question is added to the soil. In other words, $dC/dt = 0$ (indicating an equilibrium state) when $A_1 = k_1 C_1$, $A_2 = k_2 C_2$, $A_3 = k_3 C_3$, and so on. It can therefore be concluded that if the annual additions of organic matter are increased (i.e. A_1, A_2 and A_3 are increased), then the soil will respond by increasing the respective soil organic matter contents (i.e. C_1, C_2 and C_3).

An important conclusion to note from Table 11.9 is that although most of the organic matter added to the soil is in the rapidly decaying fraction; it is the very slowly decaying fractions that dominate the soil organic matter. Note also that the predicted overall concentration of organic matter in the soil (2.2% by mass) is reasonable for an arable field in the UK.

It is interesting to consider how the results in Table 11.9 would have been different if the annual addition of organic matter had been changed. For example, if the farmer's field

Worked example 11.5   Soil carbon balance at equilibrium.

Consider the case of a field of spring barley in southern England. Figure 11.1 showed an example in which the grain yield from such a field ranged from 2 to 7 tonnes hectare^{-1}. The organic matter present in the grain will, of course, be removed from the field by the farmer. Given that most of the dry-matter of a mature spring barley crop is present in the grain, then it is likely that the amount of dry-matter remaining in the field as dead leaves, stems and roots would be around 2 tonnes hectare^{-1} in an average year. This provides an estimate of the annual organic matter input to the soil.

Assume that the organic matter returned to the field decays in a similar way to the ryegrass example in Figure 11.14, in which case one might assume that 70% is fast decaying (assume $k = 2.6$ per year), 25% is slowly decaying ($k = 0.5$ per year) and the remaining 5% is very slowly decaying ($k = 0.001$ per year). These values are shown in columns 2 and 3 of Table 11.9. What would be the equilibrium soil organic matter content if the field was used for many years to grow such a crop?

At equilibrium, where $dC/dt = 0$, the equilibrium concentration of each fraction can be obtained by simple rearrangement of equation (11.17):

$$C = \frac{A}{k} \qquad\qquad (11.19)$$

Column 4 of Table 11.9 shows the results of applying equation (11.19). Column 5 expresses the equilibrium concentration in terms of the percentage by mass of the soil. This was done assuming that the organic matter is contained within the upper 30 cm of the soil profile, and that the upper 30 cm of the soil profile in a 1-hectare ($\equiv 10\,000\,\mathrm{m}^2$) field contains 4500 tonnes of soil (check this out assuming that the bulk density of soil is $1.5\,\mathrm{Mg\,m}^{-3}$; see Section 10.3).

Table 11.9 Example of calculations of the equilibrium concentrations of various organic matter fractions within a topsoil.

Fraction	Annual input (A, tonnes ha^{-1})	Turnover rate (k, year^{-1})	Equilibrium concentration (C, tonnes ha^{-1})	Equilibrium concentration (C, % by mass)
Fast	1.4	2.6	0.54	0.012
Slow	0.5	0.5	1	0.022
Very slow	0.1	0.001	100	2.2
Total	2.0		101.54	2.234

in southern UK had been mature woodland, rather than being used to grow cereals, then the organic matter input would likely have been increased by approximately a factor of 5. This higher figure is because organic matter is not being removed, and because the land surface is covered by growing vegetation for most of the year. Increasing values of A in Table 11.9 by a factor of 5 would result in the equilibrium soil organic matter contents being five times greater (i.e. ~500 t ha^{-1} or 11% soil organic matter). This suggests the possibility that converting land from arable cropping to woodland provides a means of sequestering atmospheric CO_2 into soil organic matter.

By contrast, consider land being used for growing a subsistence crop in Africa by a farmer with no resources to irrigate or apply fertilizers. In this case the productivity will be very low, and the annual input of organic matter to the soil is likely to be 10 times smaller than for the UK cereal field considered in Table 11.9 (i.e. 0.2 tonnes hectare^{-1} year^{-1}). Also, because of higher temperatures, k is likely to be about four times faster than in the UK (Figure 11.14). From equation (11.19) it is expected that the combined effect of these two factors will result in the organic matter content of the soil at equilibrium being 40 times lower than for the UK field used for growing cereals (i.e. the soil organic matter content would be about 0.05%). As noted in the discussion of Table 11.8, this depletion of organic matter can have serious implications for the supply of nutrients and the physical stability of the soil. This is one of the reasons why the traditional practice in such marginal land is to only grow crops for very few years before leaving the land fallow for a number of years so that the organic matter levels can recover.

A final point to note is that land used in the same way for a number of years tends towards an equilibrium state where the input of carbon (ultimately from photosynthesis) is balanced by the rate of decomposition. In such a state, the balance is 'carbon neutral', so that the land surface acts as neither a source nor a sink for atmospheric carbon, irrespective of whether the land is used for forest or the production of agricultural crops. It is only when land use is changed that the land surface becomes a major source or sink for carbon. Indeed, much of the terrestrial land area of the world is not in equilibrium because of the ever-changing patterns of land use. Though beyond the scope of this chapter, slightly modified versions of the model of soil organic matter dynamics given by equation (11.18) can be used to study how rapidly the soil organic matter responds to changes in land use. Typically, it is found that when land use is changed, it can take many decades before the new equilibrium state is approached.

Mathematical models similar to that described above provide a means of assessing whether such changes in land use are likely to have a significant impact upon atmospheric CO_2 concentrations.

Another interesting application of such models is to predict the effects of climate change on soil organic matter. For example, increasing temperature will tend to increase k (k increases approximately two fold with a 10°C increase in temperature). This will have the effect of decreasing soil organic matter contents, thereby releasing CO_2 into the atmosphere and possibly adversely affecting the fertility of the land.

11.6   Summary

This chapter has examined how the nature of the physical environment, including the primary process of photosynthesis and the impact of weather, generate the conditions for vegetation growth and the global carbon budget. The maintenance of the terrestrial carbon balance has implications for crop production and agricultural land use, and can be modelled using soil organic matter reaction kinetics. Modelling the role of land use and incorporating these processes into global climate models (Chapter 9) is an active area of current research with important political and social/economic consequences.

References

Barbour, M. G., Burk, J. H., Pitts, W. D., Gilliam, F. S. and Schwartz, M. W., *Terrestrial Plant Ecology*. Harlow: Addison Wesley Longman, 1999.

Bell, P. R., *Green Plants: Their Origin and Diversity*. Cambridge University Press, Cambridge: 1997.

Crawford, R. M. M., *Studies in Plant Survival*. Oxford: Blackwell, 1994.

Fitter, A. H. and Hay, R. K. M., *Environmental Physiology of Plants*, 2nd edn. London: Academic Press, 1995.

Friday, A. and Ingram, D. S., eds, *The Cambridge Encyclopaedia of Life Sciences*. Cambridge: Cambridge University Press, 1985.

Hillel, D., *Introduction to Soil Physics*. New York: Academic Press, 1982.

Hopkins, W. G., *Introduction to Plant Physiology*, 2nd edn. New York: Wiley, 1999.

Jones, H. G. *Plants and Microclimate: A Quantitative Approach to Environmental Plant Physiology*. Cambridge: Cambridge University Press, 1992.

Killham, K., *Soil Ecology*. Cambridge: Cambridge University Press, 1994.

Ksenzhek, O. S. and Volkov, A. G., *Plant Energetics*. San Diego: Academic Press, 1998.

Marshall, T. J., Holmes, J. W. and Rose, C. W., *Soil Physics*, 3rd edn. Cambridge: Cambridge University Press, 1996.

Mohr, H. and Schopfer, P., *Plant Physiology*. Berlin: Springer, 1995.

Nobel, P. S., *Physicochemical and Environmental Plant Physiology*. San Diego: Academic Press, 1999.

Raven, P. H., Evert, R. F. and Eichhorn, S. E., *Biology of Plants*, 6th edn. New York: Freeman, 1999.

Taiz, L. and Zeiger, E., *Plant Physiology*, 2nd edn. Massachusetts: Sinauer, 1998.

Tivy, J., *Biogeography: A Study of Plants in the Ecosphere*, 3rd edn. Harlow: Longman, 1998.

Discussion questions

1   Discuss the advantages and disadvantages to a plant of having erect leaves, bearing in mind the effect of leaf angle on the ability of the plant to compete with neighbours for light, and the effect of leaf angle on the efficient use of radiation to produce photosynthate.

2   What are the advantages and disadvantages of growing a C4 tropical cereal in northern Europe (rather than a C3, temperate species) for:
    (a)   The production of leaves and stems for animal fodder.
    (b)   The production of mature grain?

3   How might you assess the impact of projected global warming on the distribution of crops around the world?

4   'Removal of large areas of tropical forest is a major factor contributing to the observed increase in atmospheric CO_2 concentrations.' Discuss this assertion with reference to the carbon balance of vegetation and soils. Consider carefully the impact that changing land use has on whether the land surface in question is a net source or is a net sink for atmospheric carbon (or, alternatively, is carbon neutral).

5   What would be the ideal arrangement of leaves in a plant canopy to make most effective use of light in terms of maximizing the rate of photosynthesis per unit land area? Do any plants you know approach this ideal?

6   Discuss the impact of climate change on the growth of vegetation.

7   You have been asked to design a project to quantify the amount of carbon stored in vegetation and in soils within Europe, and to monitor how these carbon stores might have changed over the next 20 years. What do you suggest?

8   A major land use change in recent years has been to establish forests on peat moor land. How might you assess whether this is likely to result in a net sequestration of carbon, or else release carbon dioxode into the atmosphere? Bear in mind that one of the effects of planting trees will be to dry out the peat, and thereby increase its rate of degradation.

Quantitative questions

1   Using the data in Table 11.1, plot the graph of *1/time from floral initiation to heading* against *temperature* (i.e. column 3 against column 1). From your graph, estimate the base temperature and the thermal time requirement for the period from floral initiation to heading. Assume that the thermal time requirement for the period from sowing to floral initiation is 250 day°C, and that the base temperature for this phase is the same as that just calculated. Use equation (11.1) to estimate the time it would take to reach heading if the crop spent the first 20 days after sowing at a mean temperature of 13°C, and the remainder of the time at a mean temperature of 20°C.

2   Imagine you are a farmer with a contract to produce lettuces for a supermarket every 10 days during the summer. The thermal time requirement to reach maturity for the variety of lettuce you are growing is 450 day°C, with a base temperature of 5°C. Using the mean daily temperature data in Table 11.2, when should you sow the crops to produce lettuces on 8, 18, 28 June, 8 18, 28 July, 7, 17 and 27 August? A computer spreadsheet package will help with these calculations.

3   Repeat the exercise in the question above, but assuming that the mean daily temperatures have been increased by 2°C as a result of global warming. How does this change the sowing times? How do you think the change in sowing date might influence the size of the lettuces at the time of harvest?

4   Consider a case where the Sun is directly overhead (so the solar beam is vertical), and the light intensity on the surface of a horizontal leaf is $600\,W\,m^{-2}$ (i.e. $600\,J\,s^{-1}\,m^{-2}$). What would be the light intensity on the surface of a leaf inclined at 30 and 60° to the horizontal? (Hint: bear in mind that the 'shadow area' of an inclined leaf is less than the actual area. Consider a right-angled triangle that shows the relationship between the actual length of the inclined leaf (the hypotenuse) and the shadow length of the leaf (the length of the side adjacent to the angle of inclination, A). The horizontal (the adjacent) represents the length of the shadow that would be cast if the Sun was directly overhead. Simple trigonometry tells us that the ratio of the adjacent to the hypotenuse is the cosine of A. Hence the irradiance on the surface of an inclined leaf is cosine A multiplied by the irradiance on the surface of a leaf that is normal to the direction of the solar beam.)

5   Following Question 4, use equation (11.4) to calculate the rate of photosynthesis for leaves inclined horizontally and at 30 and 60° to the horizontal, assuming that α and P_{max} are those appropriate for wheat at 20°C (Table 11.3). Discuss the effect of leaf angle on the rate of photosynthesis per unit leaf area.

6   Following Question 5, calculate for each leaf inclination the amount of CO_2 taken up per second by a leaf with an area of $0.01\,m^2$. Calculate also for each leaf the Joules of solar radiation intercepted per second, based on the answer to Question 4. Now compare the leaf inclinations with respect to the amount of CO_2 taken up per Joule of radiation intercepted. You should find that the horizontal leaf is the least efficient in terms of the amount of radiation used to carry out a unit amount of CO_2 fixation. Why is this?

7   The table shows results obtained from an experiment designed to investigate the differences in water use efficiency between rain-fed maize and bean crops growing in a semi-arid region of Kenya. It shows the results of experiments in different growing seasons in which the amounts of dry-matter (W) and grain (Y) produced, and the amounts of water lost through total evaporation (i.e. transpiration plus direct evaporation from the soil surface, ET) and transpiration (T) were measured. Discuss these results with respect to the efficiency with which water is used to produce crops.

Season	Crop	Rainfall (mm)	W (kg ha⁻¹)	Y (kg ha⁻¹)	ET (mm)	T (mm)
1990	beans	253	2100	1040	250	62
	maize	257	5700	2390	263	66
1991	beans	158	450	270	163	32
	maize	158	1550	170	163	32

Chapter 12

Environmental issues for the twenty-first century

12.1 Introduction

The last quarter of the twentieth century saw a new awareness by both the public and governments that humanity may be changing the environmental balance established over the past 4.6 billion years. The dramatic observation of the ozone hole over Antarctica in 1985 provided the first clear evidence that industrialization could produce changes in the Earth's atmosphere on a gigantic scale. The increasingly erratic climate and 'freak' weather patterns experienced at the end of the twentieth century in many regions of the world: the droughts of Africa, the floods of Bangladesh, the dramatic hurricanes and typhoons which ravaged the Caribbean and western Pacific all suggested that the global climate was changing and were cited as evidence for global warming. Shrinking biodiversity; widespread deforestation; pollutants in rivers and soils; acid rain, and urban air pollution; every day the media seemed to be full of new dangers to the environment and prophecies of impending global catastrophe. In this last chapter we will look ahead into the new century and analyse the risks human beings and the planet face and what measures can be taken to reduce their impact.

12.2 Demographic change

If humanity can affect the planet, then population growth and population density will have a crucial impact on the future of the Earth's environment. Ultimately, as part of any study of the global environment, it is necessary to consider the effect of population growth.

Chapter 1 reviewed Malthusian concepts of the impact of demographic change. In 1650 the world's population was approximately 500 million, the number having doubled approximately every 1500 years since 8000 BC. In the next two centuries the world's population doubled again reaching 1 billion in 1850. Medical advances and improvements in sanitation and medicine in the developed world, resulting in a dramatic decline in deaths from disease, particularly among children, then saw the population double again in only 80 years. Then, as scientific advances extended across Asia and Africa, the population doubled again in only 40 years until, in 1990, the Earth's population was estimated at 5.3 billion (or 10% of the entire human population ever to have lived on Earth). It reached 6 billion in October 1999 and it is anticipated to be 9 billion by 2030 (Figure 12.1). Of the predicted population increase in the early twenty-first century, 95% will occur in the Third World, where resources of much needed water and agricultural land are most

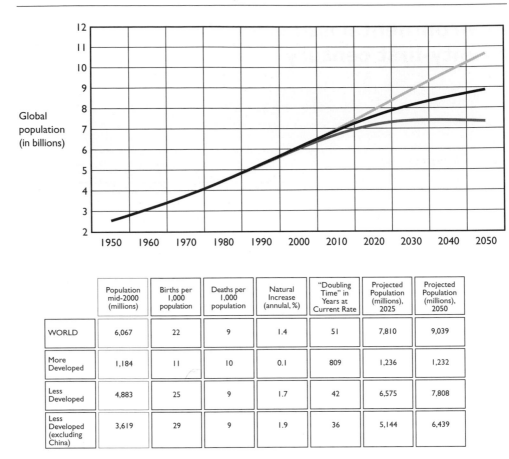

	Population mid-2000 (millions)	Births per 1,000 population	Deaths per 1,000 population	Natural Increase (annulal, %)	"Doubling Time" in Years at Current Rate	Projected Population (millions), 2025	Projected Population (millions), 2050
WORLD	6,067	22	9	1.4	51	7,810	9,039
More Developed	1,184	11	10	0.1	809	1,236	1,232
Less Developed	4,883	25	9	1.7	42	6,575	7,808
Less Developed (excluding China)	3,619	29	9	1.9	36	5,144	6,439

Figure 12.1 Global population growth 1950–2050. Trends beyond 2000 are three different predictions.

limited. Life expectancy will continue to rise, resulting in a 'greying' of the global population, with all the implications that this will have for our welfare and the quality of life. Larger populations will put increased pressure on agricultural and industrial resources both at the local and global level. For example, increased population growth requires increased power generation, which in turn will exploit the Earth's reserves of fossil fuels at an even faster rate. If we look at the world's two most populous countries, China and India, one can gauge the impact that they might have as a major driving force in the world's population growth rate. In 1991, India's population was 846.3 million. At that time the population of China and India together constituted over 40% of the global population. In 1992 the annual growth rates for China and India were 1.5 and 2% respectively so that by 1999 India's population had risen to 989 million and it is clear that within the next 30 years India's population will overtake China's.

In 1999 the world's population was increasing by 87 million a year. Limiting the ever rising number of people seeking to exploit the Earth's resources must, therefore, be of primary concern to governments world-wide in the twenty-first century or by the end of the century, in many parts of the world, natural resources will not sustain the local populations. However, of all the regulations that might be introduced to combat environmental change, legislation on birth control and family growth is likely to be the most contentious since it inevitably would be in opposition to the current philosophy of human rights and many religious beliefs. Indeed, in a democratic society, governmental interference in domestic family life is likely to be almost impossible. Only by changing peoples perceptions of the size and role of the family unit can future population growth be limited and stabilized.

12.3   Urbanization

People like to live in cities. They are social centres providing many forms of recreational activity and rapid access to work. There are, however, features of cities that can make them unhealthy and dangerous places. Increasing population-density can lead to a reduction in the quality of life, life expectations and life-expectancy. Cities are places of pollution, especially from car-exhausts and industrial plants, while noise pollution is an increasing issue for many urban citizens. Cities can also provide the environmental conditions that facilitate those physical transport processes responsible for the transmission of various infectious diseases, such as in the aerosol droplet migration that spreads tuberculosis.

As the global population has increased, so has the population of cities. Over the past five millennia rural depopulation has facilitated the rise of the city until, for the first time in history, the 1990s saw the number of people living in urban areas exceed those living in rural areas. In the twenty-first century the world will become increasingly urban. In developed countries, such as Europe, the USA and Japan, three-quarters of the population already live in urban areas, whereas in the developing countries, such as Ethiopia, only one-third of the population lives in an urban environment. In 2000 the 30 largest cities in the world had a combined population of 360 million or approximately the entire population of the world in the Middle Ages! By 2010 it is forecast that the population of these cities will have risen to 450 million, an average of 15 million per city. This introduces the concept of the mega-city and with it the prospect that billions of people will live in an artificial environment designed and conditioned purely by humanity.

Cities are not static entities. Indeed, over the course of history cities have even vanished. It is clear why some rural communities disappeared during the fourteenth century in the wake of the Black Death; what is not so easy to decipher is why whole cities, for example, in Mongolia and the Middle East, disappeared. Shortage of sustainable natural materials and/or climate change and its impact on agriculture and water supplies may have played an important role in the past and may do so in the future. It is therefore important to understand what factors allow urban environments to be sustained.

To date, much of the emphasis on environmental issues has concentrated on the rural areas and natural resources. Although these areas will, of course, remain important, in the future science must emphasize urban environments and the effects of increasing urbanization on the rest of the planet.

12.4   Sustainability

This term was advanced by the UN World Commission on Environment and Development (1987) and refers to the exploitation of the Earth's resources in a manner that does not degrade the environment and which will leave sufficient resources for future generations.

There is a widespread feeling that there is a limit to global growth, that there is a global-carrying capacity for food, utilities and the space for the world's population to develop harmoniously and productively; and that there is a limit to the reserves of fossil fuels. Nevertheless, as the global population increases a major question arises – how can more land, and the potentialities of the oceans, be brought into production, and how can agricultural productivity be increased without destroying the very soil that generates this production?

Estimates of how many people the Earth can support range from 2.5 to 40 billion. Hence, from a figure that would suggest that we have already exceeded a population the Earth can sustain for any length of time to one that suggests that we are capable of absorbing almost infinite growth in population. Thus, there is considerable debate about how to proceed in checking population growth or indeed if we need to do so.

If food production levels are maintained at the present level and the entire global population was to eat as well as citizens of the USA do now (40% more calories than are necessary to keep the body healthy), then the world would indeed only be able to support 2.5 billion, while if all the Earth's surface could ultimately be farmed the 40 billion figure could be reached. Obviously, the true figure lies somewhere between these two extremes.

The Green Revolution of the 1960s saw an increase both in world production of food and in available acreage. By 1981 the area for arable crops had reached a maximum. It has since declined. More people have a need for more living space, and this is invariably from land taken out of agricultural production. In addition, excessive working of the land adversely affects production, as does the potentialities of soil and water erosion. Twelve million acres are required annually to supplement the growing global population. The thirst for land is a factor in the destruction of the tropical rainforests and in the consequent loss of biodiversity. Allied to this requirement is the need for water. To generate 1 ton of wheat requires 1000 tons of water. At present (2001) 17 countries are experiencing scarcity of water. There is therefore the need for better irrigation and de-salination systems, and the development of crops resistant to pests and insects and which have tolerance to drought and salt.

Approximately 1.7 billion tons of grain are produced every year, which, if distributed evenly, would allow 6 billion people to eat a healthy vegetarian diet. To increase grain production either the current land must be made more productive, through further use of chemical fertilizers or genetically modified crops, or new areas of land must be developed for crop growth. A 12% increase in cultivatable lands is possible but would be expensive, and it would be beyond the resources of the developing world. Much of the virgin land being brought into cultivation is therefore being developed by the poorest people using 'slash and burn' (e.g. in the Amazonian rainforests). This quickly leads to soil erosion. Indeed, globally, more soil is lost through erosion than is being produced. There is, in fact, a net annual global deficit of soil (about 25 billion tons).

Water suitable for drinking and irrigation is only the smallest fraction of the Earth's water supplies ($<1\%$). As the population increases, the use of water will also accelerate. The world per capita use of water in 1975 was $700 \, \text{m}^3 \, \text{year}^{-1}$ or 2000 litres day^{-1} or a global rate of $3850 \, \text{km}^3 \, \text{year}^{-1}$! In 2000 this had grown to nearly $6000 \, \text{km}^3 \, \text{year}^{-1}$, in itself not a major proportion of the world's global freshwater supply. However, much of the freshwater supply is inaccessible. The relatively uninhabited Amazon basin provides one-quarter of the Earth's total runoff but in the more populated regions of the world, local demand is now exceeding local supplies. Even in the USA it is estimated that by 2020 water use may exceed surface water resources by >10%. Developing transport networks for water is very expensive and if it crosses national boundaries can be potentially divisive. Indeed, 'water wars' have been predicted in the twenty-first century.

12.4.1   *Energy resources*

A crucial factor in economic and industrial development, in post-industrial, industrial and Third World countries, is that of how are global energy requirements to be met? This factor encompasses issues of sustainability and both demographic and urban growth.

As the Earth becomes increasingly industrialized the requirement for more energy (mainly in the form of electricity) accelerates. Since 1860, the rate of energy use has increased by a factor of 30. At first this energy demand was met mainly by coal, but since 1950 by the use of oil and gas. In 1990 the global consumption of energy was esti-mated at 8730 million tonnes of oil equivalent (toe) or about 12 terawatts (1 terawatt = 1 TW = 10^{12} W). The average consumption of primary energy per person is therefore 1.66 toe or 2.2 kW per annum. However, there are great disparities in the amount of energy used per person world wide. The USA has only 5% of the Earth's popu-lation but consumes 25% of the total global energy production. Thus, while other regions of the world aspire to the wealth and affluence of the USA and hence to a higher degree of industrialization, the Earth's energy production must increase.

At present the world generates most of its energy from fossil fuels; yet these are finite resources and ultimately will be exhausted. Known reserves will allow fossil fuels to provide energy until the mid-twenty-first century after which (and before if demand con-tinues to grow) supplies will begin to come under pressure and lead to increases in their cost. Further exploration and new technology will undoubtedly allow new reserves to be found and exploited but ultimately within a century commercially viable reserves of oil and gas will have been exhausted and a return to coal is likely since there are several hundred years of coal and lignite stocks exploitable. Beyond this supplies of uranium for nuclear power may provide energy supplies for many centuries but with the risk of poten-tially disastrous nuclear accidents. Hence, there is a renewed interest in the development of renewable energy sources.

At present, renewable energies provide only 7% of the total global energy demand, mostly through hydroelectric power, but it has been proposed that by the mid-twenty-first century this may be more than doubled until 20% of the total global energy budget is met by adopting such resources. Plans are already underway, for example, in developing fuel cells, Oceanic Thermal Energy Conversion (OTEC), and in the exploitation of solar energy on a massive scale, from within deep space using satellites.

Although nuclear fusion is envisaged to be producing electrical energy within the next 40 years, fossil fuels or nuclear fission power will remain the predominant energy sources in the twenty-first century, and as they become more scarce conflicts for the control of the remaining resources may occur. Energy saving measures can, and should be introduced, introducing new concepts for the design of transport and the built environment.

12.5 Climate change, survival and health

The world has been in flux since its formation. Global environmental change is occurring now, both from natural change and the changes induced by human beings. 'The past as the key to the future' is a useful aphorism since the manner in which the Earth has evolved can provide an indication of possible future developments in the global environment in the twenty-first century.

The world's climate has always been fluctuating and as a result it has dramatically affected living organisms. Looking back over geological time it is evident that climatic variations have been marked (Figure 7.3 (c)) with mean global temperatures varying by several degrees Celsius over the past 2 million years. As recently as 6000 years ago, for example, what is now the Sahara Desert was covered in forests and lakes. Local climate changes have been proposed to explain major episodes in human history; e.g. the collapse of great civilizations, such as Sumeria and Babylon, the migration of tribes from Asia into Europe in the fourth and fifth centuries (contributory factors in the collapse of the Roman Empire), the abandoning of major cities in Arabia in the sixth century and evacuation of Viking settlements in Greenland and North America around 1400. Climate change induced by bolide collisions (such as asteroids from outer space) have been considered responsible for the extinction of several notable species.

There is now a widespread acceptance that not only is humanity excessively exploiting the Earth's natural resources, but also by its actions is beginning to alter the natural environmental balance established over the past 4.6 billion years. This is most dramatically demonstrated by the public debate over the prospect of global climate change. Industrial development, with its heavy use of fossil fuels and subsequent release of greenhouse gases, is now believed to be leading to an enhancement of the greenhouse effect, the continuing growth of which will disturb the balance of the global climate system. The most important greenhouse gas, carbon dioxide, which is responsible for half of the enhanced greenhouse effect, has increased by approximately 30% since 1860 from 280 ppm to 370 ppm.

The debate concerning the evidence for global warming remains vociferous. What is evident from the record of global mean annual temperatures during the past century (Figure 12.2) is that the global annual temperature has increased between 0.5 and 0.7°C. In fact, the 1990s was the warmest decade in the 135 years during which global temperature records have been monitored. Whether this is a part of a normal natural climatic cycle or due to global industrialization cannot have a definitive answer at the present time. During the course of the twenty-first century this crucial question will be answered since humanity will record and experience climatic change. However, to wait until climate change directly (and possibly catastrophically) influences our daily lives is nonsensical. One must utilize the knowledge of science to predict the future, not passively to wait for it, and hence to take remedial actions should they be necessary.

Climate change will have a direct medical impact on human health. A warmer climate will affect young children and the elderly, especially with heat stress. Climate change will affect biodiversity, and in particular the behaviour of insects and their role in the transmission and evolution of infectious diseases, such as malaria. Warm temperatures and photochemical smogs, especially in urban areas with high traffic density, will have a debilitating effect on the human respiratory system.

Changes in climate leading to longer periods of poor weather can also have a disastrous

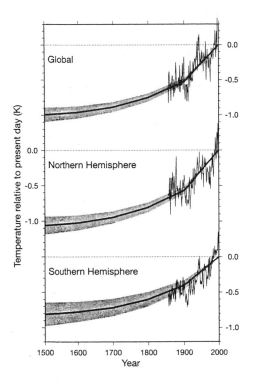

Figure 12.2 Rising global temperatures from 1500 to 2000.

effect on agriculture, especially where this involves the transmission of crop diseases. There may be an inhibition of growth and reduction of photosynthesis in certain plants, and for marine life there may be a reduction in the growth and photosynthesis of phytoplankton, which will lead to secondary effects in the marine food chain.

However, it should not be thought that the consequences of global warming are all negative. Several regions will welcome a warmer climate. For example, in Canada and Siberia energy consumption will be significantly reduced if winters are shorter and less cold. Larger CO_2 concentrations and higher global temperatures may also lead to an increase in plant growth since the photosynthesis process may be accelerated. Hence in some regions, crops previously unable to germinate and reach maturity may be introduced, adding variety to the local diet and reducing the need to import foodstuffs.

12.6   Models, predictions and uncertainties

Increasingly, scientists are of the opinion that the balance of evidence suggests that there is now a discernible effect of human influence on the global climate and that in future one must modify industrial procedures to avoid future disasters. They draw their conclusions from the successful development of climate models. Several large-scale climate models have been developed around the world and all now show similar trends and draw similar conclusions. If greenhouse gas emissions continue unchecked, then these models predict that there will be observable and significant changes in the global climate. It has been predicted that by 2025 the planet will be about 1°C warmer than at present, and up to 5.8°C higher by the twenty-second century. This change is therefore anticipated to be faster than any known natural change in the past 150 000 years.

The models also predict that the thermal expansion of the oceans and the melting of glaciers and parts of the polar ice-caps will cause the planet's average sea level to rise by at least 0.4 m by 2100. The impact will be greatest in the low-lying, highly populated areas such as Bangladesh, in the Nile delta and parts of China. The resulting economic losses could then produce vast population migrations of environmental and economic refugees.

However, global warming can also have an impact that is not readily discernible. The increased precipitation anticipated, together with increasing melt-water from the polar ice caps, will result in the dilution of the salinity that drives the global oceanic conveyor belt. The implications for such currents as the Gulf Stream could then be quite enormous, and would have a direct impact on the climate of those countries bordering the Stream, which rather than becoming warmer could become colder.

Climate models suggest that the hydrological cycle will become more active, with increasing precipitation in some parts, resulting in floods, and decreasing quantities in others, resulting in drought. Agriculture and patterns of land-use will be affected, and water as a vital resource may well lead to conflict in some regions.

Set against global trends are fluctuations at the regional level. The partial melting of the Antarctic ice sheet, the recession of the Alpine glaciers, but the extension of those in Norway and New Zealand, are just three indicators that suggest that if global warming is actually occurring then its effects will be differential.

Nevertheless, for all the calculations and modelling there are uncertainties. Even the most advanced and comprehensive climate models, using the most powerful of supercomputers, require further development. They are remarkably successful in simulating the main features of the present global climate, its monthly, seasonal and inter-annual variations. However, they will require improved knowledge of the numerous physical processes and higher spatial resolution to predict those climate changes occurring on regional and sub-regional scales as a result of human activities. As these models provide the only scientific tool for predicting changes in the global climate over the next century or more, they must be validated against observed variations in the present climate, and also against past climates, in order to assess the uncertainties involved. However, there remain larger uncertainties in predicting non-meteorological inputs to these models, such as future emissions of pollutants into the atmosphere. These will be decided by the future rates of increase of the world population, economic growth, energy consumption and governmental action.

Nevertheless, there is a natural inertia to change policy unless it can be demonstrated clearly that the change is either beneficial or of such necessity that it cannot be avoided. Introducing punitive and environmentally friendly legislation in a democratic country may, for example, lead to the government being voted out of office at the next election, while one country might, by enacting such legislation, find it makes its industry uncompetitive with other less environmentally conscious neighbours. Therefore, any government, industry or individual will require an evaluation of the *risk* of the environmental problem before enacting measures to tackle it.

12.7   Environmental risk

It is clear that humans affect the Earth, and in turn the environment can affect them. The environment is at risk from the way in which human beings behave; in turn, they are at risk from the manner in which the global environment is changing.

A distinction should be made between hazard and risk. A hazard is a situation in which a danger can be presented. It describes the nature of the danger (such as the toxicity or explosiveness of a chemical). Risk is a mathematical probability defined as 'the chance or possibility of danger, loss, injury or other adverse consequences'. Such a concept encapsulates environmental risk. There is an element of risk connected to each and all activities. Risk covers many aspects of our lives such as where we live, what type of work we do, how we spend our leisure time, and what we eat and drink. We are all at risk from many hazards in the environment. These can include hotel and home fire, industrial and domestic accident, air pollution, discotheque fire, illness, release of nuclear radiation, earthquake, and outdoor pursuits accidents. Table 12.1 lists the risk of death from various actions.

However, it is important to quantify these basic statistics. If you never take part in motor racing, water sports or play football you will not die from those activities, nor will you run the risk of violent death from dramatic weather events if you live in a region of the world where there are no hurricanes and tornadoes. Individuals may quantify the risk arising from actions that they chose to undertake. Thus individuals can decide whether the risk of developing cancer from smoking is comparable with the pleasure it gives or other likely causes of death.

At a governmental/national/international level, risk has a slightly different meaning. Governments and companies must quantify any given risk to the population and/or its employees against the resources necessary to ameliorate a specific problem. For example, a government must determine the risk to the health of its population from its industrial policies. Working days lost by employees due to illness leads to lower productivity while a sick population places great strains on any national health service. There is considerable risk connected to living in cities and industrial conurbations, particularly in relation to environmental health. Many great cities, for example Athens, Mexico City and

Table 12.1 Measures of risk.

Activity	Risk of death per year
Sports	
Motor Racing	1.2 in 1000
Football	4 in 10 000
Water sports	1.9 in 10 000
Travel	
Air	3 in 1 million
Motor Car	2 in 10 000
Natural hazards	
Hurricanes	4 in 10 million
Lightning	4 in 10 million
Tornadoes	5 in 10 million
Air pollution	1.5 in 10 000

Los Angeles, experience photochemical smogs which have a marked impact on health, and parts of the old Eastern Bloc are acknowledged environmental 'disaster zones'. The pollution from one country can be transported by winds and oceans to others, as has been the case with pollution from the north-east of England moving over to Scandinavia. Infectious diseases can now be easily transmitted through air travel.

12.7.1   Risk benefit analysis

Risk must, however, be set against benefit. Human beings do certain things either because they enjoy them or because they have to. One method of quantifying risk is through the flow chart:

Environmental audit → environmental impact assessment → environmental risk

The audit involves assessment of how inputs and outputs can affect living organisms and their environment. One can then be in a better position to quantify the results as a risk. For example in looking at toxicity the procedure can involve three stages. The initial stage involves evaluation of the extent of the hazard, which can entail scrutiny of the epidemiological data related to various levels of toxicity, such that inferences can be made regarding the effect and the manner in which it operates and manifests itself. The second stage refers to a dose–response assessment, in which the relationship between the level of the dose and the risk of being affected can be inferred. The third stage involves human exposure analysis, in which affected populations can be specified, the extent of the ranges of doses and the duration of exposure. Chemical pollution and radiation release provide two examples of the abundant contexts for environmental risk.

Both the public and the planners have to be aware of the siting of potentially hazardous facilities, such as nuclear power stations: Windscale (1957), Kyshytym in the Soviet Union (1957) and Chernobyl (1986) all experienced radiation releases into the atmosphere; or chemical factories, such as Bhopal (1984), where 30 tons of methyl isocyanate were released into the local environment, generating at least between 7000 and 10000 fatalities. Urban populations close to such centres can be 'at risk'. Inner-city mortality, in particular, can relate to pollution and poverty and raises questions not only about consumption styles, but also the impact of life quality and urban stressors such as noise.

The impact of ionizing radiations, whether artificial or natural, present major health and political issues. With the demise of the Cold War in the 1990s we could be lulled into thinking that the threat of nuclear armageddon has receded. It cannot be assumed that that will remain a permanent state of affairs. The environmental impact and, in particular, the consequences for one's health in nuclear war is unimaginable. In addition, there are the growing problems associated with ageing nuclear power stations, both in their maintenance, with their continuing seepage, the costs of their decommissioning and the disposal of their waste. Some of these problems were apparent in the incident at the Chernobyl power station (see Section 5.3).

In addition to the threat of radiation from nuclear fission there is the environmental radioactivity due to naturally occurring radon and thoron gases, which emanate from uranium- and thorium-bearing ores in the Earth's crust This is widespread in parts of Britain, particularly in the south-west where there is a greater profusion of granite. In 1991, Cornwall had the highest radon value of all English counties: $114\,\mathrm{Bq\,m^{-3}}$.

Natural radioactivity provides a rich vein for *environmental epidemiology* and occupational diseases. Uranium miners, for example, are particularly vulnerable to the long-delaying effects of irradiation which nurture lung cancers and possibly leukaemia.

12.8   What is being done?

The environment is now considered universally a very important priority for all humanity. It is high on the global political agenda and, because of the transnational nature of many environmental problems, requires international collaborative research. For most governments, with 4- or 5-year fixed-term periods of office, expediency and immediacy require realpolitik and pragmatism, with the achievement of short-term political and economic goals taking preference over the long-term strategic plans needed to tackle effectively many of the current environmental issues.

The growing appreciation of, and concern for, environmental issues has been a driving force for the development of a global environmental agenda and although it may seem that the world is slow to respond to environmental challenges it is nevertheless responding, as can be clearly seen in Table 12.2. The most dramatic example of how scientific argument has led to governmental action over a global environmental issue is the ban of CFCs that lead to ozone depletion. The global consequences of depletion of the Earth's ozone layer would be severe. The ozone layer lies in the stratosphere and depletion of the ozone layer means that more UV radiation, especially in the medium wavelength band UV-B (290–319 nm), can pass through the atmosphere and produce malignant melanomas (skin cancers), cataracts and a general decline in the effectiveness of the human immune system.

The discovery in 1985 of the seasonal depletion (up to 60%) of the ozone layer over the Antarctic – the ozone hole – caused major world-wide concern. In a remarkably short period scientists determined the basic mechanisms for such ozone depletion, identifying the chlorofluorocarbons (CFCs) as the major cause for ozone loss. Laboratory-based experiments together with satellite monitoring and knowledge of the meteorology of the

Gro Harlem Brundtland (1939–). Born in Oslo. As a medical doctor she was concerned with public health, particularly children's health. Her interest in the relationship between environment and health was a factor in her 1974 appointment as Minister of the Environment, and in 1983 she chaired the World Commission on the Environment and Development. The idea that economic considerations could facilitate the conservation of the natural environment generated ideas about sustainability, and a direct result of the 1987 Brundtland Report was the call for a major international conference on these themes. The Rio Earth Summit of 1992 resulted in the discussion of how economies could develop within environmentally-sustainable frameworks. She served three terms as Prime Minister of Norway during the 1980s and 1990s – in total 10 years. Since 1998 she has been Director-General of the World Health Organization.

Table 12.2 Escalating concern with global environmental issues.

Activity	Year
World Commission on Environment and Development set up: chaired by Gro Harlem Bruntland, Prime Minister of Norway	1983
Bruntland Report published	1987
United Nations establishes the Intergovernmental Panel on Climate Change (IPCC) to examine the scientific evidence for climate change	1988
First report of the IPCC: argues that actions against possible global warming may have to be taken	1990
Earth Summit UN Conference on Environment and Development in Rio: 165 countries are signatories to the Framework Convention on climate change: call for the stabilization of greenhouse gas emissions	1992
Population Conference in Cairo	1994
Signatories of the Rio Framework Convention debate the reduction of greenhouse gas emissions post-2000	1995
Second Report of the IPCC. It was in this report that it was stated that 'the balance of evidence suggests a discernible human influence on global climate'	1995
UN Conference on Human Settlement (Habitat II) in Istanbul This conference discussed the growth of cities with all their environmental, social and economic problems.	1996
Kyoto Conference in Japan to set targets for the reduction of greenhouse gas emissions	1997
Buenos Aires, climate change and global CO_2 emission limitation	1998
The Hague Summit on Climate Change. Meeting of national governments ends without agreement as to how to limit CO_2 and other greenhouse gas emissions	2000
Third Report of the IPCC on Climate change. Climate scientists declare that 'most of the warming is attributable to human activities'. Mean global temperatures predicted to rise between 1.4 and 5.8°C by 2100 and sea level rise by between 9 and 66 cms. Report that global warming has led to 20% decrease in global snow cover and 40% thinning of the Arctic ice cap since 1960s	2001

Antarctic led to the development of detailed computer models that could reproduce the ozone hole with remarkable accuracy. The scientific evidence was so overwhelming that effective global action has been taken to stop the production and use of ozone-destroying chemicals through the signing of the Montreal Protocol on 'Substances that Deplete the

Environmental issues for the twenty-first century   409

Ozone Layer' in 1987. This was the first global environmental treaty and one that has shown how humanity can respond to environmental risks and limit or even reverse threats to the global environment (Appendix 8). Through such a combination of well-proven science and political will many of the other major environmental challenges facing the planet may be tackled in the twenty-first century.

In democratic societies the 'green vote' cannot be ignored. Politicians are influenced both by the electorate and the scientific arguments. They have partially met the environmental challenges by introducing legislation to deal with some of the major issues. Indeed, some world politicians, like the ex-Prime Minister of Norway, G. H. Brundtland, have adopted environmental values strongly. Industry too is slowly responding to the needs for change. Conglomerates, such as Ford and BP, are becoming more environmentally conscious and are heavily involved, for example, in manufacturing a prototype hydrogen-fuelled cars and solar energy respectively. It is to be hoped that such efforts will continue and, indeed, expand during the next century.

12.9   Summary: environmental physics as an enabling science

We all have a responsibility towards the natural environment. In that respect we are guardians, safeguarding the well-being of the planet for the generations to come. Today there are many who have a sense of impending environmental crisis and we must all recognize that there are environmental problems. However, these are not insurmountable, providing that there is a unified political commitment, a willingness to take personal responsibility for the future of the planet and a firm understanding of the science involved. The solution of many of these problems provides scientists with their greatest challenges. Physics has always been concerned with providing solutions to problems, and it is hoped that this book has revealed the physical principles underpinning these challenges and described the complex interactions between the atmospheric, biological and hydrological processes which dominate the terrestrial environment. However, much remains to be discovered and many of the mechanisms described here have yet to be fully understood. In this century environmental physics will therefore make a major contribution towards ensuring that Planet Earth is a better place in which to live, develop and flourish.

Appendices

Appendix 1   Entropy

In a physical sense entropy is a measure of the disorder of a system. It should be clear what 'disorder' means. Disorder is a measure of the different ways in which the elements of a system can be distributed, and it is linked to the probability. As the number of different ways increases, so the disorder increases.

For example, toss a coin. This is an example of a mutually exclusive event in which either a head or a tail is produced. This can be depicted as a probability distribution (Figure A1.1).

Now, toss two coins – the outcomes are HH, HT, TH or TT and the distribution is shown in Figure A1.2.

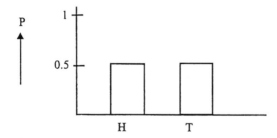

Figure A1.1 Outcomes from tossing a coin once.

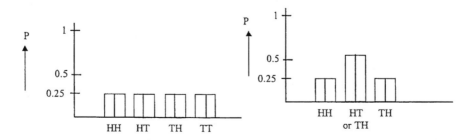

Figure A1.2 Outcomes from tossing two coins.

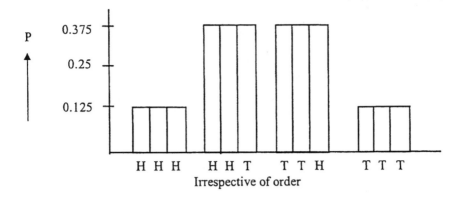

Figure A1.3 Outcomes from tossing three coins.

Tossing three coins will result in the results in Figure A1.3.

If one continues this process of tossing an increasing number of coins, the resulting distributions will provide an ever-increasing distribution of possible combinations of heads and tails.

Now, this argument can be extended to look at energy distributions, such as would be found in metabolic processes.

Entropy was first defined by Clausius in 1865, and entropy change as

$$dS = \frac{dQ}{T} \tag{A1.1}$$

implying that an entropy change results when energy is supplied to a system which is at a temperature T.

Boltzmann also defined entropy (S) in terms of the probability (W) of the number of ways in which energy distributions can be generated:

$$S = k_B \ln W \tag{A1.2}$$

where k_B is the Boltzmann constant $= 1.38 \times 10^{-23}\,\mathrm{J\,K^{-1}}$, and $\ln = \log$ to base $e = \log_e$.

W tells us that the probability of obtaining certain outcomes in a particular energy distribution depends on the number of ways it can be distributed. As Figures A1.2 and A1.3 indicate, some distributions are *more probable* than others. There is a direct relationship between probability (W) and 'disorder', and this implies that as W increases, the disorder increases. Boltzmann's equation (A1.2) helps us to understand, for example, the energy distribution of the atoms in a system when it is warmed.

It was Clausius who neatly summarized the First and Second laws as:

The energy of the whole universe is constant; its entropy, however, increases towards a maximum.

Two philosophical ideas have emerged in the development of entropy. A. Eddington suggested that there could be a relationship between the Second Law of Thermodynamics, entropy and time, i.e. an *arrow of time*, which supposedly gives a direction to time. This implies that for macroscopic closed systems time passes in the direction related to the entropy increase. The second notion, and one which may have implications for our long-term survival, is that of the *heat death of the Universe*. If the Universe is to be taken as an adiabatic closed system (i.e. one in which no heat can leave or enter the system), in which the entropy has reached its maximum possible, i.e. in which energy has been degraded into its maximum 'disorder', then the Universe would have attained a uniform temperature, in which case there would be insufficient internal energy to do any external work.

Appendix 2   Mathematics behind Newton's law of cooling

For forced convection the rate of energy flow can be mathematically modelled as:

$$\frac{dQ}{dt} = -k(T - T_0) \tag{A2.1}$$

where k is the convective energy transfer coefficient, T is the temperature at a time t and T_o is the ambient temperature.

$$\text{Since } \frac{dQ}{dt} = mc\frac{dT}{dt}, \text{ it implies that } \frac{dT}{dt} = -\frac{kA}{mc}(T - T_0) \tag{A2.2}$$

To find the equation that defines the temperature of the cooling body at any time t after cooling has commenced, we have to separate the variables of the initial equation and integrate:

$$\int_{T_{max}}^{T} \frac{dT}{(T - T_0)} = -K\int_0^t dt, \qquad \text{where } \frac{kA}{mc} = K$$

and T_{max} is the starting temperature at time $t = 0$.

$$\left\{\ln(T - T_o)\right\}_{T_{max}}^{T} = -Kt$$

$$T = T_o + (T_{max} - T_o).e^{-Kt}$$

For given wind speeds using a wind tunnel, you may wish to test the hypothesis that the equation above matches the actual results obtained.

You may also wish to use the software *DERIVE* to generate spreadsheets for the data and for calculating dT/dt.

Appendix 3   Energy consumption self-assessment

Much of the energy consumed by developed countries arises from the need to heat and light domestic properties, and with the growing number of commercial appliances in every home domestic electricity use has grown rapidly in the past 50 years. The more affluent the society, the greater its energy use.

To appreciate your energy usage in any year, it is necessary to conduct your own 'energy audit' of the home. This can be done quite easily using data you can collect or estimate for yourself:

1   First collect copies of the electricity and (if relevant) gas bills for your home in the past year (these must be complete and should cover an entire winter and summer). Such bills are usually expressed in kilowatt hours (kWh), which you should convert to J year^{-1}. Note that $1\,kWh = 3.6\,MJ$.
2   Second, draw a sketch of your home showing all the rooms and windows, estimating the area of each.
3   Write down how many people live in your home and for how many months of the year.

Energy audit of your home

Figure 3.7 showed the main routes by which the quantity of thermal energy per unit time (P) is lost from our houses. To estimate how much energy is lost (and hence how much must be produced/used to keep your home at a constant temperature) calculate:

$$P = UA(t_i - t_o) \tag{A3.1}$$

where A is the surface area and U the U-value for respective interfaces (e.g. windows, roof, walls and floor). If your home has a common wall with an adjoining property, you may assume that the energy loss through the wall is balanced by an energy inflow from your neighbour as it is likely that both properties are maintained at the same temperature. t_i is the average temperature inside the home (typically 20°C) and t_o the outside temperature may be determined from climatic data available at your local library or weather centre. Construct a table as below (see Worked example 3.3). What happens in the summer months when the outside temperature may be higher than that required inside? How might you regulate the house temperature then? How would you estimate the energy needed for such temperature regulation? Now construct a table as in A3.1

Table A3.1 Energy required to thermoregulate a home.

Interface	Area (m²)	U (Wm^{-2}K^{-1})	AU (WK^{-1})	Energy used per year
Walls				
Roof				
Floors				
Windows				
	Total energy used per year to thermoregulate your home = ? J year^{-1}			

Now consider the energy (dQ) used in providing hot water at a temperature θ_{hot} to your home:

$$dQ = mc.dT$$

where m is the mass, c the specific heat capacity (for water $c = 4200\,\mathrm{J\,kg^{-1}K^{-1}}$) and dT the temperature through which the water is heated $= \theta_{hot} - \theta_{cold}$. θ_{cold} is the initial temperature from your cold water tap, which you can measure. Construct a table as below:

Table A3.2 Energy used in heating water.

Activity	Water heated to θ_{hot}	Volume of water used	Number of times per year (for all in household)
Bath			
Shower			
Sink			
Washing machine			
Dishwasher			
Total energy used per year in heating water: $= ?\,\mathrm{J\,year^{-1}}$			

In Chapter 3, ventilation was shown to be a major source of energy use, $P = \dfrac{c_v NV(t_i - t_O)}{3600}$, where P is the energy loss/time or 'power loss' (W), c_v is the volumetric specific heat capacity of air (common working value $= 1300\,\mathrm{J\,m^{-3}K^{-1}}$), N is the number of complete air changes per hour (ach), V is the volume of the room ($\mathrm{m^3}$); t_{in} is the temperature of inside air (ºC) and t_{out} is the temperature of new air (ºC), usually from outside.

The recommended air changes per hour (see Section 3.4.1) are shown in Table 3.9. Using data in Table 3.9, complete it:

Table A3.3 Energy used in ventilation.

Room	$ach\,h^{-1}$	Volume	Energy consumed per year
Kitchen/bathroom	3		
Living rooms	2		
Bedrooms	1		
Total energy used in ventilation per year $= ?\,\mathrm{J\,year^{-1}}$			

Now consider the energy used by domestic appliances. Complete the following table:

Table A3.4 Energy used by domestic appliances.

Appliance	Electricity consumption per year (kWh)	Number of appliances	Total energy used per year
Cooker	780		
Washing machine	180		
Dishwasher	430		
Refrigerator	300		
Fridge/freezer	500		
Television	140		

In addition, an average household uses $370\,\mathrm{kWh\,year^{-1}}$ in electric lighting.

Total energy used per year $= ?\,\mathrm{J\,year^{-1}}$.

By combining the figures from Tables A3.1 to 3.4, determine the total energy used in your home in a calendar year. How does this compare with the energy bills (electric and gas)?

How do you account for any differences in your estimated values and billed values?

Appendix 4   Doppler effect

The Doppler effect is concerned with the apparent change in frequency that results from the relative motion between a source and an observer. This implies that:

1   either the source is moving towards and then receding from a stationary observer or
2   the observer is moving towards and then receding from a stationary source or
3   both observer and source are moving towards or receding from one another.

A source can be acoustic or of electromagnetic radiation (i.e. light). Let us consider the following cases:

(a) Source moving towards a stationary observer

Let the velocity of sound be v, and that of the source be v_s. If the source was stationary, then the number of waves per second reaching the observer would be unchanged, i.e. with a frequency of $f = v/\lambda$. But, as the source is moving towards the observer, the distance covered by the waves in $1\,s = v - v_s$ and the apparent wavelength of the waves passing the observer is

$$\lambda_{app} = \frac{(v - v_s)}{f}$$

This would make the apparent frequency $f_{app} = \dfrac{v}{\lambda_{app}} = \left(\dfrac{v}{v - v_s}\right).f$ (A4.1)

Because $(v - v_s) < v$ implies that as the source approaches the observer the apparent frequency increases.

(b) Source receding from a stationary observer

Now, in unit time, i.e. $1\,s$, the waves (of frequency, f) will cover a distance of $(v + v_s)$. The wavelength of these waves (λ_{app}) reaching the observer will then be

$$\lambda_{app} = \frac{(v + v_s)}{f}$$

and the apparent frequency $f_{app} = \dfrac{v}{\lambda_{app}} = \left(\dfrac{v}{v + v_s}\right).f$ (A4.2)

Because $(v + v_s) > v$, the apparent frequency will appear to decrease.

(c) Observer approaching a stationary source

With v_o being the velocity of the observer and v the velocity of sound, the source will emit waves of frequency, $f = v/\lambda$. From the observer's viewpoint, the relative velocity of the waves reaching him/her is $(v + v_o)$. The apparent frequency will then be:

$$f_{app} = \frac{(v + v_0)}{\lambda} = \frac{(v + v_0)}{v/f}$$

Thus, $f_{app} = \left(\frac{(v + v_0)}{v} \right) f$ \hfill (A4.3)

Since $(v + v_o) > v$ the apparent frequency will be greater than the source frequency.

(d) Observer receding from a stationary source

The wavelength of the waves emitted from the source, $\lambda = v/f$ and is unchanged. The relative velocity of the waves with respect to the receding observer $= (v - v_o)$. The apparent frequency will then be:

$$f_{app} = \frac{v - v_o}{\lambda} = \frac{v - v_o}{v/f} = \left(\frac{(v - v_o)}{v} \right) . f$$ \hfill (A4.4)

Since $v - v_o < v$ the apparent frequency will be less than the source frequency.

(e) Both source and observer moving towards one another

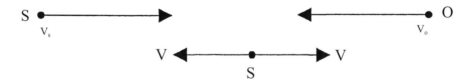

Apparent frequency, $f_{app} = \dfrac{v_{app}}{\lambda_{app}}$

Thus, v_{app} is the relative velocity of the waves with respect to the moving observer, while λ_{app} is the wavelength as perceived by the observer.

Then, $v_{app} = v + v_o$ and $\lambda_{app} = (v - v_s)/f$.

Then, $f_{app} = \left(\dfrac{v + v_o}{v - v_s} \right) . f$ \hfill (A4.5)

(f) Both source and observer receding from one another

In this case $v_{app} = v - v_o$ and $\lambda_{app} = \dfrac{v + v_s}{f}$

Therefore $f_{app} = \dfrac{v_{app}}{\lambda_{app}} = \left(\dfrac{v - v_o}{v + v_s}\right).f$ (A4.6)

Summarizing: in dealing with sound waves, when the source is moving, there will be a change in the wavelength. In contrast, when the observer is moving there will not be a change in the wavelength, but in the frequency of waves received by the observer.

Doppler effect and electromagnetic radiation

In applying the Doppler effect to light waves some comparisons with sound (i.e. acoustic) waves have to be made. Sound waves, which are mechanical and longitudinal, require a medium for their transmission and their speed is temperature-dependent. For example, as altitude increases the temperature decreases and the speed of sound decreases with height. The velocity of sound, and the velocities of observer and source can be determined relative to the medium. Light waves, in contrast, are electromagnetic waves, that do not require a medium and can be transmitted through a vacuum. The speed of light in a vacuum is constant, and this implies that it is constant for an observer irrespective of whether s/he or the source, or both, are in motion. Indeed, unlike time, which is relative, the speed of light is considered absolute – this is one of the basic tenets of Einstein's 1905 theory of special relativity.

The problem becomes compounded because both the object and the Earth are moving, and because the velocity of light is very much greater than the velocity of an object emitting the radiation. The theory of relativity incorporates the idea of relative motion, such that the apparent frequency can be expressed as:

$$f_{app} = \left(\sqrt{\dfrac{1 - \dfrac{v}{c}}{1 + \dfrac{v}{c}}}\right).f$$ (A4.7)

for an object that is both releasing radiation and receding, f_{app} would be the frequency perceived by an observer on Earth. If the observer and the source were approaching one another with a relative speed of v, v would be substituted by $-v$ in equation (A4.7).

The term 'Doppler shift' can be illustrated with reference to the manner in which stars either recede or approach the Earth. The velocity of recession of stars can be determined from examination of the spectra emitted from them. If a star is liberating electromagnetic radiation of frequency f and is moving away from the Earth with a velocity v_s, then the light waves cover a distance $(c + v_s)$ where c is the velocity of light). These waves will have an apparent wavelength $\lambda_{app} = (c + v_s)/f$. Since $\lambda = c/f$, then the shift in the wavelength will be $\lambda_{app} - \lambda = v_s.\lambda/c$.

For a receding star the spectral lines are 'shifted' towards the red region of the electromagnetic spectrum, and for an approaching star the lines are 'shifted' towards the blue region.

The *red shift* can be expressed as the fractional change in wavelength:

$$\frac{d\lambda}{\lambda} = \frac{\lambda_{app} - \lambda}{\lambda} = \frac{v_s}{c} \tag{A4.8}$$

You might like to consider the following:

- For sound waves in the fore-going discussion, the velocity of the source or the observer is much less than the velocity of sound. Consider the implications if the speed of the source is much greater than the speed of sound. *Shock waves* will be generated, with *Mach numbers* >1.
- For light waves, it has been assumed that in an astrophysical context the speed of an approaching or receding celestial object (such as a star) is much less than the speed of light. What might happen then if the speed of a hypothetical object approaches the speed of light?

Let us take the argument further. The speed of light in a vacuum is absolute, it has a fixed value of $3 \times 10^8 \, \mathrm{m\,s^{-1}}$, but what happens if now the speed of the source is greater than the speed of light? This would appear impossible to answer, but the problem can be approached by considering the case where there is a medium in which the speed of the source (v_{source}) is greater than the speed of light *in that medium*. If a charged particle travels through such a medium, *Cerenkov radiation* is released, from which the particle's speed can be determined.

The Doppler effect can, therefore, be used to ascertain the speed and direction of cloud particles or rain, through the change in wavelength of the returning signal.

Appendix 5   Pressure variation with altitude

The following is the derivation of the barometric formula (sometimes called Halley's formula) relating atmospheric pressure with altitude.

Consider a parcel of air (see Figure A5.1), in cylindrical form under isothermal conditions and at constant density. One of the exciting things about physics is that the mathematical model developed to describe a physical phenomenon depends on the assumptions that one makes about the conditions and variables involved in a given system. There are several in this derivation.

The air thins with altitude, and therefore the pressure decreases. Hydrostatic conditions show that the pressure in a fluid at a depth (h), with a density (ρ) is given by $P = h\rho g$. Now for the pressure, at any point, the pressure (P) will be the total pressure of the air above it.

The force acting downwards = the weight of the parcel = $Mg = V\rho g = A\rho g.dh$, where M is the mass of the parcel of air.

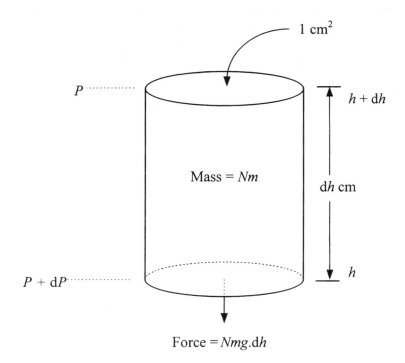

Figure A5.1 Atmospheric pressure variation with altitude.

Assuming that the parcel remains stationary, the force acting upwards will be due to the pressure difference acting on the cross-sectional (A) at both ends of the cylinder:

$$= A.\{P - (P + dP)\} = -A.dP$$

This force should equal the downward force. Thus, $A\rho g.dh = -A.dP$ with the result $dP = -\rho g.dh$. Rearranging this gives $dP/dh = -\rho g$ which is the *hydrostatic equation*.

Now, because we are dealing with an isothermal situation, the ideal gas equation can be applied, i.e. $PV = nRT = Nk_BT$, where n is the number of moles in the parcel, N is the number of molecules in the parcel and k_B is the Boltzmann constant. If we want to express the outcome in a density form the alternative expression $P = \rho RT$ can be used.

Then, $dP = -(P/RT).g.dh$

To obtain a relationship for P with height (h), this relationship is integrated using the method of separation of variables:

$$dP/P = (-g/RT).dh$$

$$\log_e (P/P_o) = -gh/RT$$

$$P = P_o.e^{-gh/RT} = P_o.e^{-kh}, \text{ where } k = g/RT$$

This can also be expressed in terms of density, in which

$$P = P_o e^{-\rho_o - gh/P_o} \tag{A5.1}$$

where ρ_o is the density of air and P_o is the pressure of the air at the Earth's surface respectively.

This is Halley's formula, and it is for an ideal atmosphere. In reality, the exponential form is not quite the case, because the temperature falls with altitude and the atmosphere's composition changes. At greater heights there is a greater proportion of gases with lower molecular masses. Finally, g decreases with altitude, though this is only 3% less at 100 km.

Appendix 6   Derivation of the lapse rate (dT/dz)

It has been shown (Appendix 5) that pressure decreases as one travels up through the atmosphere but what happens to the temperature? You may have noticed while walking in the hills the higher you climb, the cooler it becomes. To answer this question the idea of the *lapse rate* is introduced. This describes how the temperature changes with height. Consider the temperature change of a parcel of air, of unit mass, as it rises through the troposphere. As the air rises its volume changes (because of the pressure changes) while maintaining its mass. Thus, the volume of unit mass can be expressed in terms of its density, $V = 1/\rho$. This situation is an example of the First Law of Thermodynamics, which is expressed as:

$$dU = dQ + dW$$

where dU is the change of the internal energy of the air parcel, dQ the thermal energy transferred to or from the air parcel and dW the work done on or by the air parcel. Using this law and the ideal gas equation ($PV = nRT$ or $PV = mRT$) it is possible to calculate the rate of change of temperature of the air parcel as it rises through the atmosphere.

Using the ideal gas equation of the form $PV = mRT$, the First Law of Thermodynamics can be expressed as:

$$dQ = c_v + P.dV$$

where c_v is the specific heat capacity of the air at constant volume and dQ is the thermal energy supplied, $dU = mc_v.dT$ with $m = 1$ kg and $dT = 1\,°C$.

Consider a parcel of air that deforms and changes its volume but maintains its mass. The volume can be expressed in terms of the density by $V = 1/\rho$, where m is unit mass. Then,

$$dQ = c_v.dT + P.d(1/\rho) = c_v.dT + d(P/\rho) - dP/\rho$$
$$= c_v.dT + R.dT - dP/\rho = c_p.dT - dP/\rho$$

where the specific heat capacity at constant pressure $c_p = c_v + R$

If $dQ = 0$, there will be no exchange of thermal energy with the surroundings. This is the adiabatic case. Under these conditions $c_p.dT = dP/\rho$.

$$dT = dP/(c_p \times \rho) = RT.dP/P.c_p \text{ or } dT/dP = RT/P.c_p$$

By including the pressure relation with altitude (A5.1), one can obtain an expression which involves temperature and height:

$$dP/dz = -g\rho = -g.P/RT$$

Then the lapse rate, $dT/dz = dT/dP \times dP/dz = -g/c_p$ \hfill (A6.1)

In fact, the derivation above is strictly true for dry air, i.e. air which contains no water vapour. The negative sign indicates that the rising air will expand and subsequently cool, producing lower temperatures at higher altitudes. Close to the ground, i.e. well within the troposphere, this lapse rate (Γ_D) is about $-1°C$ per 100 m rise in altitude.

Appendix 7 Synoptic weather chart

An internationally agreed set of symbols describes the weather conditions recorded at every weather station (Table A7.1). These are recorded on the synoptic weather charts. The weather chart at the Earth's surface (the 'surface chart') shows the mean sea-level isobars – lines which join places having the same pressure. The resultant isobaric pattern delineates the main areas of high pressure (anticyclones), which are usually associated with fine, settled weather, from regions of low pressure (cyclones or depressions) which are usually areas associated with unsettled and often stormy conditions. The surface chart also shows positions of fronts which are zones of different air masses usually delineated from one another by having different temperature or humidity. The remaining meteorological observations (surface temperatures, humidity, distribution and type of cloud, rainfall, and visibility) cannot normally be shown continuously so are listed as spot observations on the chart according to symbols in Table A7.1. The configuration of isobars on the surface map determines the strength and direction of the wind near the ground, but wind directions are also recorded as spot values. Figure A7.1 shows a typical weather record at a surface station.

At this station the sky was nine-tenths covered in cloud (◕); with a north-westerly gale blowing at between 89 and 97 km h^{-1} (✖), a light rain was falling (9). The temperature was 15°C and the dew point was 10°C. The predominant cloud was cumulonimbus with an anvil top covering six-sevenths of the sky (⌂). The pressure was 1007.5 millibars, written as (075). By recourse to such diagrams any meteorologist can rapidly assimilate a picture of the local weather over any particular station and, hence construct a picture of the conditions along any weather front.

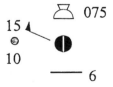

Figure A7.1 Example of the symbolic language of a weatherstation.

Table A7.1 International weather symbols.

Current weather		Sky coverage		Wind speeds		
9	light drizzle	◯	no clouds		(mph)	(kph)
9 9	steady, light drizzle	◐	one-tenth covered	◎	calm	calm
9 / 9	intermittent, moderate drizzle	◖	two-to-three-tenths covered	—	1–2	1–3
9 9 9	steady, moderate drizzle	◖	four-tenths covered	∖_	3–8	4–13
9 9	intermittent, heavy drizzle	◑	half covered	∖__	9–14	14–23
9 9 9	steady, heavy drizzle	◕	six-tenths covered	∖___	15–20	24–33
☉	light rain	◕	seven-to-eight-tenths covered	∖∖__	21–25	34–40
☉ ☉	steady, light rain	◗	nine-tenths covered	∟___	55–60	89–97
☉ ☉	intermittent, moderate rain	●	completely overcast	▙▖___	119–123	192–198
☉ ☉	steady, moderate rain	⊗	sky obscured			
☉ ☉ ☉	intermittent, heavy rain	*Low clouds*				
☉ ☉	steady, heavy rain	—	stratus			
✛	light snow	⌣	stratocumulus			
✛ ✛	steady, light snow	⌢	cumulus			
✛ ✛	intermittent, moderate snow	△	cumulus congestus			
✛ ✛ ✛	steady, moderate snow	⌂	cumulonimbus calvus			
✛ ✛ ✛	intermittent, heavy snow	⌂	cumulonimbus with anvil			
✛ ✛ ✛	steady, heavy snow	*Middle clouds*				
⊖	hail	∠	altostratus			
∩ ₒ	freezing rain	⌣⌣	altocumulus			
Ⅿ	smoke	Ⅿ	altocumulus castellanus			
)(	tornado	*High clouds*				
8	dust devils	_⌐	cirrus			
S	dust storms	∠	cirrostratus			
≡	fog	2⌣	cirrocumulus			
⌐⟨	thunderstorms					
⟨	lightning					
♗	hurricane					

Appendix 8   Environmental risk and environmental impact assessment of ozone-related disasters

The global consequences of environmental risk from ozone layer depletion could be much more severe than any natural disaster humanity has ever experienced. The ozone layer lies in the stratosphere in the upper level of the atmosphere and filters out most of the Sun's potentially harmful ultraviolet radiation from reaching the Earth's surface. However, ozone lying near the Earth's surface (i.e. in the troposphere) is a key ingredient in the formation of photochemical smog and contributes to a high risk assessment factor in analysing the state of environmental pollution and environmental safety. There is a widespread scientific concern about the considerable increase in ozone concentrations in the troposphere. The current scientific evidence indicates that stratospheric ozone is being destroyed largely by the human-made chemicals called chlorofluorocarbons (CFCs), halons, methyl chloroform and carbon tetrachloride. The nations of the world have taken effective steps in joining together to stop the production and use of ozone-destroying chemicals by signing an accord in 1987 called the Montreal Protocol on Substances that Deplete the Ozone Layer. The sequence of events related to this great cause are listed with brief details.

1928   *First CFCs, CFC-11 and CFC-12 are invented in USA*, initially as coolants for the purpose of refrigeration.

1974   *Two scientific papers* (Molina and Rowland) suggest that CFCs emitted on Earth diffuse to the upper atmosphere and breakdown there releasing chlorine atoms in the stratosphere and catalytically destroy ozone.

1975   *UNEP launches a programme of research on risks to the ozone layer.* The National Academy of Sciences launches an assessment of human impact on the stratosphere.

1977   *UNEP sets up a Coordinating Committee on the Ozone* and started *World Plan of Action on the Ozone Layer* to stimulate research. More than 30 countries joined this programme in the first phase. US government asks for warning labels to be put on bottles and spray cans with CFC containing aerosols and announces to phase out most CFC use as aerosol propellants. National Science Foundation estimates that continued release of CFCs to the atmosphere will deplete ozone by 14%.

1978   *International meeting by the developed countries* on CFC regulation recommends a significant reduction in CFC use in aerosols as a precautionary measure.

1979   A number of *developed countries started to impose legal controls* on the production and use of CFC-11 and CFC-12. National Academy of Sciences estimates ozone depletion between 16.5% and 30%, if CFC production and release continues to grow.

1980   European Community and the seven developed countries *call for an international convention to protect the ozone layer.* The US Environmental Protection Agency proposes the *first legal controls* on non-aerosols uses of CFCs. CFC manufacturers form the Alliance for Responsible CFC Policy and ask for the hard evidence of ozone depletion.

1981   *UNEP's Ninth Governing Council* proposes to begin work on the elaboration of a legal framework convention for ozone layer protection, and establishes an *ad hoc* working group of legal and technical experts for this purpose.

1982   UNEP Working Group begins to elaborate a *Framework Convention for the Protection of the Ozone Layer*, based on the draft proposal from Finland, Norway and

Sweden. NASA estimates that ozone depletion due to CFC use will be less than previously thought – between 5 and 9%.

1983   *UNEP's Coordinating Committee on the Ozone Layer* again reduces estimated eventual ozone depletion from current emission rates of CFC-11 and CFC-12, to between 3 and 5%.

1985   At a conference attended by 41 countries, the *Vienna Convention for the Protection of Ozone Layer* is adopted. The convention requires no restrictions on ozone depleting substances but allows for the future elaboration of specific controls. The resolution adopted along with the convention laid the foundation for further work on the protocol on CFC control. Two months later a scientific paper by Farmer *et al.* of the British Antarctic survey is published showing sharp seasonal depletion of the ozone layer over Antarctica – 'The Ozone Hole'.

1986   International negotiations continue on a protocol to the Vienna Convention to control CFCs. CFC-producing companies support for a 'reasonable' limit on future growth in CFC production, and estimate that at least 5 years would be needed to develop substitutes for CFC-11 and CFC-12.

1987   *Montreal Protocol on the Substances that Deplete the Ozone Layer* is adopted. The protocol requires an eventual 50% cut in the consumption of five CFCs by the end of the century and a freeze in the consumption of three halons with a 10-year grace period for developing countries to enable them to meet their basic domestic needs. During the year, some European countries legislate to restrict the use of CFCs as aerosols propellants. The controls should be reassessed at least every 4 years.

1988   A scientific *Panel on Ozone Trends* sponsored by international agencies and US national research bodies concludes that CFCs are responsible for the Antarctic Ozone Hole. A number of UNEP administered international assessment panels are created under the Montreal Protocol to review the latest information on scientific, environmental, technical and economic aspects of ozone depletion. Du Pont becomes the first CFC-producing company to announce that it will phase out CFC production. Northern Telecom, Seiko and Epson became the first multinational companies to announce phase out goals. Sweden decides to phase out by the end of 1994.

1989   *Montreal Protocol enters into force.* At the first meeting parties agree for a CFCs phase out as soon as feasible. Thirteen developed countries announce their intention to phase out the eight controlled substances by 1997. First synthesis of UNEP's Science, Environmental Effects, Technology and Economic Assessments.

1990   *Montreal Protocol calls for total CFC ban on CFCs and halons by 2000.* At the *London Meeting*, parties agree to phase out CFCs and halons by 2000 and added phase out dates for other CFCs, methyl chloroform and carbon tetrachloride. Proposals to create a *Multilateral Fund* and *Financial Assistance* to developing countries. The Government of Finland launches a fund for non-party countries.

1991   *Interim Multilateral Fund becomes operational* with a 3-year budget of US$240 million. *UNEP, UNDP* and the *World Bank* are the initial implementing agencies and later to be joined by the *UNIDO.* UNEP launches the *Ozone Action Programme.* Assessment panels operating under the Protocol conclude that even more stringent controls than those agreed by parties in 1990 are needed, including restrictions on the use of HCFCs. The panels also conclude that technologies are available to replace virtually all uses of controlled substances and that the phase-out process is less expensive than previously predicted.

1992   *Copenhagen Meeting.* Parties of the Montreal Protocol agree to speed up phase-out schedules for already controlled substances and to control new substances for developed countries – HCFCs and HBFCs. Some developed countries adopt faster timetables for phase out of controlled substances. A developing country, Mexico, announces that it is prepared, in principle, to phase out CFC use by 2000, the existing deadline for developed countries. The Multilateral Fund is officially established.

1993   Parties to the *Montreal Protocol agree not to allow any exemptions for the production of halons* beyond the 1994 phase out deadline agreed in Copenhagen, and approve a budget of US$510 million over 3 years 1994–96 for the Multilateral Fund.

1994   According to the *Ozone Secretariat Data 1994*, the developed country parties' consumption of CFCs and halons dropped by about 50% between 1986 and 1992, while developing country consumption rose for all controlled substances except halons.

1995   *Montreal Protocol assessment panels report* that the phase out is well advanced in most developed countries and that developing countries are also making good progress, though consumption of controlled substances is increasing in some. *The 1995 Nobel Chemistry Prizes* were awarded to three scientists, Paul Crutzen (The Netherlands), Mario Molina (USA) and Sherwood Rowland (USA), for their contribution in revealing the threat to the ozone layer from CFC gases by publishing an article in *Nature* in 1974. The only other *Nobel Prize* that has ever been awarded in the field of atmospheric research is to *Sir Edward Appleton in 1947.*

Appendix 9 Units and Constants

Atto (a) $= 10^{-18}$
Femto (f) $= 10^{-15}$
Pico (p) $= 10^{-12}$
Nano (n) $= 10^{-9}$
Micro (μ) $= 10^{-6}$
Milli (m) $= 10^{-3}$
Hecto (h) $= 10^{-2}$
Kilo (k) $= 10^{3}$
Mega (M) $= 10^{6}$
Giga (G) $= 10^{9}$
Tera (T) $= 10^{12}$
Peta (P) $= 10^{15}$
Exa (E) $= 10^{18}$

1 calorie $= 4.18$ joules
1 atmosphere $= 1.013 \times 10^{5}\,Pa = 1013$ millibars
$1\,kWh = 3.6\,MJ$
Conversion of Centigrade (°C) into Fahrenheit (°F): $°F = (5/9).(°F - 32)$
Conversion of Fahrenheit (°F) into Centigrade(°C): $°C = (9 \times °C/5) + 32$

Useful constants

Avogadro's number	$N_A = 6.02 \times 10^{23}\,mol^{-1}$
Universal gas constant (or molar gas constant)	$R = 8.31\,J\,K^{-1}\,mol^{-1}$
Solar constant	$S = 1353\,J\,s^{-1}\,m^{-2}\,(W\,m^{-2})$
Stefan constant	$\sigma = 5.67 \times 10^{-8}\,W\,m^{-2}\,K^{-4}$
Boltzmann's constant	$k_B = 1.38 \times 10^{-23}\,J\,K^{-1}$
Gravitational constant	$G = 6.67 \times 10^{-11}\,N\,m^2\,kg^{-2}$
Velocity of light in a vacuum	$c = 3.00 \times 10^8\,m\,s^{-1}$
Planck's constant	$h = 6.626 \times 10^{-34}\,J\,s$
Atmospheric pressure	$P_A = 1.013 \times 10^5\,Pa$
Radius of the Earth	$r = 6.37 \times 10^6\,m$

Answers to numerical questions

Chapter 2

1   $3.375 \times 10^6\,J$
2   (b) (ii) 197 g
3   (b) (i) $1.2 \times 10^2\,J\,min^{-1}$, (ii) $6.1 \times 10^2\,J\,min^{-1}$
4   (c) (i) $2.9 \times 10^2\,W$, (ii) $5 \times 10^2\,W$
5   (a) (ii) $80\,W\,m^{-2}$, (b) about 880 W
7   (a) $6.93 \times 10^{-3}\,{}^{\circ}C\,s^{-1}$, (b) 91°C, (c) $1.93 \times 10^{-4}\,kg\,m^{-2}\,s^{-1}$, (d) 102.9 $W\,m^{-2}$ is dissipated by forced convection; 10%

9   $$T = T_o + \left\{ (T_{max} - T_o)^{-1/4} + \frac{kt}{4} \right\}^{-4}$$

10   08:40 hours (i.e. 20 to 09:00 hours)

Chapter 3

2   (i) $1.21\,W\,m^{-2}\,K^{-1}$, (ii) $0.405\,W\,m^{-2}\,K^{-1}$
3   (a) 8030 W, (b) $18.25\,m^2$
4   (a) 5.868 GJ; 3.144 GJ
5   (a) 5645 kWh, (b) 4788 kWh, (c) £133.06, (d) £207.90
6   (a) 21.398 GJ, (b) 892 kg
7   44%

Chapter 4

1   2.7×10^{25} molecules m^{-3} and $1.25\,mg\,m^{-3}$
2   (a) 9300 tonnes, (b) 186 tonnes of SO_2
3   (a) $4.5 \times 10^5\,J\,km^{-1}$, (b) $2.25 \times 10^6\,J\,km^{-1}$, (c) $2.7 \times 10^7\,J\,litre^{-1}$, (d) 8640 power strokes km^{-1}
4   (a) $10^5\,J$, (b) 416 A
5   Efficiency = 42%
7   100.4 dB
8   (a) 3.3 m, (b) 3.3 cm
9   $3.16 \times 10^{-6}\,W\,m^{-2}$

Chapter 5

2   (c) (i) 6.58×10^{-7} m, (ii) 3.8×10^5 m s^{-1}, (iii) 0.39 V, (iv) 109.6 A m^{-2}
6   (b) Power developed $= 2\rho A_D . U_\infty^3 . a(1-a)^2$; power coefficient $= 4a(1-a)^2$; thrust coefficient $= 4a(1-a)$
     (c) Thrust coefficient $(C_{T,max}) = 1$ at $a = 0.5$

Chapter 6

1   Hint: determine the angular velocity and centripetal acceleration
2   101 min
3   14.5 orbits
5   100 km

Chapter 7

1   5.27×10^{18} kg
2   (a) 1536 kg m^{-3}, (b) 1.2 km
3   (a) 2.5×10^{25} m^{-3}, (b) 3.2×10^{19} m^{-3}, (c) 1.4×10^{14} m^{-3}
4   0.465 kg m^{-3}
5   More
6   (a) 483.4 m s^{-1}, (b) 1934 m s^{-1}
7   3038 W m^{-2}
8   (i) 4.96 eV (7.95×10^{-19} J)
9   5760 K
10   10 μm
11   1.65%
12   (i) 1000 mm
13   (a) 1368 W m^{-2}
14   8.2×10^{12} kg and 0.5%
15   90%
16   (b) (i) 8.5×10^3 m, (ii) 15 km
17   120 m

Chapter 8

1   (a) 0.1 N, (b) 1000 m^3
2   0.42 mm
5   119 and 278 min
6   16 pC m^{-3}
7   (a) 1250 J, (b) 12.5 K

Chapter 9

1   35.6 Pa per 10 km
2   18.5 h
3   At 30ºE, $V_g = 9.7$ m s^{-1}; at 120º, $V_g = 8.5$ m s^{-1}
5   30 m s^{-1}
6   10.66 kg s^{-1}

Chapter 10

1

Depth (cm)	Matric potential (cm)	Gravitational potential (cm)	Hydraulic potential (cm)	Water lost from layer (mm)
5	−171	−5	−176	1.2
15	−140	−15	−155	1.1
25	−125	−25	−150	1.0
35	−113	−35	−148	1.0
Approx depths of zero flux plane				
45	−101	−45	−146	0.9
55	−95	−55	−150	1.0

Evaporation = 4.3 mm, drainage <60 cm = 1.9 mm.

2   (a) 270 mm, (b) peak concentration expected at (or shortly before) 270 mm drainage, assuming that most of the water that filled the pore space is participating in carrying water

3   (a) 100-fold, (b) 10^8-fold, (c) 10 000-fold

4   (a) −226.5, −224.5 and −227.5 cm for layers 1, 2 and 3 respectively
(b) 0.1333 cm cm^{-1} (indicating upward flow) and −0.2 cm cm^{-1} (downward flow) respectively
(c) -1.89×10^{-9} m s^{-1} for the flow between layers 1 and 2 assuming the conductivity to be the average of layers 1 and 2; 4.09×10^{-9} m s^{-1} for the flow between layers 2 and 3 (assuming the conductivity to be the average of layers 2 and 3)
(d) Water moving upward from layer 2 to 1, and downward from layer 2 to 3. Hence, layer 2 gets drier

5   0.6 for short grass and 0.13 for the conifer forest

Chapter 11

1   Base temperature = 8.48°C. You may have a slightly different answer, depending on how you fitted a straight line to the data
Thermal time requirement for floral initiation to heading = 370 day°C, so thermal time requirement for sowing to heading is 620 day°C. Your answer may be slightly different, depending on how you fitted a straight line to the data
44.3 days from the date of sowing to reach heading

2   Sample results: sow on 9 April and 4 May for lettuces to be harvested on 8 and 18 June, assuming that the dates in Table 11.2 are the dates at the end of each 10-day period

3   Sample results: sow on 2 and 17 May for lettuces to be harvested on 8 and 18 June, assuming that the dates in Table 1.2 are the dates at the end of each 10-day period. The lettuces would be expected to be smaller because the growing season is shortened by about 2 weeks as a result of the 2°C increase in temperature

4   519 and 300 W m^{-2} for leaves inclined at 30° and 60° respectively

5   Photosynthetic rates are 1.04, 1.02 and 0.92 mg m^{-2} s^{-1} for leaves inclined at 0°, 30° and 60° respectively

6   CO_2 taken up per unit radiation intercepted are 0.0017, 0.002 and 0.0031 mg CO_2 J^{-1} for leaves inclined at 0°, 30° and 60° respectively. Inclined leaves are therefore much more efficient

Bibliography

Adams, J., *Risk*. London: UCL Press, 1996.

Ahrens, C. D., *Essentials of Meteorology*, 2nd edn. Belmont CA: Wadsworth, 1998.

Ahrens, C. D., *Meteorology Today: An Introduction to Weather, Climate and the Environment*, 5th edn. London: Brookes Cole, 1999.

Alexander, D. E. and Fairbridge, R. W., eds, *Encyclopaedia of Environmental Science*. Dordrecht: Kluwer, 1999.

Anderson, J. S. and Bratos-Anderson, M., *Noise: Its Measurement, Analysis, Rating and Control*. Avebury Aldershot: Technical, 1993.

Andrews, D. G., *An Introduction to Atmospheric Physics*. Cambridge: Cambridge University Press, 2000.

Andrews, J. E., Brimblecombe, P., Jickells, T. D. and Liss, P. S., *An Introduction to Environmental Chemistry*. Oxford: Blackwell, 1996.

Atkinson, B. W., *Dynamical Meteorology: An Introductory Selection*. London: Methuen, 1982.

Bader, M. J., Forbes, G. S., Grant, J. R., Lilley, R. B. E. and Waters, A. J., eds, *Images in Weather Forecasting: A Practical Guide for Interpreting Satellite and Radar Imagery*. Cambridge: Cambridge University Press, 1995.

Baird, C., *Environmental Chemistry*. New York: Freeman, 1995.

Baker, D. J., *Planet Earth: The View from Space*. Cambridge, MA: Harvard University Press, 1990.

Barbour, M. G., Burk, J. H., Pitts, W. D., Gilliam, F. S. and Schwartz, M. W., *Terrestrial Plant Ecology*. Harlow: Addison Wesley Longman, 1999.

Barrett, E. C. and Curtis, L. F., *Introduction to Environmental Remote Sensing*, 4th edn. Cheltenham: Thornes, 1999.

Barry, R. G. and Chorley, R. J., *Atmosphere, Weather and Climate*, 7th edn. London: Methuen, 1998.

Barton, R., *Outward Bound Survival Handbook*. London: Ward Lock, 1997.

Bell, P. R., *Green Plants: Their Origin and Diversity*. Cambridge: Cambridge University Press, 1997.

Benito, G. R., Baker, V. R. and Gregory, K. J., *Palaeohydrology and Environmental Change*. Chichester: Wiley, 1998.

Bigg, G. R., *The Oceans and Climate*. Cambridge: Cambridge University Press, 1998.

Blaikie, P., Cannon, T., Davis, I. and Wisner, B., *At Risk: Natural Hazards, People's Vulnerability and Disasters*. London: Routledge, 1997.

Blaxter, K., *Energy Metabolism in Animals and Man*. Cambridge: Cambridge University Press, 1989.

Boeker, E. and van Grondelle, R., *Environmental Physics*. 2nd edn. Chichester: Wiley, 1999.

Bohren, C. F. and Albrecht, B. A., *Atmospheric Thermodynamics*. Oxford: Oxford University Press, 1998.

Bonington, C., *Quest for Adventure: Remarkable Feats of Exploration and Adventure, 1950–2000*. London: Cassell, 2000.

Boubel, R. W., Fox, D. L., Turner, D. B. and Stern, A. C., *Fundamentals of Air Pollution*. 3rd edn. New York: Academic, 1994.

Boyle, G., ed., *Renewable Energy: Power for a Sustainable Future*. Oxford: Oxford University Press, 1996.

Bradbury, I., *The Biosphere*, 2nd edn. London: Belhaven, 1998.

Bradley, R. S. and Jones, P. D., eds, *Climate Since A.D. 1500*, revd. London: Routledge, 1995.

Bradley, R. S., *Palaeoclimatology: Reconstructing Climates of the Quaternary*, 2nd edn. San Diego: Harcourt/Academic, 1999.

Brady, N. C. and Weil, R. R., *The Nature and Properties of Soils*, 12th edn. Englewood Cliffs: Prentice-Hall, 1996.

Bragdon, C. R., *Noise Pollution*. Philadelphia: University of Pennsylvania Press, 1971.

Briggs, D., Smithson, P., Addison, K. and Atkinson, K., *Fundamentals of the Physical Environment*, 2nd edn. London: Routledge, 1997.

Bryant, E., *Climate Process and Change*. Cambridge: Cambridge University Press, 1997.

Burberry, P., *Environment and Services*, 8th edn. Harlow: Longman, 1997.

Burroughs, W. J., *Watching the World's Weather*. Cambridge: Cambridge University Press, 1991.

Burroughs, W. J., *Weather Cycles: Real or Imaginary*. Cambridge: Cambridge University Press, 1994.

Burroughs, W. J., *Does the Weather Really Matter?: The Social Implications of Climate Change*. Cambridge: Cambridge University Press, 1997.

Calow, P., (editor-in-chief). *The Encyclopaedia of Ecology and Environmental Management*. Oxford: Blackwell, 1998.

Camp, T. R., *Water and its Impurities*. New York: Reinhold Publishing, 1963.

Campbell, G. S. and Norman, J. M., *An Introduction to Environmental Biophysics*. 2nd edn. New York: Springer, 1998.

Campbell, J. B., *Introduction to Remote Sensing*, 2nd edn. London: Taylor and Francis, 1996.

Carraro, C., ed., *International Environmental Agreements on Climate Change*. Dordrecht: Kluwer, 1999.

Case, R. M. and Waterhouse, J. M., *Human Physiology: Age, Stress and the Environment*. Oxford: Oxford University Press, 1994.

Chalmers, J. A., *Atmospheric Electricity*. 2nd edn. Oxford: Pergamon, 1967.

Chiras, D. D., *The Natural House: A Complete Guide to Healthy, Energy-Efficient and Environmental Homes*. England: Chelsea Green, Totnes, 2000.

Clark, R. P. and Edholm, O. G., *Man and his Thermal Environment*. London: Arnold, 1985.

Corbitt, R. A., *Standard Handbook of Environmental Engineering*. 2nd edn. New York: McGraw-Hill, 1999.

Cotton, W. R. and Pielke, R. A., *Human Impacts on Weather and Climate*. Cambridge: Cambridge University Press, 1996.

Coulson, K. L., *Solar and Terrestrial Radiation, Methods and Measurements*. New York: Academic Press, 1975.

Cracknell, A. P. and Hayes, L. W. B., *Introduction to Remote Sensing*. London: Taylor and Francis, 1993.

Crawford, R. M. M., *Studies in Plant Survival*. Oxford: Blackwell, 1994.

Danson, F. M. and Plummer, S. E., *Advances in Environmental Remote Sensing*. Chichester:Wiley, 1996.

Danny Harvey, L. D., *Climate and Global Environmental Change*. Harlow, Essex: Prentice Hall/Pearson Education, 2000.

Danny Harvey, L. D., *Global Warming: The Hard Science*. Harlow, Essex: Prentice Hall/Pearson Education, 2000.

Demeny, P. and McNicoll, G., eds, *Population and Development*. London: Earthscan, 1998.

Diaz, H. F. and Markgraf, V., eds, *El Niño: Historical and Paleoclimate Aspects of the Southern Oscillation*. Cambridge: Cambridge University Press, 1994.

Diaz, H. F. and Pulwarty, R. S., eds, *Hurricanes: Climate and Socioeconomic Impacts*. Berlin: Springer, 1997.

Dinar, A. and Loehman, E. T., eds, *Water Quantity/Quality Management and Conflict Resolution*. Westport, Connecticut: Pregaer Press, 1995.

Downing, T. E., Olsthoorn, A. J. and Tol, R. S. J., eds, *Climate Change and Risk*. London: Routledge, 1999.

Drake, F., *Global Warming: The Science of Climate Change*. London: Arnold, 2000.

Drury, S. A., *Images of the Earth: A Guide to Remote Sensing*, 2nd edn. Oxford: Oxford University Press, 1998.

Dunlop, S. and Wilson, F., *Weather and Forecasting*, 2nd edn. London: Chancellor, 1998.

Dunne, T. and Leopold, L. B., *Water and Environmental Planning*. San Francisco: Freeman, 1978.

Elliott, D., *Energy, Society and Environment: Technology for a Sustainable Future*. London: Routledge, 1997.

Ernst, W. G., *Earth Systems: Processes and Issues*. Cambridge: Cambridge University Press, 2000.

Fitter, A. H. and Hay, R. K. M., *Environmental Physiology of Plants*, 2nd edn. London: Academic, 1995.

Foley, G., *The Energy Question*, 3rd edn. Harmondsworth: Penguin, 1987.

Foster, H. D., *Health, Disease and Environment*. London: Belhaven, 1992.

Fowler, C. M. R., *The Solid Earth: An Introduction to Global Geophysics*. Cambridge: Cambridge University Press, 1993.

Freeze, R. A. and Cherry, J. A., *Groundwater*. Englewood Cliffs: Prentice-Hall, 1979.

Friday, A. and Ingram, D. S., eds, *The Cambridge Encyclopaedia of Life Sciences*. Cambridge: Cambridge University Press, 1985.

Fritchen, L. J. and Gay, L. W., *Environmental Instrumentation*. New York: Springer, 1979.

Glantz, M. H., *Currents of Change: El Niño's Impact on Climate and Society*. Cambridge: Cambridge University Press, 1997.

Graedel, T. E. and Crutzen, P. J., *Atmosphere, Climate and Change*. New York: Scientific American Library, 1997.

Graedel, T. E., and Crutzen, P., *Atmospheric Change: An Earth System Perspective*. New York: Freeman, 1997.

Green, J., *Atmospheric Dynamics*. Cambridge: Cambridge University Press, 1999.

Guyot, G., *Physics of the Environment and Climate*. Chichester: Wiley, 1998.

Hall, P., *Cities and Civilization*. London: Weidenfeld and Nicolson, 1998.

Hardy, R. N., *Temperature and Animal Life*, 2nd edn. London: Arnold, 1979.

Harrison R. M. ed., *Pollution: Causes, Effects and Control*. 3rd edn. Cambridge: Royal Society of Chemistry, 2000.

Hartmann, D. L., *Global Physical Climatology*. San Diego: Academic, 1994.

Havrella, R. A., *Heating, Ventilating and Air Conditioning Fundamentals*. Englewood Cliffs: Prentice-Hall, 1995.

Haymes, E. M. and Wells, C. L., *Environment and Human Performance*. Champaign, IL: Human Kinetics, 1986.

Hibbert, C., *Cities and Civilizations*. London: Weidenfeld and Nicolson, 1996.

Hillel, D., *Introduction to Soil Physics*. New York: Academic Press, 1982.

Hobbs, J. E., Lindsay, J. A. and Bridgman, H. A., *Climates of the Southern Continents: Present, Past and Future*. Chichester: Wiley, 1998.

Hobbs, P. V., *Introduction to Atmospheric Chemistry*. Cambridge: Cambridge University Press, 2000.

Hodges, L., *Environmental Pollution*. New York: Holt, Rinehart and Winston, 1973.

Hopkins, W. G., *Introduction to Plant Physiology*, 2nd edn. New York: Wiley, 1999.

Hough, M., *Cities and Natural Process*. London: Routledge, 1995.

Houghton, J. T., *The Physics of Atmospheres*, 2nd edn. Cambridge: Cambridge University Press, 1995.

Houghton, J. T., *Global Warming: The Complete Briefing*. Cambridge: Cambridge University Press, 1997.

Howes, R. and Fainberg, A., eds, *The Energy Sourcebook: A Guide to Technology, Resources and Policy*. New York: American Institute of Physics, 1991.

Hoyt, D. V. and Schatten, K. H., *The Role of the Sun in Climate Change.* Oxford: Oxford University Press, 1997.

Huggett, R. J., *Environmental Change: The Evolution of the Ecosphere.* London: Routledge, 1997.

Ingram, D. L. and Mount, L. E., *Man and Animals in Hot Environments.* New York: Springer-Verlag, 1975.

Iribane, J. V. and Cho, H. R., *Atmospheric Physics.* Dordrecht: Reidel, 1980.

Jacob, D. J., *Introduction to Atmospheric Chemistry.* New Jersey: Princeton University Press, 1999.

Jacobson, M. Z., *Fundamentals of Atmospheric Modelling.* Cambridge: Cambridge University Press, 1999.

James, W. and Niemczynowicz, J., eds, *Water, Development and the Environment.* Boca Raton: CRC Press, 1992.

Jones, H. G., *Plants and Microclimate: A Quantitative Approach to Environmental Plant Physiology.* Cambridge: Cambridge University Press, 1992.

Jones, J. A. A., *Global Hydrology: Processes, Resources and Environmental Management,* Harlow: Longman, 1997.

Kemp, D. P., *Global Environmental Issues: A Climatological Approach,* 2nd edn. London: Routledge, 1996.

Kiely, G., *Environmental Engineering.* Boston: Irwin McGraw-Hill, 1998.

Killham, K., *Soil Ecology.* Cambridge: Cambridge University Press, 1994.

Kleiber, M., *The Fire of Life: An Introduction to Animal Energetics.* New York: Kreiger, 1975.

Kondratyev, K. Y. and Cracknell, A. P., *Observing Global Climate Change.* London: Taylor and Francis, 1998.

Ksenzhek, O. S. and Volkov, A. G., *Plant Energetics.* San Diego: Academic, 1998.

Lamb, H. H., *Climatic History and the Future.* New Jersey: Princeton University Press, 1985.

Lamb, H. H., *Climate, History and the Modern World,* 2nd edn. London: Routledge, 1995.

Langmuir, E., *Mountaincraft and Leadership,* 3rd edn. Edinburgh: Scottish Sports Council, 1995.

Lawlor, D. W., *Photosynthesis: Metabolism, Control and Physiology.* Harlow: Longman, 1990.

Leroux, M., *Dynamic Analysis of Weather and Climate: Atmospheric Circulation, Perturbations, Climatic Evolution.* Chichester: Wiley, 1998.

Linacre, E. and Geerts, B., *Climates and Weather Explained.* London: Routledge, 1997.

Littler, J. and Thomas, R., *Design with Energy: The Conservation and Use of Energy in Buildings.* Cambridge: Cambridge University Press, 1984.

Lutgens, F. R. and Tarbuck, E. J., *The Atmosphere,* 7th edn. Englewood Cliffs: Prenctice-Hall, 1998.

Lutz, W., ed., *The Future Population of the World.* London: Earthscan, 1996.

MacDougall, J. D., *A Short History of Planet Earth.* Chichester: Wiley, 1996.

MacGorman, D. R. and Rust, W. D., *The Electrical Nature of Storms.* Oxford: Oxford University Press, 1998.

Marshall, T. J., Holmes, J. W. and Rose, C. W., *Soil Physics.* 3rd edn. Cambridge: Cambridge University Press, 1999.

Mason, B. J., *The Physics of Clouds.* 2nd edn. Oxford: Clarendon, 1971.

Mason, B. J., *Clouds, Rain and Rainmaking.* 2nd edn. Cambridge: Cambridge University Press, 1975.

Mason, B. J., *Acid Rain.* Oxford: Oxford University Press, 1992.

Masters, G. M., *Introduction to Environmental Engineering and Science.* Englewood Cliffs: Prentice-Hall, 1991.

McGuffie, K. and Henderson-Sellers, A., *A Climate Modelling Primer.* Chichester: Wiley, 1999.

McIlveen, R., *Fundamentals of Weather and Climate,* 2nd edn. London: Chapman and Hall, 1992.

McIntosh, D. H. and Thom, A. S., *Essentials of Meteorology.* London: Taylor and Francis, 1969.

McLaren, R. G. and Cameron, K. C., *Soil Science: Sustainable Production and Environmental Protection.* Auckland: Oxford University Press, 1997.

McMichael, A. J., *Planetary Overload: Global Environmental Change and the Health of the Human Species.* Cambridge: Cambridge University Press, 1995.

McMichael, A. J., Haines, A., Sloof, R. and Kovats, S., eds, *Climate Change and Human Health*. Geneva: World Health Organization, 1996.

McMullan, J. T., Morgan, R. and Murray, R. B., *Energy Resources*. 2nd edn. London: Arnold, 1983.

McMullan, R., *Environmental Science in Buildings*. 4th edn. London: Macmillan, 1998.

McNeill, J., *Something New Under the Sun: An Environmental History of the Twentieth Century*. London: Allen Lane Penguin Press, 2000.

Meteorological Office, *The Observers Handbook*. London: HMSO, 1982.

Monteith, J. L. and Unsworth, M. L., *Principles of Environmental Physics*, 2nd edn. London: Arnold, 1990.

Mohr, H. and Schopfer, P., *Plant Physiology*. Berlin: Springer, 1995.

Mount, L. E., *Adaptation to the Thermal Environments: Man and his Productive Animals*. London: Arnold, 1979.

Mumford, L., *The City in History*. Harmondsworth: Penguin, 1966.

Nakayama, Y. and Boucher, R. F., *Introduction to Fluid Dynamics*. London: Arnold, 1999.

National Research Council, *Decade-to-Century Scale Climate Variability and Change: A Science Strategy*. Washington, DC: National Academy Press, 1998.

Nelissen, N., Van der Straaten, J. and Klinkers, L., eds, *Classics in Environmental Studies: An Overview of Classic Texts in Environmental Studies*. Utrecht: International Books, 1998.

Newman, E. I., *Applied Ecology and Environmental Management*. 2nd edn. Oxford: Blackwell, 2000.

Nobel, P. S., *Physiochemical and Environmental Plant Physiology*, 2nd edn. San Diego: Academic, 1999.

Ogelsby, C. H., *Highway Engineering*. New York: Wiley, 1975.

Oke, T. R., *Boundary Layer Climates*. 2nd edn. London: Routledge, 1996.

Oliver, A., *Dampness in Buildings*. 2nd revd edn by J. Dougles and J. S. Stirling. Oxford: Blackwell, 1997.

O'Neill, P., *Environmental Chemistry*. 2nd edn. London: Chapman and Hall, 1995.

O'Neill, P., *Environmental Chemistry*. 3rd edn. London: Blackie, 1998.

O'Reilly, J. T., Hagan, P., Gots, R. and Hedge, A., *Keeping Buildings Healthy: How to Monitor and Prevent Indoor Environmental Problems*. Chichester: Wiley, 1998.

Palmen, E. and Newton, C. W., *Atmospheric Circulation Systems*. New York: Academic, 1969.

Park, C., *The Environment: Principles and Applications*. London: Routledge, 1997.

Parsons, K. C., *Human Thermal Environments: The Principles and the Practice*. Taylor and Francis, 1993.

Patel, M. R., *Wind and Solar Power Systems*. London: CRC Press, 1999.

Paton, T. R., Humphreys, G.London: S. and Mitchell, P. B., *Soils: A New Global View*. London: UCL Press, 1995.

Physics Education, 5, 1991.

Physics World, 8, 1998.

Pickering, K. T. and Owen, L. A., *An Introduction to Global Environmental Issues*. 2nd edn. London: Routledge, 1997.

Pielke, R. A., *Hurricanes: Their Nature and Impacts on Society*. Chichester: Wiley, 1997.

Ponting, C., *A Green History of the World*. London: Sinclair-Stevenson, 1991.

Pugh, L. G. C. E., 'Deaths from exposure on the Four Inns walking competition', *Lancet*, 1, 1964, 1210–12.

Ramage, J., *Energy: A Guidebook*. 2nd edn. Oxford: Oxford University Press, 1997.

Raven, P. H., Evert, R. F. and Eichhorn, S. E., *Biology of Plants*. 6th edn. New York: Freeman, 1999.

Rees, W. G., *Physical Principles of Remote Sensing*. Cambridge Cambridge: University Press, 1996.

Reynolds, J. M., *Introduction to Applied and Environmental Geophysics*. Chichester: Wiley, 1997.

Riehl, R., *Introduction to the Atmosphere*. New York: McGraw-Hill, 1978.

Roberts, N., ed., *The Changing Global Environment*. Oxford: Blackwell, 1995.

Roberts, N., *The Holocene: An Environmental History*. 2nd edn. Oxford: Blackwell, 1998.

Robinson, P. J. and Henderson-Sellers, A., *Contemporary Climatology*, 2nd edn. Harlow: Longman, 1999.

Rogers, R. R. and Yau, M. K., *A Short Course in Cloud Physics*. 3rd edn. Oxford: Pergamon, 1989.

Rowell, D. L., *Soil Science: Methods and Applications*. Harlow: Longman, 1997.

Sabins, F. F., *Remote Sensing: Principles and Interpretation*, 3rd edn. New York: Freeman, 1997.

Salby, M. L., *Fundamentals of Atmospheric Physics*. San Diego: Academic Press, 1997.

Schaefer, V. J. and Day, J. A., *A Field Guide to the Atmosphere*, Boston: Houghton Mifflin, 1981.

Schnoor, J. L., *Environmental Modeling: Fate and Transport of Pollutants in Water, Air and Soil*. New York: Wiley, 1996.

Schott, J. R., *Remote Sensing: The Image Chain Approach*.Oxford: Oxford University Press, 1997.

Schowengerdt, R. A., *Remote Sensing: Models and Methods for Image Processing*. San Diego: Academic, 1997.

Scientific American, *The Oceans, Scientific American*, 9 (special issue), 1998.

Scorer, R. S., *Dynamics of Meteorology and Climate*. Chichester: Wiley, 1997.

Seeley, I. H., *Building Technology*, 5th edn. Basingstoke: Macmillan, 1995.

Seinfeld, J. H., *Atmospheric Chemistry and Physics of Air Pollution*. New York: McGraw-Hill, 1986.

Seinfeld, J. H. and Pandis, S. N., *Atmospheric Chemistry and Physics: From Air Pollution to Climate Change*. New York: Wiley, 1998.

Shaw, E. M., *Hydrology in Practice*, 3rd edn. London: Chapman and Hall, 1998.

Slaymaker, O. and Spencer, T., *Physical Geography and Global Environmental Change*. Harlow: Longman, 1998.

Smith, B. J., Peters, R. J. and Owen, S., *Acoustics and Noise Control*. 2nd edn. Addison Wesley Harlow: Longman, 1996.

Smith, K., *Environmental Hazards: Assessing Risk and Reducing Disaster*. 2nd edn. London: Routledge, 1998.

Sorensen, B., *Renewable Energy: Its Physics, Engineering, Use, Environmental Impacts, Economy and Planning Aspects*. 2nd ed. San Diego: Academic, 2000.

Sparks, D. L., *Environmental Soil Chemistry*. London: Academic Press, 1995.

Stanier, M. W., Mount, L. E. and Bligh, J., *Energy Balance and Temperature Regulation*. Cambridge: Cambridge University Press, 1984.

Stroud, M., *Survival of the Fittest: Understanding Health and Peak Physical Performance*. London: Cape, 1998.

Summerhayes, C. P. and Thorpe, S. A., eds, *Oceanography*. London: Manson, 1996.

Taiz, L. and Zeiger, E., *Plant Physiology*. 2nd edn. Cambridge, MA: Sinauer, 1998.

Taylor, R., *Noise*. Penguin, Harmondsworth 1970.

Thompson, R. D., *Atmospheric Processes and Systems*. London: Routledge, 1998.

Thompson, R. D. and Perry, A., eds, *Applied Climatology: Principles and Practice*. London: Routledge, 1997.

Thurman, H. V., *Essentials of Oceanography*. 5th edn. Englewood Cliffs: Prentice-Hall, 1996.

Tivy, J., *Biogeography: A Study of Plants in the Ecosphere*. 3rd edn. Harlow: Longman, 1998.

Turco, R. P., *Earth Under Siege: From Air Pollution to Climate Change*. Oxford: Oxford University Press, 1997.

Turiel, I., *Physics, the Environment and Man*. Englewood Cliffs: Prentice-Hall, 1975.

Turner, B. L., Clark, W. C., Kates, R. W., Richards, J. F., Mathews, J. T. and Meyer, W. B., eds, *The Earth as Transformed by Human Action: Global and Regional Changes in the Biosphere over the Past 300 years*. Cambridge: Cambridge University Press, 1993.

Twidell, J. W. and Weir, A. D., *Renewable Energy Resources*. E. and F. N. Spon, London: 1996.

Uman, M. A., *Lightning*. New York: McGraw-Hill, 1969.

Uman, M. A., *Lightning*. New York: Dover, 1983.

United Nations World Commission on Environment and Development, *Our Common Future: The Brundtland Report*. Oxford: Oxford University Press, 1987.

Vale, B. and Vale, R., *Green Architecture: Design for a Sustainable Future*. London: Thames and Hudson, 1998.

Verbyla, D. L., *Satellite Remote Sensing of Natural Resources*. New York: Lewis, 1995.

Vertebrate Structures and Functions, *Readings from Scientific American*. San Francisco: Freeman, 1974. See Part 5 for Temperature Adaptation.

Viessman, W. and Hammer, M. J., *Water Supply and Pollution Control*. 6th edn. Menlo Park: Addison Wesley Longman, 1998.

Vincent, R. K., *Fundamentals of Geological and Environmental Remote Sensing*. Englewood Cliffs: Practice-Hall, 1997.

Walker, J. F., and Jenkins, N., *Wind Energy Technology*. Chichester: Wiley, 2000.

Walker, K., *Safety on the Hills*. Skipton: Dalesman, 1995.

Wallace, J. M. and Hobbs, P. V., *Atmospheric Science: An Introductory Survey*. San Diego: Academic, 1997.

Ward, R. C. and Robinson, M., *Principles of Hydrology*. 3rd edn. Maidenhead: McGraw-Hill, 1990.

Wayne, R. P., *Chemistry of Atmospheres*. 3rd edn. Oxford: Oxford University Press, 2000.

Wellburn, A., *Air Pollution and Climate Change: The Biological Impact*. 2nd edn. Harlow: Longman, 1994.

Wells, N., *The Atmosphere and Ocean: A Physical Introduction*. 2nd edn. Chichester: Wiley, 1998.

Wheater, H. and Kirby, C., eds, *Hydrology in a Changing Environment*. Proceedings of the British Hydrological Society International Conference, Exeter. Chichester: Wiley, 1998.

White, I. D., Mottershead, D. N. and Harrison, S. J., *Environmental Systems*. London: Chapman and Hall, 1994.

White, R. E., *Pinciples and Practice of Soil Science*. 3rd edn. Oxford: Blackwell, 1997.

Wilby, R. L., ed., *Contemporary Hydrology: Towards Holistic Environmental Science*. Chichester: Wiley, 1997.

Wild, A., *Soils and the Environment: An Introduction*. Cambridge: Cambridge University Press, 1995.

Williams, J., *Geographic Information from Space: Processing and Applications of Geocoded Satellite Images*. Chichester: Praxis, 1995.

Williams, M., Dunkerley, D., De Deckker, P., Kershaw, P. and Chapell, J., *Quarternary Environments*. 2nd edn. London: Arnold, 1998.

Wilson, E. M., *Engineering Hydrology*. 4th edn. Basingstoke: Macmillan, 1990.

Wrigglesworth, J. M., *Energy and Life*. London: Taylor and Francis, 1997.

Wright, P. H. and Paquette, R. J., *Highway Engineering*. New York: Wiley, 1979.

Glossary

Absorbivity Ability of a material's capacity to absorb radiation. Expressed as the internal absorptance of the substance.

Absorptance Ability of a material to absorb radiation. Expressed as a ratio of the radiation absorbed to the radiation incident on the material.

Acid rain Precipitation that contains water with various chemicals, such as nitrogen oxide and SO_2.

Action spectrum Depicts a range of wavelengths of electromagnetic radiation in which a physiological process, such as photosynthesis, can occur.

Adenosine triphosphate (ATP) A nucleotide that plays an essential role in transmitting chemical energy in metabolic processes in living organisms.

Adiabatic change Pressure–volume change in which no thermal energy leaves or enters the system. From the First Law of Thermodynamics, $dQ = 0$. Thus, $dW = -dU$, which implies that any work done is committed through corresponding changes in the internal energy.

Aerodynamics The physics of the motion of gases (especially air) and the subsequent behaviour and motion of objects in the gas.

Aerofoil Structure designed such that its relative movement with respect to a fluid results in a lift force which is greater than a drag force. Has applications for aircraft and wind-propellers.

Aerosol Minute particulate matter, often found in the atmosphere, with diameters usually $<10\,\mu m$.

Air-conditioning Process whereby the air characteristics of a building, such as temperature, humidity and ventilation, can be adjusted.

Air-entry suction Threshold value for soil water suction beyond which there can be large losses of water from sandy soils.

Albedo Fraction, or percentage, of incident radiation that is reflected.

Alcoholic fermentation Use of micro-organisms, in the presence of oxygen, to produce ethanol from sugars by fermentation.

Altimeter Instrument used to measure the altitude of the Earth from a satellite using laser pulses or radar.

Anaerobic process Process whereby living organisms can exist without oxygen.

Anemometer Device designed to measure wind speeds.

Angle of contact Angle between the liquid in a capillary tube and the glass, where the liquid surface is just in contact with the glass, and the angle is measured through the liquid.

Angular velocity (ω)   Rate of rotation of a body (such as the Earth), $\omega = \theta/t$, and the units are in radians s^{-1}.

Anthesis   Time taken for the production of pollen by male parts of the flower to fertilize the female organs that form seeds.

Anti-cyclone   Large-scale weather system in which winds flow in the lower parts of the troposphere in a clockwise direction in the Northern Hemisphere.

Aphelion   Position of a satellite or a planet, in its orbit, when furthest away from the Sun.

Aquifer   Rocks, which are often permeable and porous, that can absorb water and let it pass. They can thus provide groundwater.

Arrow of time   Term arising from the Second Law of Thermodynamics referring to the directionality of time and suggests that time's direction occurs as the entropy of an isolated system increases.

Atmosphere   That component of the global environment consisting of the air and particles that envelop the Earth. It consists of concentric shells and extends up to about 100 km (i.e. to the thermosphere).

Atmospheric electricity   Describes the electrical nature of various atmospheric processes, such as the electrical discharges found during thunderstorms.

Atmospheric window   Sections of the electromagnetic spectrum in the infrared region, $8.5 < \lambda < 14\,\mu m$, where the radiation is not absorbed by CO_2 and water vapour.

Attractor   3D mapping that results from the locus of a point as its position changes with time.

Aurorae   Electromagnetic phenomena occurring in the thermosphere (at about 100 km altitude). The aurora borealis can be seen in the Northern Hemisphere, while the aurora australis is observed in the Southern Hemisphere. Its origin could be caused by high-velocity electrons coming from the Sun (possibly during Sun-spot activity) interacting with the Earth's magnetic field. Their resulting acceleration then ionizes atmospheric gases.

Available water capacity   Difference between the volumetric water content at field capacity and the permanent wilting point, and provides a measure of the fraction of the soil volume occupied by pores small enough to hold water against gravity, but not so small that water is held too tightly for roots to be able to extract it.

Avogadro's number (N_A)   Number of basic units (6.02×10^{23}) in one mole of a substance. For example, there are 6.02×10^{23} atoms in 4 g of helium (^4_2He).

Band theory   Theory developed in electronics that suggests that when large numbers of atoms are close together, as in the solid state, the atomic energy levels merge into bands, namely the valence, forbidden and conduction bands. The theory is useful in studying the electrical properties of a variety of materials.

Barometer   Instrument that measures air pressure.

Basal metabolic rate   Energy rate for metabolism to occur for a stationary person or animal.

Beaufort scale   Scale for wind-strengths that range from 0 (calm) to 12 (hurricane). Formulated by Admiral Beaufort in 1805.

Beer–Lambert law   Describes the exponential nature of the absorption of light through materials. For example, through the Earth's atmosphere or a leaf.

Bernoulli's theorem   Used in fluid mechanics. It is an expression of the principle of the conservation of energy, and states that for an incompressible and inviscid fluid (i.e. frictionless) the sum of the pressure, the kinetic energy per unit volume and the potential energy per unit volume, at a point on a flow-line, is constant.

Betz limit   Sometimes referred to as the power coefficient and represents the fraction of the power of the wind that can be extracted. It is a measure of the efficiency with which a rotor can extract energy from the wind.

Biodiversity   Extent of the diversity of living organisms (i.e. both flora and fauna).

Biogeochemical cycles   The motion and exchange of those chemicals, such as nitrogen, carbon, phosphorus and sulphur, through the components of the environment, that are necessary for life.

Biomass   Total mass of all living organisms, expressed as mass per unit area.

Biophotolysis   Emission of hydrogen by the metabolism of certain biological organisms.

Biosphere   That part of the global environment that includes the location of all living organisms (i.e. in the air, on the land, in the Earth, and in the oceans).

Black body   Hypothetical entity that absorbs all wavelengths of electromagnetic radiation falling on it and is capable of re-radiating. This implies that both its emissivity and absorptance are one.

Boltzmann's constant (k_B)   Ratio of the molar gas constant (R) and Avogadro's number (N_A), where $k_B = 1.38 \times 10^{-23} \, \text{J K}^{-1}$. The ideal gas equation can be expressed as $PV = Nk_BT$, where N is the number of molecules in a volume of the ideal gas.

Boundary layer   Concept developed by the German engineer L. Prandtl to describe a liquid or gaseous layer affected by a neighbouring solid or liquid surface. For example, the planetary boundary layer.

Bulk density   Measure of compaction and defined as the mass of soil solids per unit soil volume.

C3 plants   Plants in which the immediate product of photosynthesis is pyruvate (a three-carbon compound), such as wheat.

C4 plants   Plants in which the immediate product of photosynthesis is malate (a four-carbon compound), such as tropical grasses.

Calorific value   Thermal energy generated per unit mass of a substance under conditions of complete oxidation.

Calvin cycle   Process whereby the energy generated in the photosynthetic light reactions is used, in dark reactions, to produce glucose from CO_2 and H_2O.

Campbell–Stokes sunshine recorder   Device that records sunshine in relation to local time.

Candela   Unit of luminous intensity where the latter has a specified direction and the radiation frequency of the source is $5.4 \times 10^{14} \, \text{Hz}$ and the radiant intensity of $1/683 \, \text{W}$ per steradian. One steradian is the unit of solid angle and equals the angle at the 'centre of a sphere subtended by a part of the surface equal in area to the square of the radius'.

Carbon cycle   Biogeochemical cycle through which carbon moves through the global environment. The oceans and the atmosphere, for example, act as carbon sinks.

Carnot cycle   Cycle of operations for a hypothetical heat engine, and is named after S. Carnot (1796–1832).

Catalytic converter   Device built into a vehicle's exhaust system and designed to convert toxic chemicals into less harmful products. Through the use of the three-way converter hydrocarbons and CO are oxidized to form CO_2 and nitrogen oxides to N_2.

Cavity wall   Term used in building engineering to describe a structure of two vertical brick walls separated by a space, either filled with air or other insulating material, such as foam.

Centripetal force   Force acting on a body following a circular path. The force, and the radial acceleration, act toward the centre of the circle.

Chaos   Type of behaviour characteristic of a deterministic system that appears to be random. The idea was first developed by the American meteorologist E. Lorenz.

Chernobyl   Location of the nuclear power station in the Ukraine in which there was a major release of radiation in 1986.

Chlorofluorocarbons (CFC)   Organic compounds containing chlorine, fluorine and methyl (CH_3) groups, and which are held responsible for the deterioration of stratospheric ozone.

Chlorophyll   Pigmentation in plants that makes them green, and is essential for absorbing light in photosynthesis.

Chloroplasts   Organelles that contain chlorophyll, and which are essential for photosynthesis.

Clausius–Clapeyron equation   Shows how saturation vapour pressure varies with temperature.

Climate   Long-term weather patterns prevailing for a given region over several decades.

Climate modelling   Application of mathematical models and computer simulations to determine short-term weather predictions and long-term climate forecasting.

Clo   Quantity of insulation provided by indoor clothing for a sedentary person in thermally comfortable conditions. 1 clo unit $= 0.155°C\,m^2\,W^{-1}$, where the skin temperature is 32°C and environmental temperature is 21°C.

Clone crop   Crop developed from cells genetically identical and produced from a 'parent' by mitotic division, which is asexual.

Cloud condensation nuclei   Parts of aerosols that provide the focii for the development of cloud droplets.

Coalescence   Part of the process of cloud formation in which developing water droplets collide with other droplets and combine.

Coefficient of performance   Term used for a heat pump and is expressed by the ratio of the thermal energy that can be transferred by the system to the electrical energy supplied.

Composite materials   Materials constructed out of two or more layers.

Compression ratio (r)   Term relating to the internal combustion engine. It is the ratio of the maximum petrol–air mixture volume contained by the cylinder to the minimum volume produced by the compression stroke.

Condensation   Opposite process to evaporation, in which water is formed from water vapour. The process describes a phase change in which energy is dissipated.

Conduction   Method of thermal energy transfer which, in the case of metals, is due to the mobility of free conduction electrons throughout the crystal lattice.

Continuity equation   Equation used in global circulation models of climate forecasting to describe the mass conservation of moving air masses.

Convection   Method of thermal energy transfer through the motion of a fluid arising from a temperature difference and therefore changes in density.

Coriolis force   Apparent force ($F_c = 2\omega v \sin \phi$) acting on moving air and water systems, such as winds and waves, due to the rotation of the Earth. Named after the French mathematician G. de Coriolis (1792–1843). See also Coriolis parameter.

Coriolis parameter   The term $2\omega\sin\phi$, where ω is angular velocity and ϕ is angle of the latitude.

Crop coefficient   Proportionality constant in the inverse relationship between the amount of dry matter produced per unit of water transpired in plants and the daytime saturation deficit of the air during the growing season.

Crop factor   Measure of the amount of leaves covering a land surface. It is a coefficient multiplied by the potential evaporation in the Penman equation.

Cryosphere   That part of the global environment that includes the colder regions of the Earth, such as the polar regions and glaciers.

Cyclone   Large-scale weather system characterized by low pressures and low tropospheric winds rotating in an anti-clockwise direction in the Northern Hemisphere.

Dalton's law of partial pressures   Pressure of a mixture of gases that is the sum of the partial pressures of the individual gases present.

Darcy's law   Describes the rate of water flow that shows that the volume flow rate per unit area is proportional to the gradient in soil water potential.

Dark reaction   Mechanism that can occur either in darkness or in light in which the chemical energy generated in a light reaction is utilized in manufacturing organic compounds.

Decibel   Unit used in acoustics to measure differences in power levels and differences in sound pressure levels. It is not a unit of loudness.

Dehydration   Loss of water due to perspiration particularly while engaged in strenuous activities.

Demography   Scientific study of populations. It involves the structure of populations, fertility, mortality, migration and the impact of disease.

Demographic transition   Change between a period of high birth- and death-rates to one of low birth- and death-rates. This was first seen in Britain between 1780 and 1960, and has been characteristic of nearly all industrialized nations.

Depression   Extra-tropical cyclone consisting of a low-pressure system.

Desertification   Process by which land assumes desert-like characteristics, sometimes as a result of environmental degradation.

Dew point temperature   Temperature to which air can be cooled at a constant pressure so that saturation takes place. It is often followed by condensation.

Dipole   Molecular structure consisting of two opposite but about equal charges separated by a short distance.

Dissociation   Break up of molecules or ions into components by electron or photon impact.

Diurnal patterns   Daily 24-h environmental patterns.

Doppler effect   Apparent change in the frequency of a source of acoustic or light waves as a result of the relative motion between the source and the observer.

Dry deposition   Variety of acid precipitation in which dry and acidic particles are deposited.

Eccentricity   Measure of the ellipticity, i.e. the shape of the orbit of the Earth as it spins around the Sun.

Ecosystem   Biologically stable system characterized by the relationship between living and non-living organisms and their environment.

Einstein   One Einstein is the total energy required to initiate a photosynthetic reaction for one mole of a substance.

Electromagnetic induction Process developed by M. Faraday in which an electrical current can be generated by the relative motion between a coil and a magnetic field.

Electromagnetic radiation Radiation consisting of oscillating electric (E) and magnetic (B) fields, travelling perpendicularly to one another through space and time, at the speed of light, and in phase with one another.

Electromagnetic spectrum Spectrum of electromagnetic waves, extending from very high frequency short wavelength cosmic rays to very low frequency and long wavelength radio waves, all travelling at the speed of light.

Ellipticity Term used in the Milankovitch hypothesis to describe the change in the elliptical shape of the Earth's path around the Sun. The cycle of this change has a period of about 100 000 years.

El Niño Periodic patterns in sea-surface temperature along the eastern Pacific Ocean, in particular characterized by warm currents of water from the equator flowing south along the Peruvian coast. Is held responsible for global changes in weather patterns.

El Niño Southern Oscillation (ENSO) El Niño-associated changing patterns of pressure with air flow reversals in the Pacific.

Emissive power Total amount of radiation released per unit area per second from a black-body radiator.

Emissivity Emittance of a body as a fraction of the value for a black body at the same temperature.

Emittance Energy emitted per second from a unit area of the Sun's photosphere or a source of radiant energy.

Endothermic reaction Occurs when thermal energy is absorbed.

Energy budget equation Expression of the various energy transfer modes as an energy audit for the body.

Energy flux Rate at which energy flows (i.e. in $J s^{-1}$ or W), usually normal to a surface.

Enhanced greenhouse effect Enhancement of the natural greenhouse effect mainly through the impact of anthropogenic emissions of greenhouse gases.

Enthalpy Thermodynamic state function that describes the heat content of a system.

Entropy Measure of the disorder of a system. Can be defined mathematically as $dS = dQ/T$.

Environmental audit Account or assessment of how the inputs into and outputs from a system can affect living organisms and their environment.

Environmental epidemiology Branch of epidemiology that examines environmental factors in patterns of disease.

Environmental impact assessment Examination of the possible impact of human behaviours on the environment, and how the environment can act on human beings.

Environmental physics Branch of physics that examines environmental processes and issues in the global environment through a study of the interacting relationship between the atmosphere, hydrosphere, geosphere and biosphere. A central theme is the relationship between living organisms and their environment.

Environmental (or natural) radioactivity Radioactivity due to naturally occurring ores (such as uranium) in the natural environment.

Environmental risk Risk to living organisms emanating from the environment.

Epidemiology Study of the origin, prevalence, virulence and distribution of diseases.

Equation of continuity   Term used in fluid dynamics to describe the relationship between cross-sectional area, density and velocity at a certain point in a tube of flow and the corresponding variables at a later point in the same tube.

Erythema   Medical condition characterized by a reddening of the skin.

Escape velocity   Minimum velocity required by an object to escape from a body's gravitational field (i.e. its force of attraction). To escape from the Earth's gravitational field requires an escape velocity of $11\,000\,\mathrm{m\,s^{-1}}$.

Evaporation   Process whereby a liquid is transformed into a vapour through a phase change.

Evapotranspiration   Summation of evaporation from the land and transpiration from plants.

Exosphere   Outermost region of the Earth's atmosphere, existing from about 400 km.

Exothermic reaction   Reaction in which thermal energy is released.

Extinction coefficient   Measure of a medium's ability to absorb or reflect radiation, rather than allowing it to be transmitted.

Ferrel cell   Part of the tricellular model of atmospheric circulation that exists between 30° and 60° north and south of the equator. Air ascends at about 60° and descends at latitudes of about 30°, after which it moves towards the poles. Named after W. Ferrel (1856).

Field capacity   Upper limit to the amount of water that soil can hold in the long-term.

First Law of Thermodynamics   A formulation of the Principle of the Conservation of Energy and expressed as $\mathrm{d}Q = \mathrm{d}U + \mathrm{d}W$, where $\mathrm{d}Q$ is the energy transferred to or from the system, $\mathrm{d}U$ is the change in internal energy and $\mathrm{d}W = P.\mathrm{d}V$ = work done by the system.

Finite energy   Energy sources of which there are finite (i.e. exhaustible) supplies.

Fission   Fragmentation of an atomic nucleus due to collision by another particle, such as a neutron in nuclear fission.

Fixation   Process involving the absorption of a gas into the organic compounds of living organisms.

Fluid dynamics   Study of the forces acting on fluids and the resulting motion.

Fluid mechanics   Branch of physics that deals with fluids in motion (fluid dynamics) and at rest (fluid statics).

Food chain   Energy transfers and motion of biochemical material through biological communities.

Forced convection   Process of energy transfer in which a fluid flows past an object, which has been heated, and that cools as a result.

Fossil fuels   Combustible material, carbon-based, geologically deposited and originally biological.

Fourier's law of thermal conduction   Mathematical model for the rate at which energy flows during thermal conduction.

Fraunhofer lines   Absorption lines occurring in the spectrum of the solar photosphere.

Front   Moving mass of rain or clouds over a large area and characterized by elongated boundaries that separate air masses from different sources and different properties. The theory was developed by the Norwegian meteorologist V. Bjerknes (1862–1951).

Fuel cell   Device that generates electrical energy from the chemical reaction between oxygen and hydrogen.

Fusion   Process whereby lighter atomic nuclei are 'fused' together, under high pressures and temperatures, to form heavier ones with the liberation of energy.

Genera   Taxonomy or classification of organisms into groupings which have characteristics in common and which are different from those of others.

Geosphere   Part of the global environment that comprises both the surface and internal regions of the solid Earth.

Geostationary orbit   A satellite in such an orbit has the same angular velocity and period as that of the Earth. Therefore, if it is at a distance of 42 180 km from the Earth's centre, the satellite will orbit above the same point of the Earth.

Geostrophic balance   For a parcel of air in the Earth's atmosphere to be in a steady state, the pressure gradient force has to be balanced by the Coriolis force.

Geostrophic wind   Wind flowing horizontally which is at 90° to the Coriolis force and the pressure gradient force.

Geosynchronous orbit   Similar to a geostationary orbit, except that it has an angle of inclination to the equatorial plane.

Geothermal energy   Energy source derived from deep in the Earth's crust.

Gibbs free energy   A thermodynamic state function that refers to the energy available for work, especially at the cellular level.

Global Circulation Models (GCM)   Means by which global weather events and climate change can be monitored and predicted by computer simulation techniques and mathematical modelling.

Global oceanic circulation   Global conveyor movement of water, with the upper layers affected primarily by winds and the lower layers influenced by temperature, density and salinity. In the transport of mass and energy, oceans are essential components of the global climate.

Global Positioning System (GPS)   Method of locating a point on the Earth's surface to a very accurate order by the use of three or more satellites.

Global warming   Increase in the mean global temperature in, at least, the past 100 years, due to the enhanced greenhouse effect resulting from the anthropogenic-induced increase in the emission of greenhouse gases.

Glycolysis   Initial phase in which respiration can produce energy.

Graphite moderator   Material used in nuclear fission reactors to regulate the speeds of neutrons so that the chance is increased of neutron capture and, therefore, that fission will occur.

Great Smog of 1952   Air pollution occurring in London in 1952, caused by a combination of fog and smoke from coal-fired domestic and industrial heating, and weather conditions. Caused in excess of 4000 deaths, especially of those with respiratory problems.

Greenhouse effect   Natural warming effect of the Earth's surface due to those gases in the atmosphere that absorb and re-radiate in the infrared. It is due to the atmosphere's ability to allow transmission of short-wave solar radiation but not the longer-wave terrestrial radiation.

Greenhouse gases   Gases in the atmosphere, such as CO_2, methane, water vapour, nitrogen oxides and trace gases, that absorb terrestrial infrared radiation and generate the greenhouse effect.

Ground heat flux   Rate at which thermal energy is conducted into the ground.

Gulf Stream   Current of warm sea water, with its origin in the Gulf of Mexico that passes the eastern coast of America and then turns north-east towards Britain and Norway.

Hadley cell That part of the tricellular model of atmospheric circulation that lies between the equator and 30° north and south, in which air ascends in the Inter Tropical Convergence Zone and descends in the subtropical regions anti-cyclonically. First developed by G. Hadley in 1735.

Hagen–Poiseuille equation Developed independently by J. Poiseuille and G. Hagen to describe the flow of liquids through pipes under laminar flow conditions. It describes the volume flow rate of a fluid through a tube in terms of the pressure difference between the ends of the tube and its geometry.

Half-life Time taken for the mass or the count-rate of a radioactive species to be reduced by half.

Heat Measure of the energy that can be transferred to or from a system as a result of a temperature difference.

Heat engine Device that transfers thermal energy into useful mechanical energy as a result of temperature difference between a heat source and a heat sink.

Heat pump Device (opposite to a heat engine) in which work can be done to transfer thermal energy from a source at a lower temperature to a sink at a higher temperature.

Heat of reaction Enthalpy change that can be assessed by the amount of energy generated or absorbed in a reaction.

Hectare Area of a surface, such as a field, measuring 100 m square, i.e. with an area of 10 000 m².

Homeostatis Describes the ability of a biological system to maintain stability and equilibrium in the face of change.

Homeotherms Living organisms capable of maintaining a constant body temperature through a thermoregulation mechanism orchestrated by the hypothalamus. Examples are humans, birds and mammals.

Humidity Amount of water vapour in the atmosphere.

Hurricane Weather pattern found in the Caribbean and the North Atlantic characterized by a very low pressure system which can have wind speeds >118 kph. See also Typhoon.

Hydraulic conductivity Measure of the ability of the soil to conduct water. It represents the rate of water flow that would occur due to gravity alone.

Hydraulic head Term used in soil physics that expresses pressures and suctions in units of 'hydraulic head', where a unit hydraulic head is the pressure exerted by the weight of a unit height of water column, where 1 cm hydraulic head exerts a pressure of 100 Pa.

Hydrodynamic dispersion Term used in the leaching of solutes through soil profiles that shows that solute pulses tend to disperse as they move down the profile by the leaching process.

Hydroelectric power Generation of electrical energy from the gravitational potential energy of water usually stored in reservoirs, and released to flow and cause the rotation of turbines.

Hydrogen bond Weak chemical bond connecting hydrogen atoms with other atoms, such as oxygen in the water molecule.

Hydrological cycle Depicts the interconnections of water in the global environment and involves the movement of water within the Earth–Atmosphere coupling system, commencing with evaporation, especially from the oceans, followed by condensation into clouds and falling as precipitation.

Hydrostatic equation Mathematical formulation of the pressure at a given point within the body of a fluid. It can, for example, describe the relationship between atmospheric pressure and height ($dp/dz = -g\rho$).

Hygrometer Instrument that measures humidity; quite often relative humidity.

Hydrolysis Process in which water is involved in a chemical reaction.

Hydrosphere That part of the global environment that encompasses water, be it in the form of oceans, rivers, lakes or groundwater.

Hyperthermia Raising of the core body temperature. Characterized by the following increasing levels of heat stress: sweating and vasodilation, heat cramp, heat exhaustion and heat stroke.

Hypothermia Lowering of the normal core body temperature. Characterized by mild, moderate, severe and acute stages.

Ideal gas A gas that obeys the gas laws (such as Boyle's, Charles's, Gay-Lussac or Pressure Law, the ideal gas equation, Dalton's law of partial pressures and Avogadro's Law). Its internal energy (U) depends only on the absolute temperature and not on its volume. Obeys the assumptions of the kinetic theory of gases.

Ideal gas equation Mathematical model that describes the behaviour of an ideal gas, namely $PV = nRT$, in terms of its pressure, volume and temperature, where n is the number of moles and R the universal molar gas constant.

Illuminance Measure of the intensity of an illumination, and expressed as the luminous flux per unit area acting on a surface.

Impaction Process of removing atmospheric particulates by collision and adhesion to surfaces.

Incompressible fluid Fluid in which the density remains constant.

Infra-red radiation Part of the electromagnetic spectrum lying beyond the red-end of the visible spectrum with wavelengths lying between 0.7 and 1000 µm.

Internal combustion engine Heat engine in which the combustion of a petrol–air mixture provides mechanical work done in giving rotational motion to the wheels.

Internal energy Energy that an ideal gas has by virtue of the sum of the kinetic energies of all its atoms or molecules. For an ideal gas, the internal energy depends on the temperature only.

Inter-Tropical Convergence Zone (ITCZ) Low-latitude and low pressure zone within which air converges from opposite directions near the Earth's surface.

Ionizing radiation Radiation capable of ionizing the material through which it passes.

Ionosphere Layer of the upper atmosphere, from about 50 to 1000 km, consisting of ionized air caused by radiation from space.

Irradiance Rate of flow of radiant energy through unit area at 90° to the beam of solar radiation, at any point (unit $= W\,m^{-2}$). Sometimes referred to as the solar energy flux density.

Irreversible process Process where a system is not in a state of equilibrium at all times.

Isobaric change Change under conditions of constant pressure.

Isothermal change Pressure–volume change under conditions of constant temperature, in which the change in internal energy of the gas (dU) is 0, and in which any energy supplied to the system is used in doing work, i.e. $dQ = dW = P.dV$

Isotope   Elements that have the same number of protons but different nucleon number, i.e. with different number of neutrons.

Jetstream   Streams of high-velocity air masses in the higher sections of the troposphere.

Kepler's laws   Describe the mechanics of planetary motion.

Kinetic energy   Energy of a body by virtue of its motion.

Krebs cycle (or citric acid cycle)   Sets of cyclical biochemical processes essential to the metabolism of living organisms.

Laminar   Slow-moving fluid in which there is a velocity differential between the maximum velocity of flow at the centre of a pipe and at the boundary where it is a minimum. In a pipe's tube of flow the laminar flow is composed of streamlines in which the fluid particles in each streamline have the same velocity and there is no exchange between successive layers.

Landsat   Polar-orbiting satellite that completes an orbit in approximately 99 min.

Lapse rate   Rate with which the temperature decreases with increasing atmospheric height.

Latent heat   Physical process whereby energy is either gained or lost by a system during a phase change, without a change in temperature.

Latent heat of vaporization   Energy required to bring about a change of state (or phase change) of unit mass of liquid into vapour, without a change of temperature.

Leaching   Process of the passage of water through a substance, such as the soil, and its taking in dissolved particles with it.

Leaf area index   Area of leaf per unit land surface as an indicator of vegetation growth.

Legionnaire's disease   Medical condition, identified in 1976, consisting of a form of pneumonia caused by bacteria that may live in water and the air-conditioning systems of buildings.

Light reaction   Stage in the photosynthetic process that involves the photolysis of water by sunlight and the evolution of oxygen.

Light response curve   Measure of the irradiance in the photosynthetically active region of the spectrum.

Loam   Permeable soil, consisting of silt, sand and clay.

Lumen   Unit of luminous flux. Defined as the luminous flux radiated by a point source of 1 candela intensity in a cone of solid angle 1 steradian.

Luminous flux   Rate of flow of radiant energy.

Luminous intensity   Luminous flux radiated per unit solid angle by a source in a specified direction.

Lux   Measure of illuminance, and defined as the illumination provided by 1 lumen over unit area (m^2).

Magnetosphere   Layer around the Earth consisting of charged particles influenced by the Earth's magnetic field.

Manometer   Instrument for measuring fluid pressure, or differences in fluid pressure.

Mass-energy equivalence   Defined by the mathematical model $\Delta E = \Delta mc^2$, and showing that matter and energy can be interchangeable.

Maxwellian speed distribution   Maxwell–Boltzmann distribution of molecular speeds from which the most probable speed, the mean speed and the root-mean-square speed can be determined.

Melanoma   Form of skin cancer characterized by malignant growths.

Mesopause   At about 80 km above the Earth's surface, the region between the mesosphere and thermosphere.

Mesosphere   Part of the upper atmosphere that extends beyond the stratosphere.

Metabolism   Sum total of the biochemical reactions taking place in the cells of living organisms and which involves catabolism and anabolism.

Meteorological Optical Range (MOR)   Length of path in the atmosphere required to reduce the luminous flux in a collimated beam from a monochromatic light source to 5% of its original value.

Meteorology   Science of the atmosphere and atmospheric processes, particularly where the weather and climate are concerned.

Meteosat   Meteorological satellite.

Milankovitch hypothesis   Theory developed by M. Milankovitch to explain the astronomical cycle of ice ages. He argued that variations in the three factors relating the geometry between the Earth and the Sun (i.e. ellipticity, tilt and precession) produced these cycles.

Moisture content (or water-release curve)   For a given soil, the relationship between the water content of the soil and the soil water suction.

Mole   Gram-equivalent of a substance. It is the amount of substance that contains the number of units, such as atoms, or molecules or ions, as there are in $12\,g$ of the isotope $^{12}_{6}C$. The number is equal to the Avogadro number, N_A. For example, oxygen is diatomic and an oxygen atom is represented as $^{16}_{8}O$. Thus, the gram equivalent is $32\,g$ and the molar mass is $0.032\,kg$.

Monsoon   Significant tropical and subtropical seasonal fluctuations in which masses of air in the low troposphere are accompanied by strong winds and heavy rainfall.

Navier–Stokes equations   Set of non-linear partial differential equations that can be used in weather prediction.

Near infrared   That part of the infrared region close to the red end of the visible spectrum.

Nebule   Unit of opacity, such that a filter of opacity of 100 nebules transmits 1/1000 of the incident light.

Neutrino   Elementary subatomic particle. Assumed to be massless and chargeless.

Newtonian mechanics   Branch of physics enshrined in Newton's laws of motion.

Newton's law of cooling   Rate at which an object cools is directly proportional to its excess temperature above the ambient temperature. The law is governed by conditions of forced convection.

Newton's Second Law   Rate of change of momentum of a body is directly proportional to the force applied to it. Mathematically expressed as $F = ma$, where a is linear acceleration, and acts in the direction of the force.

Occlusion   Process whereby a cold front interacts with a warm front.

Oceanic thermal expansion   Expansion of the oceans either through natural solar heating or through global warming.

Organelles   Structures in the cells of living organisms with specific functions.

Oscillating air column   Concrete structure that captures the energy of flowing and ebbing sea water and then converts it into electrical energy.

Oxygen isotopic ratios   Used in deciphering past climate change involving variations in the ratio of the oxygen isotopes O^{18}/O^{16} with temperature.

Ozone   Triatomic gas composed of three oxygen atoms. Stratospheric O_3 absorbs a significant proportion of incoming solar ultraviolet radiation.

Ozone depletion   Reduction of stratospheric ozone.

Ozone hole   Thinning of the ozone layer in the stratosphere which lies above Antarctica.

Ozone layer (sometimes referred to as the ozonosphere)   Region of the Earth's atmosphere containing ozone that exists between 15 and 50 km above the Earth's surface (i.e. within the stratosphere), with a maximum concentration between 20 and 26 km above the Earth's surface.

Palaeoclimatology   Science of ancient climates.

Particulates   Suspended particles, including dust, in the atmosphere caused by, for example, emissions from vehicles and volcanic eruptions.

Pascal   SI unit of pressure. 1 Pascal $= 1\,N\,m^{-2}$.

Passive solar heating   Process of the collection and storage of solar radiation by buildings.

Pathology   Scientific study of diseases.

Penman equation   Assesses how radiation, wind speed, air temperature and humidity affect the rate of evaporation expected from a wet surface freely supplied with water.

Perihelion   Position of a satellite or a planet, in its orbit, when it is nearest to the Sun.

Permanent wilting point   Lower limit to the amount of water that can be extracted by plant roots.

pH $= -\log_{10}[H^+]$. Represents the negative value of the logarithm of the hydrogen ion concentration. It is useful in determining the alkalinity or acidity of a substance. The pH scale is from 0 to 14, with 0–7 as acid, 7 as neutral and 7–14 alkaline.

Phase change   Change of state, e.g. solid to liquid or liquid to vapour.

Phonon   Refers to the lattice vibrations that generate acoustic standing waves of high frequency in the thermal conduction in some solids.

Photochemistry   Branch of chemistry that involves those chemical reactions facilitated by the absorption or release of electromagnetic radiation.

Photochemical smog   Smog generated by photochemical action on the products of combustion (such as the oxides of nitrogen and hydrocarbons from cars).

Photodissociation   Breakup of molecules by solar radiation.

Photoelectric effect   Process of releasing electrons from metals by the absorption of electromagnetic radiation.

Photo-ionization   Ionization of a substance by solar radiation.

Photon   Quantum or discrete packet of energy (such as in electromagnetic radiation) modelled mathematically as $E = h\upsilon$, where E is the energy associated with the photon, h is Planck's constant and υ is the frequency of the radiation. The idea of the photon is used to explain the particulate nature of light.

Photosynthesis   Process occurring in the cells of green plants in which water and CO_2 are absorbed and, under the mediation of solar energy and chlorophyll, produce oxygen and glucose.

Photovoltaic effect   Process by which solar radiation can be converted into electrical energy.

Phytoplankton   Plankton that consists of plants and microscopic organisms, such as algae, living in the surface layers of oceans and lakes, and which undergo photosynthesis.

Planck's constant   Constant, h, given in the expression for the energy of a photon $E = h\upsilon$, and which has the value of $6.63 \times 10^{-34}\,J\,s$.

Planetary boundary layer   Sometimes referred to as the atmospheric boundary layer.

Refers to the lower troposphere (i.e. the bottom 500 m) affected by the Earth's surfaces, both oceanic and land.

Plasma   Ionized gas.

PM$_{10}$ Particulates of dimension $\leqslant$10 μm.

Poikilotherms   Living organisms whose thermoregulation and, therefore, body temperature are environmentally determined. Examples include reptiles.

Polar cell   That part of the tricellular model of atmospheric circulation that exists between 60°, north and south, and the poles.

Polar orbiting satellite   Satellite that orbits the Earth with an inclination approximately 90° to the equatorial plane. For example, NOAA (American National and Oceanographic and Atmospheric Administration) is used to monitor weather patterns.

Polar stratospheric clouds   High-altitude clouds (possibly with ice particles) formed by very low temperatures in the stratosphere during winter above the poles.

Polar vortex   Region of winds flowing past the poles at high levels in the atmosphere.

Pollution   Contamination of the natural environment, especially the air, water and soil, by chemical and biological sources.

Pores   Spaces that exist between soil particles.

Porosity   Fraction of the soil occupied by pores.

Post-industrial society   Term coined by the American political scientist D. Bell. It refers to the type of society in which the primary (agricultural) and secondary (industrial manufacturing) sectors have been superseded by the tertiary sector, in which information, finance and services have become the paramount sources of capital.

Potential energy   Energy that a body has by virtue of its position, i.e. its height above a reference line.

Potential evaporation   Evaporation that would be expected if the land surface was continually wet, i.e. freely evaporating, and depends on the prevailing meteorological conditions.

Precession   'Wobble' of the Earth as it spins on its axis (rather like a spinning top).

Pressure   Force acting on a unit area of a surface, and caused by molecular bombardment. Units are Pascals.

Pressure gradient force   Force acting on a parcel of air due to its siting within a pressure gradient.

Primary energy   Energy derived from the original source, such as fossil fuels or wind energy.

Principle of the conservation of energy   Total energy of a system is conserved, i.e. it remains constant. This implies that it cannot be created or be destroyed, but can be transferred.

Pyrolysis   Decomposition of chemical compounds due to high temperatures, sometimes in the absence of air.

Quantum theory   Theory developed by M. Planck (1900) to explain the spectral curves resulting from black body emission.

Radar (radio detection and ranging)   Navigational and detecting system of distant bodies through the reflection of electromagnetic radiation, which can have wavelengths between 0.8 and 30 cm.

Radar altimetry   Method of determining heights using radar techniques.

Radiation   Method of energy transfer from one point to another through the transmission of electromagnetic waves, such as those from the Sun to the Earth through Space.

Radiation balance   Difference between the quantity of radiation incident on the Earth's surface and that emitted by the Earth as infrared radiation.

Radiative forcing   A perturbation to the Earth–atmosphere energy balance, such as a change in the solar output or the concentration of atmospheric CO_2. The climate system tries to respond to this forcing to redress the balance. Positive radiative forcing will warm the surface, while negative forcing will cool it.

Radioactive nuclide   Nuclear species, defined by its proton and neutron number, which is radioactive.

Radiometer   Instrument that measures solar energy fluxes.

Radiosonde   Balloon-carrying device that 'senses' the troposphere and the lower part of the stratosphere, by measuring the temperature, pressure, humidity with the use of a radar reflector.

Rain gauge   instrument that measures the extent of rainfall.

Relative humidity   Ratio of the mass of moisture in the air at a specific temperature to the maximum mass of moisture that the air could contain at that temperature.

Renewable energy   Sources of energy derived from the natural environment that are not finite.

Remote sensing   Method by which distant objects and surfaces can be observed with sensors.

Residence time   Mean time that a particle spends in a specific part of a system which is in dynamic equilibrium.

Respiration   Biochemical process in which energy is liberated from organic compounds for behaviours that require energy.

Reynolds number   Dimensionless quantity used to determine the type of fluid flow for a given velocity of a fluid and measured by the ratio of inertial to viscous forces.

Risk   Mathematical probability of the chance of any adverse condition.

Rural depopulation   Process whereby rural populations decline, invariably through the migration of its population, especially of the younger elements, to the cities.

Salter's duck   Oscillating mechanical system designed to 'capture' wave energy and through a hydraulic system to convert it into electrical energy.

Sankey diagram   Diagram showing the proportions of the energy transferred in a given system, with their sources and losses. Originated with the Irish military engineer R. Sankey.

Saturated vapour pressure   Vapour pressure of water vapour in an air sample that contains the maximum amount of vapour possible at that temperature.

Saturation deficit   Measure of the amount of water vapour required to be added to the air for the air to become saturated with water vapour. It can also be defined as the difference between the actual vapour pressure of the air and the saturated vapour pressure at the same temperature.

Savannah   Grassland found in the tropics.

Scale height   Altitude above the Earth's surface at which the pressure is reduced by a factor of e^{-1} ($= 0.37$).

Seasat   Satellite using radar to measure the height of the ocean surface.

Second Law of Thermodynamics   Energy flows from a higher to a lower level, and cannot of itself go in the opposite direction, unless acted upon by an external agency

(Clausius). It can also be expressed in terms of the change in entropy, through the relationship $dS = dQ/T$, since the Second Law defines the direction of a chemical or physical change. The process will go in the direction associated with an increase in entropy.

Sensible heat   Thermal energy that can be 'sensed' (e.g. with a thermometer) and which can be transferred into the air through conduction and convection, causing a change in temperature.

Sidereal day   Time interval between the motion of a point across a specified meridian.

Sinks   Reservoirs occurring in nature in which certain materials can be stored as part of the circulation of them through the global environment, e.g oceans are a primary sink for carbon.

Smog   Fog polluted with smoke.

Soil erosion and degradation   Removal of the top layer of the soil, through wind and water. Degradation implies a reduction in the soil quality.

Soil water balance   Measure of the difference between the gains and losses that must equal the change in the amount of water stored in the soil profile.

Soil water potential   Measure of the potential energy of the soil water.

Solar constant   Rate of incoming solar radiation flow at the top of the atmosphere ($=1353\,\mathrm{W\,m^{-2}}$).

Solar emittance   Rate of solar radiation per unit area acting on a surface.

Solar energy   Energy emanating from the Sun, mostly in the ultraviolet and visible part of the spectrum.

Solar tracking   Process of synchronizing the 'motion' of the Sun, with that of a solar collector or solar cell. This implies that the angle between the direct solar radiation and the plane of the solar device is 90º.

Solar wind   Flow of ionized particles (principally electrons and protons) from the Sun.

Sonar (sound navigation ranging)   Means of finding underwater objects using ultrasonic beams.

Specific heat capacity   Quantity of thermal energy required to raise the temperature of a substance of unit mass by 1ºC or 1K. Note that a change of 1°C is the same as a change of 1K. Measured in $\mathrm{J\,kg^{-1}\,K^{-1}}$.

Steady-state conditions   Conditions reached when energy flows through a material and in which the two temperatures on either side of the material are constant and the rate of energy flow in is equal to the rate of energy flow out.

Stefan–Boltzmann law   Mathematical model describing the total amount of energy released by a black body radiator.

Stokes's law   Involves fluid resistance. For example, if a ballbearing is falling, under gravity, through a fluid of high viscosity, then the viscous drag force acting on the bearing, once terminal velocity (v_t) has been reached, will be given by $F_v = 6\pi r \eta v_t$ where r is the bearing's radius, and η is the coefficient of viscosity of the medium.

Stomata   Apertures in the surfaces of leaves that allow the passage of gases.

Stevenson screen   Ventilated box-like device used to measure the temperature of the air, without the effects of radiation.

Stratosphere   Region of the atmosphere, above the Earth's surface, between the tropopause (at 10km) and the stratopause (at 50km).

Streamlines   Parallel laminar flow-lines in fluids.

Substrate   Chemical compound subjected to enzymatic activity.

Suction   Amount by which water pressure is reduced below atmospheric pressure.

Sun-synchronous orbit   Satellite that processes around the Earth at the same angular speed as the Earth goes round the Sun, with the satellite always passing over a particular point over the Earth at the same local time.

Sunspots   Dark regions in the Sun's outer region, the photosphere, resulting from a temperature collapse from 6000 to about 4000 K. The sunspot cycle is approximately 11.5 years. Explanations have been suggested, including the influence of powerful magnetic fields on the convection currents that drive the energy transfer from the Sun's interior to the exterior.

Surface tension   In the context of soil physics, it refers to the magnitude of the force per unit length between the rim of the water meniscus and the pore wall.

Sustainability   Process of attempting to meet the demands of the present generation while not depleting resources or degrading the environment for future generations.

Synoptic weather chart   Chart depicting meteorological observations, such as pressure and wind, at the same time.

Synthetic aperture radar   Satellites that use large 'synthetic' apertures to provide high-resolution imagery in the microwave region.

Temperature   Measure of the kinetic energy of all the particles of a system.

Temperature gradient   Ratio between the temperature difference across a body (such as a pane of glass) and the thickness of the material.

Temporary threshold shift   Temporary loss of hearing, which recovers after 1 or 2 days following exposure to excessive noise.

Tensiometer   Instrument used to measure the soil water suction.

Terminal velocity   Velocity achieved by an object moving through a medium when the resultant force acting on it is zero. Occurs when the force acting downwards is balanced by the upward forces of the viscous drag of the medium and the Archimedian upthrust, and there is no longer any acceleration.

Texture triangle   Triangular diagram showing the distribution of particle size in soils through, for example, the relationship between the percentages of sand, silt and clay, and the soil textural class.

Thermal conductivity   Rate of thermal energy flow per unit area of a material per unit temperature gradient.

Thermal infrared   Part of the infrared region of the electromagnetic spectrum with wavelengths between 3 and 20 μm.

Thermal insulation   Material that reduces energy transfer between internal and external environments of either a body or a building.

Thermal resistance   Coefficient used to describe the opposition to thermal transfer offered by a specific component of a building structure.

Thermodynamics   Branch of physics concerned with the relationship between energy transferred, the internal energy of the system and the useful work done. It is expressed in the form of the First Law of Thermodynamics.

Thermodynamic state function   Characteristic and descriptive of the thermodynamic state of a system. For example, temperature (T) or entropy (S).

Thermoregulation   Mechanism by which homeotherms and poikilotherms maintain a core body temperature in response to environmental temperatures.

Thermonuclear solar fusion   Fusion where the lighter element, hydrogen, is transformed into the heavier element, helium, under conditions of very high temperature (up to 20 million °C) and pressures (90 000 Pa) with the release of energy in the form of electromagnetic radiation.

Third Law of Thermodynamics (Nernst Heat Theorem)   Relates to the entropy of a system as its temperature decreases towards absolute zero. At absolute zero, the entropy of the system is assumed to be zero.

Threshold of hearing   Weakest sound that the 'average' person can detect.

Threshold of pain   Strongest sound that the human ear can tolerate.

Thunderstorm   Storm with lightning, thunder and, invariably, heavy rainfall. The thunder is due to the rapid heating and expansion of the air as a result of the electrical discharge which generates pressure (acoustic) waves.

Tilt   Inclination of the axis of the Earth with respect to its plane of orbit (the ecliptic) around the Sun.

Tip speed ratio   Term used in wind energy engineering and defined as the ratio V/U, where v is the speed of the propeller blade tip and u is the wind speed.

Tokomak   Type of nuclear fusion reactor that employs magnetic confinement to contain the very hot plasma.

Torque   Turning effect of a moment.

Trade winds   Winds flowing in the low troposphere from a subtropical high towards the intertropical convergence zone.

Transducer   Device that converts energy from one form into another, such as a solar cell in which solar energy (electromagnetic radiation) is converted into electrical energy.

Transmission coefficient   Fraction of irradiance that plant leaves intercept and which is transmitted.

Transmittance   Measure of an object's ability to allow the transmission of electromagnetic radiation and is obtained from the ratio of the luminous flux transmitted compared with the incident luminous flux.

Transpiration   Integral part of the circulation of water in plants from the roots to its shoots. It refers to the passage of water vapour from the leaves of plants into the local environment. The process also brings about evaporative cooling.

Transport phenomena   Physical processes that govern flow, such as the energy transfer in thermal conduction and in the mass transfer as in winds in global convection.

Tropopause   Region of the atmosphere that lies above the troposphere but below the stratosphere.

Troposphere   Layer of the atmosphere extending from the Earth's surface to approximately 10 km above it.

Turbulence   Type of fluid flow characterized by eddies, vortices and swirls and in which there is mixing between successive layers.

Turbulent transfer   Mechanism for transporting water vapour and sensible heat between land surface and the overlying air, through eddies and convection, which sweep away humid air and replace it with drier air from above.

Typhoon   Term used instead of 'hurricane' in the China Sea and the western North Pacific Ocean.

Ultraviolet radiation   High-energy section of the electromagnetic spectrum with wavelengths extending from 10 to 400 nm between X-rays and the violet end of the visible region.

U-value (or thermal transmittance coefficient)   Term used by building engineers to describe the total thermal energy transfer of an element of a construction. It is the rate at which energy flows through $1\,m^2$ of material with a temperature difference of $1\,K$.

Van Allen belts   Two layers of charged particles, mostly protons and electrons, surrounding the Earth and which lie in the equatorial plane.

Vapour   State of matter that consists of the gaseous state.

Vapour pressure   Partial pressure caused by water vapour.

Ventilation loss   Loss of warm air from a building and its replacement by cooler air, which needs to be heated.

Viscosity   Refers to fluid friction and is a measure of the resistance to flow. The greater the coefficient of viscosity of the fluid, the greater will be the viscous drag on any object passing through it.

Visibility   Greatest distance at which an object can be seen and recognized in daylight.

Visual extinction coefficient   Fractional reduction in luminous flux of a collimated beam in unit distance.

Volumetric specific heat capacity   Specific heat capacity measured in terms of volume.

Walker cells   Cells that lie in the vicinity of the equator and circulate in the vertical at $90°$ to the Hadley cells, and are associated with the sinking air over the eastern equatorial Pacific, and the ascent over the western equatorial Pacific.

Water potential   Measure of the potential energy of the soil water, and refers to the ability of water to cross a semipermeable membrane in osmosis.

Water release curve   Graph showing how the volumetric water content of a soil decreases as the suction increases.

Wave energy   Type of renewable energy in which the energy of waves is extracted to produce electrical energy.

Wave particle duality   Idea that particles (such as electrons) and electromagnetic radiation can behave as waves or particles, but not simultaneously.

Weather   Conditions in the atmosphere (particularly the troposphere), and characterized by the extent of solar radiation, pressure, air temperature, precipitation, wind, humidity, and cloud cover.

Wien's law   Relationship for the spectral curves of black body radiators at various temperatures in which the maximum wavelength for the curve at a given temperature times the temperature (i.e. $\lambda_{max} \times T$) is a constant.

Wind   Moving mass of air caused by the pressure differences arising from the unequal heating of the Earth's surface and the global convection that results.

Wind chill factor   Effect that wind velocity has in reducing the prevailing air temperature. It has important implications for living organisms, especially humans, as it can facilitate a fall in body temperature from increased energy loss that can result in hypothermia.

Windsonde   Devices using balloons to record wind speed and wind direction.

Wind turbine   Transducer for transforming wind energy into electrical energy.

Work function   Minimum energy required to release an electron from its location in the crystal lattice of a metal. Units $= J$ or eV.

Zero'th law   Defining statement for temperature which states that if in a thermally isolated system a body A is in thermal equilibrium with body B and, in turn, B is in thermal equilibrium with body C, then A and C must also be in thermal equilibrium.

Index

Absorbivity 226
Absorption spectrum 213, 217,218,373
Acid rain 10, 107, 114,133, 397 *see* Pollution
    Damage to fish 116
Action spectrum 219,373
Active solar heating 145
Adenosine diphosphate (ADP) 22, 26
Adenosine triphosphate (ATP) 22, 26, 374
Adiabatic change 118, 119
Adiabatic condition 412, 420, *see* Lapse rate
Adiabatic system *see* Entropy
Advanced very high resolution radiometer
    (AVHRR) 192
Aerodynamic drag 97
Aerodynamics 155
Aerofoil 155 *see* Bernoulli's theorem
Aerosol 6, 107, 231 *see* Particulates
    Aerosol droplets and disease 399
    Aerosol, atmospheric 266
    Aerosols from volcanic eruptions 224
Agriculture 4,14 ,319, 402–04
Air conditioning 42, 74–75 *see* Ventilation
Air masses 275
Air pollution 10, 397 *see* Smog
Air temperature 242, 243, 255
Air-entry suction 334
Albedo 224, 226, 262, 307
Alcoholic fermentation 172 *see* Biofuels
Alpha particle 134, 140 *see* Radioactivity
Altimeter 436 *see* Radar altimeter
Ammonia cycle 209
Anabolism 21
Anaerobic process 101,172 *see* Biofuels
Anemometer 247
Aneroid barometer 245
Angular frequency 94, 183
Angular velocity 278, 279
Antarctic polar vortex 222, 224 *see* Ozone Hole
Antarctic, melting of ice sheet of 404
Antarctica 297, 300
Anthesis 365

Anticyclone 257, 262, 275, 281, 291, 297, 311
Aphelion 200, 203
Aquifers 175, 344
Arrow of time 412 *see* Entropy
Atmosphere xviii, 3, 7, 78, 232, 275, 300, 363
Atmospheric electricity 269 *see* Thunderstorms
Atmospheric pressure 209, 242
Atmospheric window 182, 228
    Electromagnetic spectrum 182
    Infrared region 182
    Visible region 182
Attractor 437 *see* Lorenz attractor
Aurora Australis 206, 215, Plate 5
Aurora Borealis 206, 215
Aurorae 13, 141, 206, 215
Avogadro's number 137, 372–73

Band theory 149–50
Barometer 245
Beaufort scale 246–47
Beaufort, F. 246
Beer-Lambert law 213, 379
Benzene 116, 119
Bernoulli, D. 156, 262
Bernoulli's theorem 155–58
Beta particles 134 *see* Radioactivity
Betz limit 160, 162
Bhopal (1984) 406
Biochemical processes 23, 26, 172
Biodiversity 6, 397, 400, 402
Biogeochemical cycles 12, 14
Biological material as energy sources 172
    Agrochemical fuel extraction 173
    Biochemical processes 172
    Thermochemical processes 172
Biomass 121, 131, 133, 193
Biomass and biofuels 171–73
Biosphere xviii, 3, 7, 11, 14, 78, 232, 319
Black body (or full body) radiator 33, 226, 371
Black body curves 35–36
Black Death 399

Boltzmann, L. 20, 37, 411 *see* Entropy
Boltzmann's constant 20, 36, 411, 419
Boundary layer 39, 43–44, 60
Brundtland, G.H. 407, 409
Brundtland Report (1987) 407, 408
Building Regulations 61, 71
Building Research Establishment 62, 71, 85
Building Services Engineers, Chartered
　　Institution of 62, 71
Built environment 7, 9, 56
Bulk density 321–2

C3 plants 376, 380–2, 384–6
C4 plants 376, 380–2, 384–6
Calvin cycle 375
Campbell-Stokes sunshine recorder 249–50,
　　250–251
Candela 102
Capillarity 14
Carbohydrates 22, 372, 374, 382
Carbon balance 382, 383, 388
Carbon cycle 209,363, 383
Carbon dioxide 116, 208, 228–30, 232, 235,
　　402
Carbon dioxide emissions 6, 11, 229 *see* Energy
　　use and *see* Global Warming
Carbon monoxide 10, 106, 116, 121
Carbon store, terrestrial 388
Carnot cycle 119
Carnot, S. 18–19
Catalytic converter 112, 116, 119, 121
Cataracts 218, 407
Cavity wall 63
Centripetal force 162, 165, 183
　　see Turbulent flow
Chaos 14, 162
　　Chaos in weather forecasting 258
Charged droplets 270–71
Chernobyl (1986) 138–39, 406
Chlorofluorocarbons (CFCs) 11, 114, 221, 224,
　　229–30, 407, 423 *see* Ozone Hole
Chlorophyll 193, 372–74
Chloroplasts 372–73, 383
Chromosphere 140, 198, 215
Clausius-Clapeyron equation 266
Clean Air Acts (1956, 1968, 1970) 121
Climate 7, 12, 198, 242
Climate change 13, 180, 232, 363,402–3 *see*
　　Global Warming and Sea Level Rise
Climate modelling 14, 232, 403
Clo 43
Cloud physics 262–67
Cloud seeding 268
Clouds 12, 198, 233, 236, 242, 261, 421, Plate
　　11
　　Clouds, Types of 265

Coal 109, 132–33, 401
Coalescence 266–67
Coefficient of performance 75,160–6
Collision efficiency 266 *see* Cloud Physics
Combined heat and power 176
Compression refrigeration cycle 75–76
Condensation 60, 76–77, 207, 267
　　Condensation (or hygroscopic) nuclei 266
　　Condensation in buildings 82–84
Condensation nuclei 107 *see* Cloud Physics
Conduction 8, 23, 27, 40, 60–61,144, 225, 351
Conductive deafness 125
Conservation of energy 261
Constants, useful 426
Continuity equation 439
Convection 8, 23, 27, 31, 40, 45, 60, 62–63, 67,
　　144, 204, 285, 351–52
　　Convection coefficients 62
　　Convection currents 154
　　Convection, extratropical 300
Convective energy transfer coefficient 32, 412
Coriolis effect 261, 276, 278–80
　　Coriolis parameter 279
Coriolis, G.G. 278
Creutzen, P. 221, 425
Croll, J. 201
Crop coefficient 386
Crop diseases 403
Crop factor 357–58
Cryosphere 11, 232
Cyclone (or depression) 243, 257, 262, 280,
　　421 *see* Weather

Dalton's law of partial pressures 78
Darcy's law 39, 339–40
Decibel 124 *see* sound
Deforestation 6, 180, 397 *see* Climate Change
Dehydration 8, 47
Demographic change 4, 6, 397–99, 401
Demographic transition 5
Depression *see* cyclone
Desertification 2, 6, 180 *see* Climate Change
Developmental phase 365
Dew point 78, 79, 300,421
Dew point temperature 79, 300
Dipole moment 263
DNA 218
Doldrums 290
Doppler, C. 190
Doppler effect 11, 189–90, 415–18
　　Doppler shift 417
　　Electromagnetic radiation 417
　　Red shift 418
　　Sound waves 418
Double glazing 64
Dry deposition 114

Dynamic viscosity 162, 324

Earth as a black body 226,
Earth's atmosphere, composition of 181, 207
Earth's atmosphere, structure of the 204, 205
Earth's energy balance 224–25
Earth's ionosphere 214
Earth's magnetic field 206, 216
Eccentricity 200, 202
Edison, T. 103, 130
Efficiency index 65
Efficiency of a car engine 119–20
Efficiency of energy use in buildings 65
Einstein, A. 190
Einstein, an 373, 440
Einstein's mass-defect equation 134, 140
El Chichon 109, 224
El Nino 12, 314–15
El Nino Southern Oscillation (ENSO) 314
Electric field 269, 271
Electrical power transmission 95–96
Electromagnetic induction 93–95
Electromagnetic spectrum 33–34, 182, 373, 418
    and see Atmospheric windows
Electron-volt 136
Ellipticity 202
Emissive power 36, 226
Emissivity 37, 62
Emittance 199
Endothermic reaction 18, 374
Energy audit 413
Energy balance 26, 33
    Energy balance of land surfaces 348
Energy budget equation 40
Energy consumption 93, 412–15
Energy efficient buildings 9, 85, 86, 144 see
    Built Environment
Energy efficient car, 97–98
Energy efficient lamp 103–4
Energy loss from a body 23, 27
Energy losses from buildings 65, 66
Energy resources 401 see Fossil Fuels and
    Renewable Energy
Energy transfer and thermoregulation 26, 47
Energy use and CO_2 emissions 86 see Global
    Warming
Enhanced greenhouse effect 14, 229, 402 see
    Global Warming
Enthalpy 8, 18, 24, 25, 441
Entropy 8, 18–20, 24–25, 410–12
    First and Second laws of thermodynamics
    411
    Arrow of time 412
Entropy in terms of probability 20, 411
Environmental health 7, 405
Environmental impact assessment 406, 423–25

Environmental Physics xviii, 3, 6, 7, 15, Plate 1
Environmental risk 6, 405–07, 423–25
Epidemiology 105, 441
Equation of continuity 156, 160–62
Erectophile plants 380–82
Erythema 218
Escape velocity 210
Evaporation 8, 12, 15, 23, 27, 38–40, 78, 319,
    343, 353, 359, 365
    Evaporation from land surface
    347–60,355–57, 357–60
    Evaporation, effect of temperature, humidity
    and windspeed on 351, 353
Evapotranspiration 346, 442
Exosphere 207, 442
Exothermic reaction 18, 219, 374
Extinction coefficient 379

Faraday, M. 93, 130
Feedbacks in global warming 232
Ferrel cell 286–87
Fick's law 39
Field capacity 326, 333, 335, 342
First and Second laws of thermodynamics 411
    see Entropy
First law of thermodynamics 18, 23, 25,118–19,
    261, 371, 420
First order kinetics and degradation of organic
    matter 390–92
Fission 134–39
Flat-plate solar collector 145–46
Fluid dynamics 13, 14, 154, 162, 262
Fluid flow and convection 32–33
Forced convection 32, 412
Forces acting on air masses 276
    Coriolis force 276
    Frictional force 276
    Gravitational force 276
    Pressure gradient force 276
Fossil fuels 106, 131–33, 229, 371, 398, 401,
    402
Fourier, J. 28
Fourier's law of thermal conduction 28–30, 39,
    175, 339
Fraunhofer lines 212 see Solar Spectrum
Frequency 95
Frictional forces on air masses 276, 279, 291
Front 243, 257, 262, 275–76, 421
Fuel cells 401
Fusion 443 see Nuclear Fusion

Gamma radiation 134 see Radioactivity
Gas 133, 401
Geosphere 3, 7, 11
Geostationary orbit 182
Geostrophic balance 280, 291

Geostrophic wind 280–82
Geosynchronous orbits 184
Geothermal energy 133, 142, 174–75 *see*
    Renewable Energy
Gibbs equation 25
Gibbs free energy 8, 24–26, 375
    Gibbs free energy and the human body 26
Glaciers 264
    Glaciers, melting and recession of 235, 404
Global Circulation Models 14, 260–62
Global circulation patterns 286 *see* Winds
Global convection 285
Global humidity patterns 300–301
Global humidity patterns, Charts of 304–305
Global population growth 5, 398
Global precipitation patterns, Charts of 312–13
Global pressure fields 287–90
Global pressure fields, Charts of 288–89
Global surface temperatures, Change of 231
Global temperature fields, Charts of 298–99
Global temperatures, Rising of 402 *See* Global
    Warming and Carbon dioxide
Global village 194
Global warming 2, 6, 11, 13, 87, 114, 171,
    229–35, 363, 388, 397, 402, 404, Plate 10
Global weather monitoring network 253–54
Global weather patterns 275
Global wind fields 292–93
Global wind patterns 290–97
Glycolysis 22, 375
Gravitational force 276
Great Storm of 1987 256
Greenhouse effect 198, 227–29, 388
    Greenhouse effect on other planets 235
Greenhouse gas emissions 232–33, 403 *see*
    Carbon dioxide and Methane
Greenhouse gases 11, 14, 134, 171, 213, 228,
    230, 402 *see* Carbon dioxide and Methane
Ground heat flux 347–48
Gulf Stream 201

Hadley cell 286–87, 297, 309
Hadley, G. 285, 286
Hagen-Poiseuille law 324 *see* Poiseuille's law
Hague Summit on Climate Change (2000) 408
Hailstones 269
Hair hygrometers 248
Half-life 391
Halley's formula 418, 420
Hartley Band 217 *see* Ozone
Health and Safety at Work 71
Hearing loss 125 *see* Noise Pollution
Heat capacity 32, 42
Heat engine 10, 18, 275
Heat pump 10, 75
Heat stress 8, 42, 47, 402

Heat stroke 47, 48
Heliosphere 207
Heterogeneous reactions 223
Heterosphere 207
Homeostasis 4
Homeotherms 17, 27
Homosphere 207
Hottel-Whillier-Bliss equation 146
Humidity 12, 39, 71, 78, 80–81, 242, 255, 257,
    275, 307, 365, 385, 421
    Humidity measurement 247
    Humidity mixing ratio 306
Hurricane 12, 281, 294, 311, 405, Plate 12 *see*
    Typhoon
Hydraulic conductivity 339–40, 343, 346, 357
Hydraulic head 324, 330,337
Hydrocarbons 112, 116, 121
Hydroelectric power 131, 133, 142, 163, 401
    see Renewable Energy
Hydrogen bond 263
Hydrogen fuel cells 11, 173–74, 176
Hydrological cycle 7, 11, 208–09, 264–65, 319,
    363–64, 404
Hydrolysis 21, 26
Hydropower 158 *see* Hydroelectric power
Hydrosphere xviii, 3, 7, 11, 264
Hydrostatic equation 245, 261, 419
Hydrostatic pressure 335
Hygrometers 248
Hyperthermia 48
Hypothermia 8, 42, 45–46, 48, 50

Ice crystals 268
Ice lattice, Structure of 263, 268
Ice nuclei 267
Ideal gas 445
Ideal gas equation 261, 419–20
Illuminance 102, 192
Image processing 186
    NOAA strip 186–87
    Pixels 186, 193
    Resolution 186–87
    Swath width 187
Incompressible fluid 156, 161
Industrial Revolution 1, 5, 10, 90, 130, 163
Infectious diseases 399, 402, 406
Infrared radiation 33, 144, 182, 211, 213, 225,
    228, 230, 235
Insects and disease 402
Insulation 9,27,42, 66, 82
Interference (or perturbation) factor 161
Intergovernmental Panel on Climate Change
    (1988) 408
    First Report (1990) 408
    Second Report (1995) 408
    Third Report (2001) 408

Internal combustion engine 117, 174
Internal energy 18, 25, 284, 420
Interstitial condensation 82 *see* Condensation
Inter-Tropical Convergence Zone 291, 294, 301, 306–07, 309–10
Ionizing radiation 406
Ionosphere 206, 214–216, 269 *see* Aurorae
Isobars 257, 283, 290, 421 *see* Pressure gradient
Isothermal change 21, 418
Isotherms 257, 297
Isovolumetric change 119

Jetstream 294, 297
Joule 136
Joule, J.P. 18, 371

Kelvin, Lord 174
Kepler, J. 183,267
Kepler's laws 11,184
Kinetic energy 24, 154, 168, 170, 210, 216, 275
Krebs cycle 22
k-values 60–61
Kyoto Conference (1997) 388, 408

Laminar flow 44, 162
Land breeze 285
Landsat 192, 194, Plate 6
Lapse rate 210, 420–2 *see* Temperature
Latent heat 12, 244, 261, 263–66, 297, 353
    Latent heat flux 347–48
    Latent heat of fusion 21, 263
    Latent heat of vaporization 38, 225, 263, 266, 347, 353, 446
Lawrence, T.E. 49
Laws of thermodynamics 17, 23, 60
Leaching of contaminants 346–47
    Leaching of solutes through soils and contamination of groundwater 344
Leaf area index 371
Legionnaire's disease 75 *see* Air-conditioning
Light pollution 103, 192–93
Light response curves 375–6
Lightning 271–72, 405 *see* Thunderstorms
Lithosphere xviii, 7, 11, 14, 232
London Smog (1952) 10, 112 *see* Smog and Pollution
Lorenz attractor 259, 260
Lumen 102, 103
Luminous efficacy 103
Luminous flux 102, 252
Luminous intensity 102

Mach numbers 418 *see* Doppler effect
Magnetic confinement 141–42
Magnetosphere 207, 446
Malthus, T. 4, 397

Manometer 328, 337
Mass extinction and climate change 13, 402
Mass transport into Inter-Tropical Convergence Zone 291
Maxwellian distribution of molecular speeds 211
Mayer, J.R. 18, 371
Mega-city 10, 399
Mercury-in-glass thermometer 244
Mesopause 206
Mesosphere 206, 207
Metabolic rate 24, 42
Metabolism 8, 17, 21, 133, 371, 374–75
Meteorological Office 258
Meteorological Optical Range 252 *see* Visibility
Meteorological satellites 183, 191
    Advanced very high resolution radiometer 192
    Light pollution 192–93
    Meteosat 185, 191, 255, 447
    NOAA weather satellites 192
Methane (natural gas) 133, 229, 230 *see* Greenhouse gases
Mid-tropospheric temperatures, Charts of 302, 303
Mid-tropospheric wind patterns, Charts of 294–296
Milankovitch–Croll hypothesis 14, 201–204
Moisture characteristic 58, 60, 78–80, 333
Moisture content 80, 81
Molina, M. 221, 425
Monsoon 290, 294
Montreal Protocol (1987) 224, 408–09, 423–24
Mother of pearl clouds 208
Mount Pinatubo 109, 224, 231 *see* Volcanic activity
Mutagenesis 218

National and Oceanographic and Atmospheric Administration (NOAA) 184, 186
Natural convection 32 *see* Air conditioning and Energy Transfer
Natural ventilation 72
Navier-Stokes equations 262
Nebule 253 *see* lighting
Nerve deafness 125 *see* Hearing loss
Neutrino 139, 140
Newton, I. 182–83, 269
Newton's law cooling 8, 32–33, 62,412
Newton's law of gravitation 183, 210
Newton's laws of motion 11, 261, 278
Newton's second law 161, 262
Nitric oxide 208
Nitrogen dioxide 107, 208
Nitrogen oxide 10, 107, 116, 121 *See* Pollution
Nitrous oxide 229–30 *see* Ozone Hole

NOAA strip 186–87
NOAA weather satellites 192
Noctilucent clouds 208
Noise 122
    Decibels 124
    Hearing loss 125–26
    Human ear 123
    Noise pollution 122
    Threshold of hearing 123
    Threshold of pain 123
Nuclear energy 131
Nuclear fission 134, 401
Nuclear fusion 139–42,198, 401
    Nuclear fusion reactors 141–42
Nuclear power 134, 401

Oceanic conveyor belt 404, Plate 2
Oceanic Thermal Energy Conversion 176, 401
Oceanic thermal expansion 234, 404
Oceans 232, 234, 260
Ohm's law 39, 96
Oil 133, 401 see Fossil Fuels
Oscillating air column 168–69
Otto cycle 119–20
Ozone (tropospheric) 230
Ozone 106, 112, 206, 208, 217, 228
Ozone chemistry 219,
Ozone depletion 2, 6, 11, 13, 114, 208,218 221,
    224, 297, 300, 407,423
Ozone hole 11, 13, 220, 222, 397, 407, 423
    Plate 8
Ozone loss in the Antarctic 222 see Polar
    vortex
Ozone loss in the Arctic 224
Ozonosphere 219, 224

Palaeoclimatology 13, 232
Parabolic concentrator 145, 146–47
Particulates 6, 10, 105, 106, 109–12, 116 see
    Aerosols
Passive solar heating 144
Penman equation 353, 357, 448
Penman, H. xviii, 353–54
Penman-Monteith equation 354
Percentage saturation 80
Perihelion 200, 203
Permanent wilting point 326, 333, 335
pH 114, 364 see Acid rain
Phase change 7, 21, 38
Photo-absorption cross section 213, 217
    Photo-absorption spectrum of ozone 217
Photochemical smog 111–12, 402, 406, 423 see
    also Smog
Photodissociation 219–20
Photoelectric effect 148–49, 371
Photoelectric equation 149

Photo-ionization 207, 214 see Ionosphere
Photon 147, 149, 372
Photosphere 140, 198, 211
Photosynthesis 12, 14, 26, 36, 133, 171, 198,
    207, 236, 262, 357, 363–65, 371–75,
    382–84, 388, 390, 394, 403
Photosynthesis and effect of temperature
    376–77
Photosynthesis and light 372,381
Photosynthesis and vegetation canopy
    376–83
Photovoltaic cells 147, 176 see Renewable
    Energy
Pixels 186, 193 see Image Processing
Planck's constant 36, 149, 213, 372, 448,
Planck's law 213
Planophile plants 380–82
Plant development 365
    Plant development and effect of temperature
    366
Plant distribution, Impact of global warming on
    369–70
Plant growth and factors affecting development
    364–65, 370
PM_{10} 111,134 see Aerosols and Pollution
Poiseuille, J. 325
Poiseuille's law see Hagen-Poiseuille law
Polar cells 286
Polar ice caps 264, 404
Polar orbiting satellite 184, 449
Polar stratospheric clouds 223–24
Polar vortex 300
Pollution 6, 208,220,397. 399
Population growth 10, 397–99,408
Pores 320, 449
    Pore radius and rate of flow 324–25
    Pore size distributions 333–34
Porosity 326, 449
Potential energy 24, 26, 164, 166, 168–69, 284
Potential evaporation 327, 339, 353–54
Potential gradient 29
Power coefficient 161
Power of a moving mass of fluid 154
Precipitation 261, 265, 310, 319, 342, 343
    Precipitation measurement 248
    Precipitation radar 249, 310–13
Presbyacusis 126 see Hearing loss
Pressure 12, 255
    Pressure as a function of altitude 209,418–20
    Pressure gradient 155, 261, 284, 324,
    Pressure gradient force 276–77, 449
    Pressure gradients and winds 280
    Pressure measurement 245
Primary energy 65, 131, 449
Principle of the conservation of energy 18, 158,
    371

Probability *see* Entropy
Pyrolysis 172, *see* Biofuels

Quantum theory 149, 371–72, 449
   Quantum theory and black body radiation 34

Radar 187–90
   Altitude of satellite 189
   Principle of radar 188
   Synthetic radar 189
   Wavelengths used in radar 188
Radiation 8, 11–12, 23, 27, 33, 40, 60, 62, 144, 285
Radiation, laws of 12, 261 *see* Stefan-Boltzmann law
Radiative equilibrium 225–26,
Radioactive waste 137,450
Radioactivity 134
Rain gauge 249, 450
Rainfall 12, 242, 365, 421 *see* Precipitation
Rate of turnover 209
Red shift 190, 418 *see* Doppler effect
Relative humidity 78, 80–81, 247–48, 352
Remote sensing 11, 180
   Remote sensing data, applications of 191
Renewable energy 7, 11, 85, 142
   Renewable energy sources 133, 401
Residence time 208, 224
Resolution 186, 187, 194
   Resolution of satellite images 185
Respiration 374, 382
Reynolds number 163
Richardson. L.F. 14, 258–60
Rio Treaty 224, 407, 408
Risk 404, 405
Roaring Forties 290, 294, 306
Rothamsted Experimental Station 354, 387
Rowland, F.S. 221, 425

Salter's duck 168–69
Satellite orbits 182–84
   Geostationary orbits 182
   Geosynchronous orbits 184
   Meteorological 183
   Newton's law of gravitation 183
   Sun-synchronous orbits 184–85
Satellites 11, 181, 194,401
Saturated air 78, 266
Saturated vapour pressure 79
Saturation deficit 81, 352–54, 385–86
Saturation vapour pressure 12, 80, 266
Scale height 210
Scott's Antarctic expedition (1911) 46–47
Sea breeze 285, 300
Sea level rise in 21st century 234, 404
Sea surface temperatures 191, Plate 7

Seasat, use of radar 188
   Height of ocean surface 188
   Mapping elevation of polar ice-sheets 188
   Mapping topography of seabed 188
Second law of thermodynamics 18–21, 24–25, 65, 90, 411
Semiconductors 27, 149
Sensible heat 319, 348, 352–53, 354, 363
   Sensible heat flux 351–52, 347–48
Sewage and its treatment 100–102
Shear stress 163, *see* Laminar flow
Shock waves 418, *see* Doppler effect
Sidereal day 183
SimCity 92
Smog 111–14, 451
   See also London Smog and Pollution
Snow crystals 267
Soils 14, 319–20
Soil erosion 319, 343, 400
Soil particle size, Classification of 320
Soil pores, Distribution of water in 330
Soil porosity 322
Soil profiles, Carbon stored in 389, Plate 13
Soil texture 320–21, 326
Soil water balance 333, 358–360
   Deep drainage 343
   Evaporation 343
   Precipitation 342–43
   Transpiration 343
   Surface runoff 343
Soil water content 323, 340–41,351
Soil water potential 337–39
Soil water suction 328–30, 333–34
Soils, Temperature gradient in 348
Soils, Water holding capacities of 326
Solar (magnetic) storm 215–16
Solar cells 36, 147, 149, 150–52 *see* Renewable energy
Solar constant 199, 201, 226, 235
Solar cycles and climate change 198, 201 *see* Croll-Milankovitch
Solar emission spectrum 218
Solar emittance 199
Solar energy 11, 12, 133, 142–43, 198, 319, 401 *see* Renewable energy
Solar flux 211
Solar irradiance 145–47, 199, 201–02, 375
Solar output 198, 201
Solar powered cars 122
Solar radiation 13, 198, 206, 211, 213, 236, 262, 365
Solar spectrum 211–13
Solar wind 140, 216, 451
Sonar 451
Sound levels 124 *see* Hearing
Sound waves 417 *see* Doppler effect

Specific heat capacity 32, 119, 144, 285, 353, 413–14, 420
Stefan's constant 37
Stefan-Boltzmann law 8, 36–38, 62, 226, 261
Stevenson screen 244
Stokes' law 267, 451
Stomata 264, 357, 384, 385
Stratosphere 206, 207, 217, 220, 255, 257, 407
    see Ozonosphere
Suction 330
    Suction and pore radius 331–33
    Suction and radius of curvature 331
Suction gradient 335
Sulphur dioxide 10, 106–07, 112, 133, 208 see
    Acid rain
Sulphuric acid 107
Sun 139, 235
Sunspot activity 140, 198, 201
Sun-synchronous orbits 184, 185
Surface runoff 327, 343
Surface tension 330, 332
Surface weather chart 257
Survival in cold climates 42–47, Plate 3
Survival in hot climates 47–50
Sustainability 6, 399–40
Synoptic weather chart 256, 421–22
Synoptic weather station network 254
Synthetic Aperture Radar 185, 189, 190

Temperature 12, 18, 25, 275, 365
    Temperature as a function of altitude 209
    Temperature distribution of the body 23
    Temperature fields 17, 29, 39, 297
    Temperature of Earth from Vostock ice core
    233
Tensiometer 335–37
Terrestial radiation 198, 213, 224, 236 see
    Greenhouse Effect
Texture triangle 320–21
Thermal circulation 284–85
Thermal conductivity 28, 43, 348, 351
Thermal gradient 23, 297
    Thermal gradients and winds 284
Thermal resistance 61, 64, 146
Thermal time 367–70
Thermodynamics 14, 26
    Thermodynamics and the human body 22
    Thermodynamics, laws of 8, 12
Thermography 33, Plate 4
Thermoregulation 8, 41
Thermosphere 206–07
Third law of thermodynamics 19–20
Threshold frequency 148–49
Threshold of hearing 123
Threshold of pain 123
Thunderstorms 13, 270

Tidal power 11, 164–66 see Renewable energy
Tip-speed ratio 158–59, 453
Tornadoes 405 see Hurricanes and Cyclones
Toxicity 406
Trade winds 286, 291, 301, 310, 314
Transmission bands 213
Transmission coefficient 37
Transpiration 15, 264, 343, 359, 384–85
Transport phenomena 7, 11, 39
Troposphere 12, 204, 207, 209, 220, 242, 245, 255, 266, 294, 297, 300, 310, 314, 421
Turbulent flow 44, 162, 226
Turnover rate 390, 393
Typhoons 311 see Hurricanes and Cyclones

Ultrasound 190
Ultraviolet radiation 11, 103, 206, 211, 213–15, 221, 407, 423 see Ozone
Universal gravitational constant 183, 210
Uranium 132, 134–35 see Nuclear Fusion
Urban environment 7, 10, 90
Urban pollutants 2, 105, 106
U-value 61, 63–64, 67–8, 413

Van Allen belts 206, 454 see Magnetosphere
Vapour barrier 83–84
Vapour pressure 78, 82
Vegetation canopy 363, 370
    Erectophile plants 380
    Gross photosynthesis of a 380
    Gross photosynthesis and light intensity 381
    Planophile plants 380
    see Effect of leaf index on transmission of
        light
    see Photosynthesis
    see Transmission of light
Ventilation 42, 66, 68, 71–2 82, 414 see Air-
    conditioning
Viscosity 44
Visibility 243, 251–52, 421
Visual extinction coefficient 252
Volcanic activity 12, 105, 107, 110, 231, 266,
    Plate 9
Volumetric specific heat capacity 68, 414

Walker cells 315 see El Nino
Water consumption 99–100
Water molecule, Structure of 263
Water potential gradient 340, 341
Water release curve 329, 333–34
Water resources 363
Water stress 329
Water transport 14
Water vapour 12, 76, 78, 81, 207, 228–29, 301, 307
Water vapour transport 261

Wave energy 11, 166–71 *see* Renewable Energy
Wave-particle duality 372
Weather 7, 12, 242
Weather forecasting 14, 257
Weather prediction 243, 260
Weather satellites 255, 307
Weather station 242
Wet and dry bulb thermometer 79, 248, 300
Wien's law 36, 261, 454
Wind chill factor 44, 45, 244
Wind energy 11, 15, 131, 133, 142, 158
Wind farm 152, 153

Wind measurement 246
Wind patterns 198, 307
Wind speed 242, 255, 365
Wind systems 276
Winds 12, 236, 255, 275, 277
World energy demand 131
World energy supplies 132
Wren, C. 84

X-rays 206

Zeroth law 25